Graduate Texts in Mathematics 245

Graduate Texts in Mathematics

Graduate Texts in Mathematics bridge the gap between passive study and creative understanding, offering graduate-level introductions to advanced topics in mathematics. The volumes are carefully written as teaching aids and highlight characteristic features of the theory. Although these books are frequently used as textbooks in graduate courses, they are also suitable for individual study.

For further volumes:
http://www.springer.com/series/136

Rubí E. Rodríguez • Irwin Kra • Jane P. Gilman

Complex Analysis

In the Spirit of Lipman Bers

Second Edition

Springer

Rubí E. Rodríguez
Facultad de Matemáticas
Pontificia Universidad Católica de Chile
Santiago, Chile

Irwin Kra
Department of Mathematics
State University of New York at Stony Brook
Stony Brook, NY, USA

Jane P. Gilman
Department of Mathematics
and Computer Science
Rutgers University
Newark, NJ, USA

ISSN 0072-5285
ISBN 978-1-4899-9908-5 ISBN 978-1-4419-7323-8 (eBook)
DOI 10.1007/978-1-4419-7323-8
Springer New York Heidelberg Dordrecht London

Printed on acid-free paper

Springer is part of Springer Science+Business Media (www.springer.com)

For Victor, Eleanor and Bob
and to the memory of
Mary and Lipman Bers

Preface to Second Edition

The second edition contains significant new material in several new sections. We have also expanded sections from the first edition (improving, we expect, the exposition throughout) and included new figures and exercises for added clarity, as well as, of course, corrected errors and typos in the previous version.

Among the most important changes are:

- We have expanded and clarified several sections from the first edition.
- We have significantly enlarged the exercise sections. Some of the problems are routine, others challenging, and some require knowledge of other subjects usually covered in various first-year graduate courses. The problems are listed in more or less random order as far as their difficulty.
- In both editions of this text, we use the approach to integration based on differential forms. In an alternative approach differential forms are a by-product of work on integration of functions motivated by ideas from standard treatments of integral calculus. That is the approach that Bers took in courses that he taught; it is also the approach used by Ahlfors. Of course, either of the two approaches are equally valid and lead to the same major results. In this second edition, we provide an appendix that outlines this alternative path to the main results.
- New sections on Perron's method for solving the Dirichlet problem, Green's function, an alternative proof of the Riemann mapping theorem, and a description of the divisor of a bounded analytic function on the disc via infinite Blaschke products are included.
- We prove the Bers theorem on isomorphisms between rings of holomorphic functions on plane domains.
- A section on historical references prepared by Ranjan Roy has been added.

In addition, the following items related to our work might be of interest to those reading this volume.

- An answer manual for the exercises prepared by Vamsi Pritham Pingali is available, to instructors using the book for a course, from the publisher.
- An electronic version of the book is available from the publisher.

- One of the authors (IK) has created and maintains a section about the book on his web site. It contains, among other things, updated information and errata. http://www.math.sunysb.edu/~irwin/bookcxinfo.html
  The other authors may also have information about the book on their web sites.

It is our pleasure to thank Ranjan Roy for producing and allowing us to include in this volume his historical note. We are grateful to our colleagues and students who pointed out places for improvement in the first edition and in drafts of the second one. Among them: Bill Abikoff, Robert Burckel, Eduardo Friedman, Bryna Kra, Peter Landweber, Howard Masur, Sudeb Mitra, Lee Mosher, Robert Sczech, and Jacob Sturm. It is still true, of course, that errors and shortcomings may remain in the final version of this edition and these are entirely our responsibility.

Spring 2012

New York, NY, USA — Irwin Kra
Santiago, Chile — Rubí E. Rodríguez
Newark, NJ, USA — Jane P. Gilman

Preface to First Edition

This book presents fundamental material that should be part of the education of every practicing mathematician. This material will also be of interest to computer scientists, physicists and engineers.

Because complex analysis has been used by generations of practicing mathematicians working in a number of different fields, the basic results have been developed and redeveloped from a number of different perspectives. We are not wedded to any one viewpoint. Rather we will try to exploit the richness of the development and explain and interpret standard definitions and results using the most convenient tools from analysis, geometry and algebra. Complex analysis has connections and applications to many other subjects in mathematics, both classical and modern, and to other sciences. It is an area where the classical and the modern techniques meet and benefit from each other. We will try to illustrate this in the applications we give.

Complex analysis is the study of complex valued functions of a complex variable and its initial task is to extend the concept of differentiability from real valued functions of a real variable to these functions. A complex valued function of a complex variable that is differentiable is termed *analytic*, and the first part of this book is a study of the many equivalent ways of understanding the concept of analyticity. The equivalent ways of formulating the concept of an analytic function are summarized in what we term the fundamental theorem for functions of a complex variable. In dedicating the first part of this book to the very precise goal of stating and proving the fundamental theorem we follow a path in the tradition of Lipman Bers from whom we learned the subject. In the second part of the text we then proceed to the leisurely exploration of interesting consequences and applications of the fundamental theorem.

We are grateful to Lipman Bers for introducing us to the beauty of the subject. The book is an outgrowth of notes from Bers's original lectures. Versions of these notes have been used by us at our respective home institutions, some for more than 20 years, as well as by others at various universities. We are grateful to many colleagues and students who read and commented on these notes. Our interaction with them helped shape this book. We tried to follow all useful advice and correct, of course, any mistakes or shortcomings they identified. Those that remain are entirely our responsibility.

Newark, NJ, USA — Jane P. Gilman
New York, NY, USA — Irwin Kra
Santiago, Chile — Rubí E. Rodríguez

Acknowledgement

The first author was supported in part by Fondecyt Grant # 1100767. The third author was supported in part by grants from the National Security Agency, from the Rutgers University Research Foundation, and from Yale University while a visiting Fellow.

Contents

Standard Notation and Commonly Used Symbols

Standard Terminology

Term	Meaning
LHS	Left-hand side
RHS	Right-hand side
Deleted neighborhood of z	Neighborhood with z removed
CR	Cauchy–Riemann equations
$\subset$	Proper subset
$\subseteq$	Subset, may not be proper
d	Euclidean distance on $\mathbb{C}$
ρ_D	Hyperbolic distance on D
MMP	Maximum modulus property
MVP	Mean value property
$\mathbf{C}^k_{\mathbb{R} \text{ or } \mathbb{C}}(D \text{ or } [a,b])$	k-times differentiable real- (or complex-) valued functions on the domain D (or the interval $[a,b]$)
iff	If and only if
$A - B$	The complement of B in A $\{a \in A; a \notin B\}$
pdp	Piecewise differentiable path
$\int_{\vert\tau - z_0\vert = R}$	Integration over the path $\gamma(t) = z_0 + Re^{2\pi \imath t}, 0 \leq t \leq 1$
$\emptyset$	The empty set

A List of Symbols

Term	Meaning
$\mathbb{Z}$	Integers
$\mathbb{Q}$	Rationals
$\mathbb{R}$	Reals
$\mathbb{C}$	Complex numbers
$\widehat{\mathbb{C}}$	$\mathbb{C} \cup \{\infty\}$
$\imath$	The square root of -1
$\imath\mathbb{R}$	The imaginary axis in $\mathbb{C}$
$\Re z$	Real part of z
$\Im z$	Imaginary part of z
$z = x + \imath y$	$x = \Re z$ and $y = \Im z$
$\bar{z}$	Conjugate of z
$r = \lvert z\rvert$	Absolute value of z
$\theta = \arg z$	An argument of z
$\mathrm{Arg}\, z$	The principal argument of z
$z = re^{\imath\theta}$	$r = \lvert z\rvert$ and $\theta = \arg z$
u_x, u_y	Real partial derivatives
$u_z, u_{\bar{z}}$	Complex partial derivatives (of real-valued functions)
$\frac{\partial f}{\partial x}, \frac{\partial f}{\partial y}$	Real partial derivatives (of complex-valued functions)
$\frac{\partial f}{\partial z}, \frac{\partial f}{\partial \bar{z}}$	Complex partial derivatives (of complex-valued functions)
∂R	Boundary of set R
$\lvert R\rvert$	Cardinality of set R
$\mathrm{cl}R$	Closure of set R
$\mathrm{int}R$	Interior of set R
$X_{\mathrm{condition}}$	The set of $x \in X$ that satisfy *condition*
$\nu_\zeta(f)$	Order of the function f at the point ζ
$i(\gamma)$	Interior of the Jordan curve γ
$e(\gamma)$	Exterior of the Jordan curve γ
$f\vert_B$	Restriction of the function f to the subset B of its domain
$U(z,r) = U_z(r)$	$\{\zeta \in \mathbb{C};\ \lvert\zeta - z\rvert < r\}$
$\mathbb{D}$	$U(0,1)$
$\mathbb{H}^2$	$\{z \in \mathbb{C};\ \Im z > 0\}$
S^1	$\{z \in \mathbb{C}; \lvert z\rvert = 1\}$
$S^2 \cong \mathbb{C} \cup \{\infty\}$	Riemann sphere

Chapter 1
The Fundamental Theorem in Complex Function Theory

This introductory chapter is meant to convey the need for and the intrinsic beauty found in passing from a real variable x to a complex variable z. In the first section we "solve" two natural problems using complex analysis. In the second, we state what we regard as the most important result in the theory of functions of one complex variable, which we label the fundamental theorem of complex function theory, in a form suggested by the teaching and exposition style of Lipman Bers; its proof will occupy most of this volume. The next two sections of this chapter include an outline of our plan for the proof and an outline for the text, respectively; in subsequent chapters we will define all the concepts encountered in the statement of the theorem in this chapter. The reader may not be able at this point to understand all (or any) of the statements in our fundamental theorem or to appreciate its depth and might choose initially to skim this material. All readers should periodically, throughout their journey through this book, return to this chapter, particularly to the last section, that contains an interesting account of part of the history of the subject.

1.1 Some Motivation

1.1.1 Where Do Series Converge?

In the calculus of a real variable one encounters two series that converge for $|x| < 1$ but in no larger open interval (on the real axis):

$$\frac{1}{1+x} = 1 - x + x^2 - \cdots + (-1)^n x^n + \cdots$$

and

$$\frac{1}{1+x^2} = 1 - x^2 + x^4 - \cdots + (-1)^n x^{2n} + \cdots .$$

R.E. Rodríguez et al., *Complex Analysis: In the Spirit of Lipman Bers*, Graduate Texts in Mathematics 245, DOI 10.1007/978-1-4419-7323-8_1,

It is natural to ask why these two series that are centered at the origin have radius of convergence 1. The answer for the first one is natural: the function $\frac{1}{1+x}$ has a singularity at $x = -1$, and so the series certainly cannot represent the function at this point, which is at distance 1 from 0. For the second series, the answer does not appear readily within real analysis. However, if we view $\frac{1}{1+x^2}$ as a function of the complex variable[1] x, then we again conclude that the series representing this function should have radius of convergence 1, since that is the distance from 0 to the singularities of the function; they are at $\pm\imath$.

1.1.2 A Problem on Partitions

A natural question in elementary additive number theory is the following: Is it possible to partition the positive integers $\mathbb{Z}_{>0}$ into finitely many (more than one) infinite arithmetic progressions with distinct differences? The answer is **no**. It is obviously possible to construct such partitions if some differences are allowed to be equal. So assume that the differences are different and to the contrary that

$$\mathbb{Z}_{>0} = S_1 \cup S_2 \cup \cdots \cup S_n,$$

where $n \in \mathbb{Z}_{>1}$, and for $1 \le i \le n$, S_i is an arithmetic progression with initial term a_i and difference d_i, for $1 \le i < j \le n$, $S_i \cap S_j = \emptyset$, and $1 < d_1 < d_2 < \cdots < d_n$. Then

$$\sum_{i=1}^{\infty} z^i = \sum_{i \in S_1} z^i + \sum_{i \in S_2} z^i + \cdots + \sum_{i \in S_n} z^i,$$

and each series converges for $|z| < 1$. Summing the above geometric series we see that

$$\frac{z}{1-z} = \frac{z^{a_1}}{1-z^{d_1}} + \frac{z^{a_2}}{1-z^{d_2}} + \cdots + \frac{z^{a_n}}{1-z^{d_n}} \text{ for all } z \text{ with } |z| < 1. \tag{1.1}$$

Choose a sequence of complex numbers[2] $\{z_k\}$ of absolute value less than 1 with $\lim_{k\to\infty} z_k = e^{\frac{2\pi\imath}{d_n}}$. Then

$$\lim_{k\to\infty} \frac{z_k}{1-z_k} = \frac{e^{\frac{2\pi\imath}{d_n}}}{1-e^{\frac{2\pi\imath}{d_n}}}$$

[1]In the subsequent parts of this text (except the historical remarks) we usually use z, w, and c, among others, but not x or y (which usually denote the real and imaginary part of z) to denote a complex variable.

[2]Notation for the polar form of a complex number is established in Chap. 2.

and

$$\lim_{k\to\infty} \frac{z_k^{a_i}}{1-z_k^{d_i}} = \frac{e^{\frac{2\pi\imath a_i}{d_n}}}{1-e^{\frac{2\pi\imath d_i}{d_n}}} \quad \text{for } i = 1, 2, \ldots, n-1,$$

(all these quantities are finite) while $\lim_{k\to\infty} \frac{z_k^{a_n}}{1-z_k^{d_n}}$ does not exist. This is an obvious contradiction to (1.1).

1.1.3 Evaluation of Definite Real Integrals

Some integrals are difficult, perhaps impossible, to evaluate using methods usually studied in undergraduate calculus courses. Examples are

$$\int_{-\infty}^{\infty} \frac{\mathrm{d}x}{1+x^4} \text{ and } \int_0^{2\pi} \frac{\mathrm{d}\theta}{2+\sin\theta}.$$

Using the residue theorem (see Sect. 6.6), it is quite easy to evaluate them.

1.2 The Fundamental Theorem of Complex Function Theory

Theorem 1.1. *Let $D \subseteq \mathbb{C}$ denote a domain (an open connected set), and let $f = u + \imath v : D \to \mathbb{C}$ be a complex-valued function defined on D. The following conditions are equivalent (here u and v are real-valued functions of the complex variable $z = x + \imath y$, with x and y real):*

(1) *The complex derivative*

$$f'(z) \text{ exists for all } z \in D \qquad \text{(Riemann);}$$

that is, the function f is holomorphic on D.

(2) *The functions u and v are continuously differentiable and satisfy*

$$\frac{\partial u}{\partial x} = \frac{\partial v}{\partial y} \text{ and } \frac{\partial u}{\partial y} = -\frac{\partial v}{\partial x}. \qquad \text{(Cauchy–Riemann: CR)}$$

Alternatively, the function f is continuously differentiable and satisfies

$$\frac{\partial f}{\partial \bar{z}} = 0. \qquad \text{(CR–complex form)}$$

(3) *For each simply connected subdomain $\widetilde{D}$ of D there exists a holomorphic function $F : \widetilde{D} \to \mathbb{C}$ such that $F'(z) = f(z)$ for all $z \in \widetilde{D}$.*

(4) *The function f is continuous on D, and if γ is a (piecewise smooth) closed curve in a simply connected subdomain of D, then*

$$\int_\gamma f(z)\,\mathrm{d}z = 0.$$

((1) $\Longrightarrow$ (4): Cauchy's theorem; (4) $\Longrightarrow$ (1): Morera's theorem)

An equivalent formulation of this condition is: The function f is continuous on D and the differential form $f(z)\,\mathrm{d}z$ is closed on D.

(5) *If $\{z \in \mathbb{C} : |z - z_0| \le r\} \subseteq D$ with $r > 0$, then*

$$f(z) = \frac{1}{2\pi\imath} \int_{|\tau - z_0| = r} \frac{f(\tau)}{\tau - z}\,\mathrm{d}\tau \qquad \text{(Cauchy's integral formula)}$$

for each z such that $|z - z_0| < r$.

(6) *The nth complex derivative*

$$f^{(n)}(z) \text{ exists for all } z \in D \text{ and for all integers } n \ge 0.$$

(7) *If $\{z : |z - z_0| \le r\} \subseteq D$ with $r > 0$, then there exists a unique sequence of complex numbers $\{a_n\}_{n=0}^{\infty}$ such that*

$$f(z) = \sum_{n=0}^{\infty} a_n\,(z - z_0)^n \qquad \text{(Weierstrass)}$$

for each z such that $|z - z_0| < r$. Furthermore, the series converges uniformly and absolutely on every compact subset of $\{z : |z - z_0| < r\}$. The coefficients a_n may be computed as follows.

$$a_n = \frac{1}{2\pi\imath} \int_{|\tau - z_0| = r} \frac{f(\tau)}{(\tau - z_0)^{n+1}}\,\mathrm{d}\tau \qquad \text{(Cauchy)}$$

and

$$a_n = \frac{f^{(n)}(z_0)}{n!}. \qquad \text{(Taylor)}$$

(8) *Choose a point $z_i \in K_i$, where $\bigcup_{i \in I} K_i$ is the connected component decomposition of the complement of D in $\mathbb{C} \cup \{\infty\}$, and let $S = \{z_i ; i \in I\}$. Then the function f is the limit (uniform on compact subsets of D) of a sequence of rational functions with singularities only in S.*

(Runge)

1.3 The Plan for the Proof

We prove the fundamental theorem by showing the following implications.

$$(1) \Leftrightarrow (2) \Rightarrow (3) \Rightarrow (4) \Rightarrow (5) \Rightarrow (6) \Rightarrow (1);$$
$$(5) \Rightarrow (7) \Rightarrow (1) \Leftrightarrow (8).$$

It is of course possible to follow other paths through the various claims to obtain our main result. For the convenience of the reader, we describe where the various implications are to be found. At times the reader will need to slightly enhance an argument to obtain the required implication.

(1) $\Leftarrow$ (2): Corollary 2.41.
(1) $\Rightarrow$ (2): Theorem 2.33 and Corollary 5.8.
(1) $\Rightarrow$ (3): Theorem 4.61 and Corollary 4.52.
(3) $\Rightarrow$ (4): This is a trivial implication. See Lemma 4.14 and the definitions preceding it.
(4) $\Rightarrow$ (5): Theorems 5.12 and 5.2.
(5) $\Rightarrow$ (6): The proof of Theorem 5.5.
(6) $\Rightarrow$ (1): This is a trivial implication.
(5) $\Rightarrow$ (7): The proof of Theorem 5.5.
(7) $\Rightarrow$ (1): Theorem 3.19.
(1) $\Rightarrow$ (8): Theorem 7.37.
(8) $\Rightarrow$ (1): Theorem 7.2.

In standard texts, typically each of these implications is stated as a single theorem. The tag words in parentheses in the fundamental theorem are the names or terms that identify the corresponding results. The forward implication (1) $\Rightarrow$ (n) would be the theorem: "If f is a holomorphic function, then condition (n) holds," where $n \in \{2, 3, 4, 5, 6, 7, 8\}$. For example, (1) $\Rightarrow$ (2) would be stated as "If f is holomorphic, then the Cauchy–Riemann equations hold". The organization of all these conditions (potentially 56 theorems—some trivial) into a single unifying theorem is the hallmark of Bers's mathematical style: clarity and elegance. Here it provides a conceptual framework for results that are highly technical and often computational. The framework comes from insight that, once articulated, will drive the subsequent mathematics and lead to new results.

While the organization of the results into one unifying theorem is a distinctive characteristic of Bers's mathematics, the treatments we give for some of the topics are not necessarily the ones that Bers used when he taught the course. In particular, his approach to integration theory did not normally start with differentials, and his emphasis on homotopy and homology were minimal. Currently, most graduate students will often either have studied some topology and some differential geometry before they take a complex analysis course or will be taking courses that cover these topics concurrently. Such students may skip over these parts of the text or skim them as a review. Additionally, our choice of topics beyond the fundamental theorem and hyperbolic geometry has been guided by the tastes of the various authors. Some important theorems we have omitted or not treated in detail include the Picard Theorem 6.9, whose proof Bers sometimes included in his courses.

1.4 Outline of Text

Chapter 2 contains the basic definitions. It is followed by a study of power series in Chap. 3. Chapter 4 contains the central material, the Cauchy theory, of the subject. We prove that the class of analytic functions is precisely the same as the class of functions having power series expansions, and we establish other parts of the fundamental theorem. Many consequences of the Cauchy theory are established in the next two chapters. Some readers may skip or skim parts of chaps. 2 and 3.

In the second part of the text we proceed to the leisurely exploration of interesting ramifications and applications of the fundamental theorem. It starts with an exploration of sequences and series of holomorphic functions in Chap. 7, that also contains Runge's theorem on approximations of holomorphic functions by rational functions. The Riemann mapping theorem (RMT) and the connection between function theory and hyperbolic geometry are the highlights of Chap. 8. The next chapter deals with harmonic functions, including a discussion of the Dirichlet problem and an alternative proof of the RMT. Zeros of holomorphic functions are discussed in the last chapter, which contains a study of the ring of holomorphic functions on a fixed domain, infinite Blaschke products, and an introduction to special functions. The latter is the beginning of the deep connections to classical and modern number theory.

1.5 Appendix: Historical Notes by Ranjan Roy

The statements of the propositions comprising the fundamental theorem of complex function theory, given above, are accompanied by the names of the mathematicians to whom those results are attributed. The history of the discoveries of these and other mathematicians can not only provide interesting footnotes in a complex analysis course but can also complement our understanding of the substance of the topic. The study of the development of complex analysis is complicated but also enhanced by the fact that several results are named for mathematicians whose work was anticipated by earlier mathematicians. For example, Theorem 10.21, credited to the Danish mathematician J. Jensen (1859–1925) who discovered it in 1899, was proved by C. Jacobi (1804–1851) for polynomials with real coefficients in his 1827 paper, *Ueber den Ausdruck der verschiedenen Wurzeln einer Gleichung durch bestimmte Integrale*, and, as noted by E. Landau, the proof carries over to the general case. We also note that the mathematicians of past centuries often used a notation different from ours. For instance, for Gauss, Cauchy, and Weierstrass,

x or y denoted complex variables; again, Gauss and Weierstrass employed the symbol i, whereas Cauchy used $\sqrt{-1}$ for "the" square root of -1. In presenting the specific details of their researches, it is often helpful to work within their notational perspective, as in these notes.

C.F. Gauss (1777–1855) appears to be the first mathematician to state Cauchy's (integral) theorem–indeed, to have a clear conception of integration in the complex domain. In his letter of December 18, 1811 to his friend F.W. Bessel, Gauss discussed the meaning of $\int \phi x.\mathrm{d}x$ for $x = a + bi$. He explained that, just as the realm of real numbers may be conceived of as a line, the set of complex numbers may be viewed as a two-dimensional plane with abscissa a and ordinate b. He then defined the integral over any (rectifiable) curve as the sum of infinitesimals $\phi(x).\mathrm{d}x$, where $\mathrm{d}x$ was an infinitesimal increment along the curve. Gauss wrote that the value of $\int \phi x.\mathrm{d}x$ remained the same along the two paths as long as $\phi x \neq \infty$ for all points x in the region between the two paths. To contrast this result with one for which $\phi x = \infty$ inside the region, he defined $\log x$ by the integral $\int \frac{1}{x}\,\mathrm{d}x$ starting at 1 and ending at $x \neq 0$. If the curve from 1 to x circumscribed the origin, then each circuit would add $\pm 2\pi i$. He noted that this helped explain why $\log x$ was a multivalued function.

Though Gauss promised to publish these remarkable theorems and their proofs at an appropriate occasion, he never did so; such instances of delayed publication provide further obstacles in our study of the mathematical past. Thus, it remained for A.L. Cauchy (1789–1857) to gradually work out the theory of complex integration. In 1814, he set out to rigorously establish some earlier results on real definite integrals of Euler, Laplace, and Legendre. They had obtained such results by formally replacing real with complex parameters within the integrals. But Cauchy proved a formula, equivalent to Cauchy's theorem for a rectangular contour, and established a 1781 result of Leonhard Euler (1707–1783) on the Gamma function. Note that Euler's result went unpublished until 1794.

As early as 1729, Euler had discovered the Gamma function as an infinite product:

$$\Gamma(m+1) = \frac{1^{-m} \cdot 2^{m+1}}{m+1} \cdot \frac{2^{-m} \cdot 3^{m+1}}{m+2} \cdot \frac{3^{-m} \cdot 4^{m+1}}{m+3} \cdots, \quad m \geq 0; \tag{1.2}$$

then, in 1730, he showed that the infinite product (1.2) equaled the Gamma integral: $\int_0^1 (-\log x)^m\,\mathrm{d}x$. Soon afterwards, Euler defined e and the exponential function e^t, and, in fact, when he returned to the study of the Gamma integral late in life, he took $x = \mathrm{e}^{-t}$ and $m = n - 1$ to get

$$\Gamma(n) = \int_0^{\infty} t^{n-1}\mathrm{e}^{-t}\,\mathrm{d}t. \tag{1.3}$$

Observe that (1.3) is equivalent to formula (10.17), except that Euler took $n > 0$ to be real. We note that the notation $\Gamma(n)$ was introduced in 1811 by Legendre and the modern notation for the limits of integration was first given by Fourier in 1818.

In his 1781 paper, Euler used (1.3) to get (in modern notation)

$$\int_0^\infty x^{n-1}\,\mathrm{e}^{-kx}\mathrm{d}x = \frac{\Gamma(n)}{k^n}, \quad \text{with } k > 0.$$

He then boldly took $k = p + q\sqrt{-1}$ with $p > 0$ and obtained

$$\int_0^\infty x^{n-1}\,\mathrm{e}^{-px}\,\cos qx\,\mathrm{d}x = \frac{\Gamma(n)\cos n\theta}{f^n} \tag{1.4}$$

$$\int_0^\infty x^{n-1}\,\mathrm{e}^{-px}\,\sin qx\,\mathrm{d}x = \frac{\Gamma(n)\sin n\theta}{f^n}, \tag{1.5}$$

where $f = (p^2 + q^2)^{\frac{1}{2}}$ and $\tan\theta = \frac{q}{p}$. He then deduced a number of remarkable and novel integrals as particular and limiting cases. For example:

$$\int_0^\infty \frac{\cos x}{\sqrt{x}}\,\mathrm{d}x = \sqrt{\frac{\pi}{2}}, \qquad \int_0^\infty \frac{\sin x}{\sqrt{x}}\,\mathrm{d}x = \sqrt{\frac{\pi}{2}},$$

$$\int_0^\infty \mathrm{e}^{-px}\,\frac{\sin qx}{x}\,\mathrm{d}x = \theta, \qquad \int_0^\infty \frac{\sin x}{x}\,\mathrm{d}x = \frac{\pi}{2}.$$

Thus, by assuming k complex, Euler came upon results he regarded as amazing and he expounded on them in some detail, sensing them to be potentially very useful.

Cauchy also found Euler's results on the Gamma function significant, and the first part of his 1814 paper, *Mémoire sur les intégrales définies*, contained a proof of a particular case of his integral theorem, thereby providing a rigorous foundation for Euler's results. Cauchy started with the integral $\int f\,\mathrm{d}y$, where $y = M + N\sqrt{-1} = M(x,z) + N(x,z)\sqrt{-1}$, x, and z real. He noted that

$$\frac{\partial}{\partial x}\int f(y)\,\mathrm{d}y = f(y)\,\frac{\partial y}{\mathrm{d}x}, \quad \frac{\partial}{\partial z}\int f(y)\,\mathrm{d}y = f(y)\,\frac{\partial y}{\partial z}. \tag{1.6}$$

Setting $f(y) = P + Q\sqrt{-1}$, he obtained

$$f(y)\,\frac{\partial y}{\partial x} = \left(P\,\frac{\partial M}{\partial x} - Q\,\frac{\partial N}{\partial x}\right) + \left(P\,\frac{\partial N}{\partial x} + Q\,\frac{\partial M}{\partial x}\right)\sqrt{-1} \equiv S + T\sqrt{-1}, \tag{1.7}$$

$$f(y)\,\frac{\partial y}{\partial z} = \left(P\,\frac{\partial M}{\partial z} - Q\,\frac{\partial N}{\partial z}\right) + \left(P\,\frac{\partial N}{\partial z} + Q\,\frac{\partial M}{\partial z}\right)\sqrt{-1} \equiv U + V\sqrt{-1}. \tag{1.8}$$

From (1.6)–(1.8), and taking $\frac{\partial}{\partial z}\frac{\partial}{\partial x}\int f(y)\,\mathrm{d}y = \frac{\partial}{\partial x}\frac{\partial}{\partial z}\int f(y)\,\mathrm{d}y$, he had

$$\frac{\partial S}{\partial z} + \frac{\partial T}{\partial z}\sqrt{-1} = \frac{\partial U}{\partial x} + \frac{\partial V}{\partial x}\sqrt{-1}.$$

Equating the real and imaginary parts, Cauchy derived

$$\frac{\partial S}{\partial z} = \frac{\partial U}{\partial x}, \quad \frac{\partial T}{\partial z} = \frac{\partial V}{\partial x}. \tag{1.9}$$

Note that, by taking $y = x + \imath z$, Equations (1.9) become the Cauchy–Riemann equations. We remark that d'Alembert had found these equations in the course of his researches in fluid mechanics; similar equations had appeared in the work of Lagrange and Euler. But Cauchy was the first to understand the significance of these equations in distinguishing analytic from nonanalytic functions, though he came to this insight late in his life. To arrive at the particular case of his integral theorem, Cauchy next integrated the equations in (1.9) with respect to x and z within finite limits to obtain two equations, expressible in modern form as a single formula:

$$\int_C f(y)\,dy = 0, \tag{1.10}$$

where C represents a rectangular curve (or even a class of curvilinear quadrilaterals), though we note that Cauchy did not employ geometric language in his 1814 paper. Later in this paper, he extended this result to an infinite interval, and then, using the substitution $y = (p + \imath q)x$, he transformed the integral $\int_0^\infty f(y)\,dy$ to an integral along a ray through $p + \imath q$. Euler's results (1.4) and (1.5) followed as a corollary by taking $f(y) = y^{n-1}\,e^{-y}$.

Cauchy's 1814 basic approach was to conceive of and express the real and imaginary parts of $\int f(z)\,dz$ as separate: $\int P dx - Q dy$ and $\int Q dx + P dy$. But his 1825 paper, *Mémoire sur les intégrales définies prises entre des limites imaginaires*, took a much more geometric point of view. He considered points in the complex plane and discussed contours within the complex plane. He gradually began to regard $\int f(z)\,dz$ as an entity in itself. In his 1825 paper, Cauchy presented a proof of his integral theorem (1.10) for a more general curve C and for functions he called finite and continuous. To his mind, such functions were continuously differentiable. He showed that the value of the integral from a point A to a point B did not change for two neighboring curves. Then by a homotopy type of argument, he extended the result to two non-neighboring curves. This argument is not rigorous by modern standards. Indeed, Cauchy himself may have been unsure of his reasoning; the paper contains his attempt to reformulate it in terms of the calculus of variations. And later, in his 1846 *Sur les intégrales qui s'étendent à tous les points d'une courbe fermée*, he expressed the integral as

$$\int_C f\,dz = \int_C (P dx - Q dy) + \imath \int_C Q dx + P dy, \tag{1.11}$$

where C was a closed curve within which there was no singularity of $f(z)$. He applied Green's theorem, Theorem 4.20,

$$\int_C P\mathrm{d}x + Q\mathrm{d}y = \pm \iint_D \left(\frac{\partial Q}{\partial x} - \frac{\partial P}{\partial y}\right) \mathrm{d}x\, \mathrm{d}y$$

(where the plus sign was chosen if C had a positive orientation and D represented the region within C) to each of the two real integrals. The Cauchy–Riemann equations in the form $\frac{\partial P}{\partial x} = \frac{\partial Q}{\partial y}$ and $\frac{\partial P}{\partial y} = -\frac{\partial Q}{\partial x}$ then showed that the two integrals were equal to 0.

In fact, it appears that Green's theorem was first stated in Cauchy's 1846 paper. It is possible that this work was inspired by the 1828 paper by George Green (1793–1841) on the application of mathematics to electric and magnetic phenomena. Cauchy did not prove Green's theorem; Riemann presented the first proof in his famous dissertation of 1851. Also in 1851, Cauchy noted that the Cauchy–Riemann equations implied analyticity; he assumed that the partial derivatives were continuous. We also note that Weierstrass was aware of the proof of Cauchy's integral theorem by means of Green's theorem as early as 1842. All these proofs of Cauchy's integral theorem assumed the continuity of $f'(z)$. In 1883, E. Goursat (1858–1936) retained this assumption in his discovery of a new proof of the theorem for the case of a rectangular contour. But he soon realized that he did not require continuity, so that when he republished the proof in 1900, in the first issue of the *Transactions of the American Mathematical Society*, he removed this condition.

To deal with functions with singularities inside the closed curve, Cauchy's 1814 paper had a result amounting to a particular case of the residue theorem, Theorem 6.17. But it was in his 1826 paper, *Sur un nouveau genre de calcul analogue au calcul infinitésimal*, that he formally defined a residue for a function with a pole. Note that the idea of an essential singularity was not then known. He further showed that residues could be applied to the evaluation of definite integrals. In 1830, leaving France for political reasons, Cauchy became professor of theoretical physics at the University of Turin. There he published an important paper on the power series expansions of analytic functions, significant portions of which were reprinted with some changes in his *Exercices d'analyse* of 1841. He first proved Cauchy's integral formula

$$f(x) = \frac{1}{2\pi} \int_{-\pi}^{\pi} \frac{\overline{x}\, f(\overline{x})}{\overline{x} - x} \mathrm{d}p, \tag{1.12}$$

where $\overline{x} = X\mathrm{e}^{p\sqrt{-1}}$ and $|x| < X$; note that $\overline{x}$ did not represent the complex conjugate. He began the proof by showing that

$$\int_{-\pi}^{\pi} f(\overline{x})\, \mathrm{d}p = 2\pi f(0). \tag{1.13}$$

For this purpose, Cauchy observed that $D_X f(\overline{x}) = \frac{1}{X\sqrt{-1}} D_p f(\overline{x})$; he then integrated with respect to X from 0 to X and with respect to p from $-\pi$ to π. The left-hand side simplified to

$$\int_{-\pi}^{\pi} (f(\overline{x}) - f(0))\, \mathrm{d}p = \int_{-\pi}^{\pi} f(\overline{x})\, \mathrm{d}p - 2\pi f(0),$$

and the right-hand side became zero, yielding (1.13). This in turn implied that when $f(0) = 0$ he had $\int_{-\pi}^{\pi} f(\overline{x})\, \mathrm{d}p = 0$. He then replaced $f(\overline{x})$ by $\overline{x}\,\frac{f(\overline{x})-f(x)}{\overline{x}-x}$, taking $|x| < X$, to arrive at

$$\int_{-\pi}^{\pi} \frac{\overline{x} f(\overline{x})}{\overline{x} - x}\, \mathrm{d}p = \int_{-\pi}^{\pi} \frac{\overline{x} f(x)}{\overline{x} - x}\, \mathrm{d}p$$

$$= f(x) \int_{-\pi}^{\pi} \left(1 + \frac{x}{\overline{x}} + \frac{x^2}{\overline{x}^2} + \cdots\right) \mathrm{d}p = 2\pi f(x).$$

This proved the integral formula (1.12). To obtain the Maclaurin series (a particular case of the Taylor series), he expanded $\frac{\overline{x}}{(\overline{x}-x)}$ in (1.12) as a geometric series

$$f(x) = \frac{1}{2\pi} \int_{-\pi}^{\pi} \frac{f(\overline{x})}{1 - \frac{x}{\overline{x}}}\, \mathrm{d}p = \frac{1}{2\pi} \int_{-\pi}^{\pi} f(\overline{x}) \left(1 + \frac{x}{\overline{x}} + \frac{x^2}{\overline{x}^2} + \cdots\right) \mathrm{d}p.$$

The general term in the series was given by

$$\frac{1}{2\pi} \int_{-\pi}^{\pi} x^n \frac{f(\overline{x})}{\overline{x}^n}\, \mathrm{d}p = \frac{x^n}{2\pi} \int_{-\pi}^{\pi} \frac{f(\overline{x})}{\overline{x}^n}\, \mathrm{d}p.$$

We mention that, although he could easily have done so, Cauchy did not establish the existence of the derivatives of $f'(x)$. Nevertheless, Cauchy applied repeated integration by parts to write

$$\frac{1}{2\pi} \int_{-\pi}^{\pi} \frac{f(\overline{x})}{\overline{x}^n}\, \mathrm{d}p = \frac{1}{2\pi n} \int_{-\pi}^{\pi} \frac{f'(\overline{x})}{\overline{x}^{n-1}}\, \mathrm{d}p = \frac{1}{2\pi(1 \cdot 2 \cdots n)} \int_{-\pi}^{\pi} f^{(n)}(\overline{x})\, \mathrm{d}p.$$

Note that, by (1.13), the last integral could be set equal to $\frac{f^{(n)}(0)}{n!}$; Cauchy therefore argued that the function $f(x)$ could be expanded as a Maclaurin series for those values of x whose modulus was less than the least value for which $f(x)$ was not continuously differentiable. More generally, he could have obtained the coefficient as $\frac{f^{(n)}(a)}{n!}$ by expanding the series in powers of $x - a$. This was and is known as the Taylor expansion of $f(x)$; Brook Taylor (1685–1731) proved this result for real functions in 1712 and published it 3 years later in his *Methodus Incrementorum.*

We should perhaps not be surprised, however, that the first explicit statement of the Taylor series and the particular case of the Maclaurin series were given by the father of calculus, Isaac Newton (1642–1727). He presented these results in unpublished portions of his *De quadratura curvarum* of 1691–92, in connection with series solutions of algebraic differential equations. We must also note that in a 1671 letter to John Collins, James Gregory (1638–1675) wrote down power series expansions of functions such as $\tan x$ and $\sec x$ by computing the derivatives of these functions. This implies that he was aware of the Taylor expansion 40 years before Taylor.

The Laurent series given in Theorem 6.1, a generalization of the Taylor series, was derived by P. A. Laurent (1813–1854), a French military engineer. In 1843, he published a short note on the series expansion of a function analytic in an annulus. He generalized Cauchy's integral formula to the annulus and deduced that the function had a series expansion, provided that infinitely many terms with negative powers of $x - a$ were included. Cauchy reported to the French Academy on this work, calling Laurent's result a new theorem but mentioning that it could be deduced from his own 1840 work on the mean values of a function.

Surprisingly, Karl Weierstrass (1815–1897) had already proved Laurent's theorem in his 1841 paper, *Darstellung einer analytischen Function einer complexen Veränderlichen, deren absoluter Betrag zwischen zwei gegebenen Grenzen liegt.* In this paper, Weierstrass expanded a function $F(x)$, for x inside an annulus around the origin, into a series of the form

$$A_0 + A_1 x + \cdots + A_\nu x^\nu + \cdots + A_{-1} x^{-1} + \cdots + A_{-\nu} x^{-\nu} + \cdots = \sum_{\nu=-\infty}^{\infty} A_\nu x^\nu .$$

In addition to the conditions of continuity and finiteness, the function was required to satisfy the condition that the difference of the two quotients

$$\frac{F(x+hk) - F(x)}{hk} \quad \text{and} \quad \frac{F(x+k) - F(x)}{k},$$

where x was in the annulus and h was any complex number, became infinitely small when k became infinitely small. This condition would be satisfied by the class of continuously differentiable functions. Thus, a function analytic in the annulus had a Laurent series expansion. Weierstrass expressed the coefficient A_ν as an integral around a circle in the annulus with center at the origin. From this he deduced that if $|F(x)| \leq M$ on $|x| = a$, then $|A_\nu| \leq Ma^{-\nu}$. This implied that if F remained bounded in a neighborhood of the origin, then $|A_{-\nu}| \leq Ma^\nu$. He then let $a \to 0$ to derive $A_{-\nu} = 0$ so that

$$F(x) = A_0 + A_1 x + A_2 x^2 + \cdots .$$

Weierstrass's paper containing this theorem was not published until 50 years later when it appeared in his collected works. It is clear, however, that he did his work independently, since he became aware of Cauchy's results much later. It seems that the Laurent expansion was never mentioned in Weierstrass's lectures. Weierstrass may have been dissatisfied with his proof of the Laurent expansion because it employed integration, whereas he regarded power series as conceptually more fundamental than differentiation or integration. Weierstrass's 1876 proof of the Casorati–Weierstrass theorem (Theorem 6.8) did not employ Cauchy's integral formula; the proof was not too elegant but very interesting. Weierstrass's valuable insight was to use power series to define the analyticity of a function; this perspective became especially valuable when analysts began to take p-adic or other noncomplex numbers as the underlying field of constants. We note that the Casorati–Weierstrass theorem was first published in 1868 by F. Casorati (1835–1890), who used integration in an annulus. However, Schwarz reported to Casorati that he had been present when Weierstrass discussed this theorem in his 1863 Berlin lectures. Then again, in 1868, the Russian complex analyst Y. Sokhotskii (1842–1927) employed Cauchy's integral formula to independently prove the Casorati–Weierstrass theorem in his master's thesis at St. Petersburg University; thus, the theorem is also named the Casorati–Sokhotskii–Weierstrass theorem.

Bernhard Riemann (1826–1866) took the Cauchy–Riemann equations as the basis of his theory of analytic functions. He was primarily interested in the geometric properties of these functions, that is, in their orientation-preserving and conformal (or angle-preserving) properties. In his 1851 dissertation, Riemann defined simple connectivity of a domain and outlined a proof of the proposition that every simply connected proper subdomain of the complex plane could be biholomorphically (conformally) mapped onto the unit disc. This is the RMT, Theorem 8.20. The proof was not quite correct and also depended upon the unproven Dirichlet principle. It required the efforts of such outstanding mathematicians as H. A. Schwarz, H. Poincaré, D. Hilbert, P. Koebe, and C. Carathéodory to put Riemann's ideas and intuitions on a solid basis. Poincaré famously remarked that Riemann's method was a method of discovery, whereas that of Weierstrass was a method of proof.

Schwarz gave a proof of the RMT in his 1869 paper, *Zur Theorie der Abbildung*. The paper begins with a proof of Schwarz's lemma for a biholomorphic function f, mapping the unit disc onto a region U and satisfying $f(0) = 0$. This lemma was then forgotten until it was rediscovered in 1905 by Carathéodory, who applied it to give a short proof of Landau's extension of Picard's Theorem 6.9. Landau's extension states that if

$$f(z) = a_0 + a_1 z + a_z z^2 + \cdots, \quad a_1 \neq 0$$

is holomorphic and does not take the values 0 and 1 in $|z| < R$, then R is bounded by a number that depends on the first two coefficients, a_0 and a_1. The result as given in Theorem 5.34 is due to Carathéodory. He reported that his initial unpublished proof of this theorem depended on the Poisson integral formula (9.3). But he showed this work to Schmidt, famous for his fundamental contributions to functional analysis, who provided a superior argument. This proof, utilizing the maximum principle (Corollary 5.31), is presented in this and other modern texts. Carathéodory discovered other applications of this lemma and in 1912 he named it after Schwarz.

Morera's theorem, contained in the fundamental theorem, was first proved in a paper of 1886, *Un teorema fondamentale nella teo rica delle funzioni di una variabile complessa*. The Italian mathematician, G. Morera (1856–1909), supplied another way to define analyticity; he proved that any locally integrable complex function was analytic. This is also known as the converse of Cauchy's theorem. In another paper of 1886, *Sulla rappresentazione delle funzioni di una variabile com plessa per mezzo di espressioni analitiche infinite*, Morera used his theorem to reprove Weierstrass's convergence theorem for compactly convergent series, established in his *Zur Functionenlehre* of 1880.

Chapter 2
Foundations

The first section of this chapter introduces the complex plane, fixes notation, and discusses some useful concepts from real analysis. Some readers may initially choose to skim this section. The second section contains the definition and elementary properties of the class of holomorphic functions—the basic object of our study.

2.1 Introduction and Preliminaries

This section is a summary of basic notation, a description of some of the basic algebraic and geometric properties of the complex number system, and a disjoint collection of needed facts from real analysis (advanced calculus). We remind the reader of some of the formalities behind the standard notation which we usually approach quite informally. Not all concepts used as prerequisites are defined (among these are neighborhood, connected, path-connected, arc-wise connected, and compact sets); we assume that the reader has been exposed to them.[1]

We start with some standard **notation**:

$$\mathbb{Z}_{>0} \subset \mathbb{Z} \subset \mathbb{Q} \subset \mathbb{R} \subset \mathbb{C} \subset \widehat{\mathbb{C}}.$$

Here $\mathbb{Z}$ represents the integers, $\mathbb{Z}_{>0}$ the positive integers,[2] $\mathbb{Q}$ the rationals (the integer n is included in the rationals as the equivalence class of the quotient $\frac{n}{1}$), and $\mathbb{R}$ the reals. Whether one views the reals as the completion of the rationals or identifies

[1]The reader may want to consult J. R. Munkres *Topology (Second Edition)*, Dover, 2000, or J. L. Kelley, *General Topology*, Springer-Verlag, 1975 as well as definitions in Chap. 4.

[2]In general $X_{\text{condition}}$ and $\{x \in X; \text{condition}\}$ will describe the set of all x in X that satisfy the indicated condition.

R.E. Rodríguez et al., *Complex Analysis: In the Spirit of Lipman Bers*, Graduate Texts in Mathematics 245, DOI 10.1007/978-1-4419-7323-8_2,

them with Dedekind cuts (we will not use these concepts explicitly), their most important property from the perspective of complex variables is the *least upper bound property*; that is, that every nonempty set of real numbers that has an upper bound has a least upper bound.

The inclusion of $\mathbb{R}$ into the complex numbers $\mathbb{C}$ needs a bit more explanation. It is specified as follows: for z in $\mathbb{C}$, we write $z = x + \imath\, y$ with x and y in $\mathbb{R}$, where the symbol $\imath$ represents a square root of -1; that is, $\imath^2 = -1$. With these conventions we can define addition and multiplication of complex numbers using the usual rules for these operations on the reals[3]: for all $x, y, \xi, \eta \in \mathbb{R}$,

$$(x + \imath y) + (\xi + \imath \eta) = (x + \xi) + \imath(y + \eta)$$

and

$$(x + \imath y)(\xi + \imath \eta) = (x\xi - y\eta) + \imath(x\eta + y\xi).$$

The *real numbers*, $\mathbb{R}$, are identified with the subset of $\mathbb{C}$ consisting of those numbers $z = x + \imath\, y$ with $y = 0$; the *imaginary* numbers, $\imath\mathbb{R}$, are those with $x = 0$. For $z = x + \imath\, y$ in $\mathbb{C}$ with x and y in $\mathbb{R}$ we write $x = \Re z$, the *real part* of z, and $y = \Im z$, the *imaginary part* of z. Geometrically, $\mathbb{R}$ and $\imath\mathbb{R}$ represent the *real* and *imaginary axes* of $\mathbb{C}$, viewed as the complex plane and identified with the cartesian product $\mathbb{R}^2 = \mathbb{R} \times \mathbb{R}$ (see Fig. 2.1).

The complex plane may be viewed as a subset of the *complex sphere* $\widehat{\mathbb{C}}$, which is $\mathbb{C}$ compactified by adjoining a point, known as *the point at infinity*, so that $\widehat{\mathbb{C}} = \mathbb{C} \cup \{\infty\}$. The space $\widehat{\mathbb{C}}$ is also called *the extended complex plane* or the *Riemann sphere*. This last name comes from identifying the points on the unit sphere in $\mathbb{R}^3$, with the exception of the *south pole*, the point $(0, 0, 1)$, with the points in the complex plane under what is known as stereographic projection; the point $(0, 0, 1)$ is identified with ∞. See Exercise 3.20 for the details.

We now describe some basic algebraic and geometric properties of the complex numbers.

For $z = x + \imath\, y$, with x and y real numbers, the complex number

$$\overline{z} = x - \imath\, y$$

is called the *complex conjugate* of z. Note that then

$$\Re z = \frac{z + \overline{z}}{2} \text{ and } \Im z = \frac{z - \overline{z}}{2\imath}.$$

One easily verifies the following basic

[3]With these operations $(\mathbb{C}, +, \cdot)$ is a field.

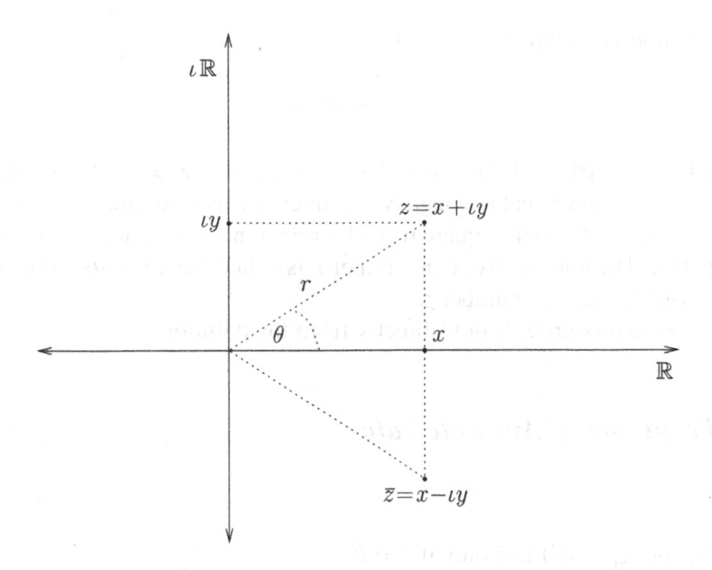

Fig. 2.1 The complex plane; rectangular and polar representations, conjugation

2.1.1 Properties of Conjugation

For z and $w \in \mathbb{C}$,

(a) $\overline{z+w} = \overline{z} + \overline{w}$
(b) $\overline{zw} = \overline{z}\,\overline{w}$
(c) $\overline{z} = z$ if and only if $z \in \mathbb{R}$
(d) $\overline{\overline{z}} = z$

There is a simple and useful **geometric interpretation of conjugation**: it is represented by mirror reflection in the real axis; see Fig. 2.1.

From a slightly different point of view, conjugation may be seen as a self-map of $\mathbb{C}$, denoted by $^{-}$.

$$^{-} : \mathbb{C} \to \mathbb{C}.$$

Then its properties (a) through (d) may be restated as follows:

(a) $^{-}$ preserves the sum of complex numbers.
(b) $^{-}$ preserves the product of complex numbers.
(c) $^{-}$ fixes precisely the real numbers.
(d) $^{-}$ is an involution of $\mathbb{C}$; that is, when composed with itself, it gives the identity map on $\mathbb{C}$.

It is not hard to show that any self-map of $\mathbb{C}$ satisfying these properties coincides with complex conjugation; see Exercise 2.19.

Another important map, $z \mapsto |z|$ or

$$| \, | : \mathbb{C} \to \mathbb{R}_{\geq 0}$$

is defined by $r = |z| = (z\bar{z})^{\frac{1}{2}} = (x^2 + y^2)^{\frac{1}{2}}$, where $z = x + \imath\, y$. Note that $z\bar{z} = x^2 + y^2$ is always a real nonnegative number; we use the usual convention that unless otherwise specified the square root of a real nonnegative number is chosen to be nonnegative. The nonnegative real number r is called the *absolute value* or *norm* or *modulus* of the complex number z.

The following properties follow directly from the definition.

2.1.2 Properties of Absolute Value

For z and $w \in \mathbb{C}$,

(a) $|z| \geq 0$, and $|z| = 0$ if and only if $z = 0$
(b) $|zw| = |z|\,|w|$
(c) $|\bar{z}| = |z|$
(d) $|\Re z| \leq |z|$, and $\Re z = |z|$ if and only if $z = x \in \mathbb{R}_{\geq 0}$
(e) $|\Im z| \leq |z|$, and $\Im z = |z|$ if and only if $z = \imath\, y$ with $y \in \mathbb{R}_{\geq 0}$

2.1.3 Linear Representation of $\mathbb{C}$

As a vector space over $\mathbb{R}$, we can identify $\mathbb{C}$ with $\mathbb{R}^2$. Vector addition agrees with complex addition, and scalar multiplication by real numbers ($\mathbb{R} \times \mathbb{C} \to \mathbb{C}$) is the restriction of complex multiplication ($\mathbb{C} \times \mathbb{C} \to \mathbb{C}$).

This identification provides a powerful geometric interpretation for many results on complex numbers. One example is provided by conjugation, which can be viewed as the $\mathbb{R}$-linear map of $\mathbb{R}^2$ (with basis $1 = (1, 0)$ and $\imath = (0, 1)$) that sends 1 to 1 and $\imath$ to $-\imath$. Another instance is provided by the next geometric interpretation of the following:

2.1.4 Additional Properties of Absolute Value

From the Pythagorean equality, $r = |z|$ is the (Euclidean) distance in the plane from z to the origin; see Fig. 2.1.

Furthermore, for z and $w \in \mathbb{C}$, the following properties hold.

(f) $|z + w|^2 + |z - w|^2 = 2(|z|^2 + |w|^2)$.
(g) $|z + w| \leq |z| + |w|$. Equality holds whenever either z or w is equal to 0. If $z \neq 0$ and $w \neq 0$, then equality holds if and only if $w = az$ with $a \in \mathbb{R}_{>0}$.
(h) $||z| - |w|| \leq |z - w|$.

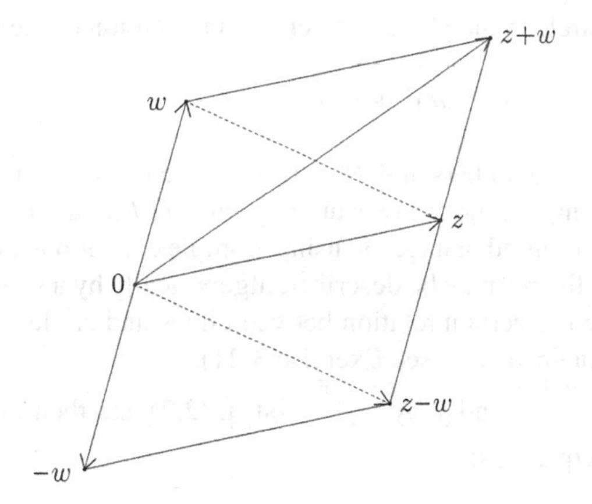

Fig. 2.2 Vector sums

Equality (f) is sometimes called the *parallelogram law*: the sum of the squares of the lengths of the diagonals in a parallelogram is equal to the sum of the squares of the lengths of its sides, see Fig. 2.2. This equality can be proven directly from the definition of absolute value and properties of the complex conjugation we have already stated:

$$|z \pm w|^2 = (z \pm w)(\overline{z} \pm \overline{w}) = |z|^2 \pm 2\,\Re(z\,\overline{w}) + |w|^2 . \tag{2.1}$$

Inequality (g) is called the *triangle inequality*: the length of a side of a triangle is at most equal to the sum of the lengths of the other two sides, with equality if and only if the triangle is degenerate (z and w lie on the same ray); see Fig. 2.2.

The triangle inequality (or rather its squared version) follows from (2.1) and the previous properties of the absolute value, by observing that $\Re(z\,\overline{w}) \leq |z\,\overline{w}| = |z|\;|w|$ and using the conditions for equality given in property (d) for the absolute value.

Through the identification of $\mathbb{R}^2$ with $\mathbb{C}$ given above, we can use real or complex notation to describe geometric shapes in the plane. As we show next, sometimes the use of complex notation simplifies the description of the objects under study.

2.1.5 Lines, Circles, and Half Planes

Any line in the plane $\mathbb{R}^2$ with orthogonal coordinates $x$and y is given by an equation of the form

$$a\,x + b\,y + c = 0, \tag{2.2}$$

with a, b, and c real numbers, and a and b not both equal to zero.

Similarly, any circle in the plane is given by an equation of the form

$$(x-d)^2+(y-f)^2-R^2=0, \tag{2.3}$$

with d, f, and R real numbers and $R>0$. In this case, we can read off from the equation that the center of the circle is at the point (d, f), and its radius is R.

We will now see an advantage of using complex notation: both of the above types of geometric figures may be described algebraically by a single equation, thus implying that there is a certain relation between lines and circles on the plane (this relation will be explained later: see Exercise 3.21).

Replacing x by $\dfrac{z+\overline{z}}{2}$ and y by $\dfrac{z-\overline{z}}{2\imath}$ first in (2.2) and then in (2.3), we obtain the following two equations:

$$B\,z+\overline{B}\,\overline{z}+c=0\,,\ \text{ with } B=\frac{a-\imath\,b}{2}\neq 0, \tag{2.4}$$

and

$$|z|^2+(-d+\imath\,f)\,z+(-d-\imath\,f)\,\overline{z}+d^2+f^2-R^2=0, \tag{2.5}$$

or, equivalently,

$$|z-E|=R\ \text{ with } E=d+\imath\,f. \tag{2.6}$$

We claim that both equations (2.4) and (2.5) are special cases of

$$A\,|z|^2+B\,z+\overline{B}\,\overline{z}+C=0\,, \tag{2.7}$$

with A and C real numbers, B complex, $A\geq 0$, and $|B|^2>AC$.

Indeed, if $A=0$ then (2.7) becomes (2.4), which is equivalent to (2.2), whereas if $A>0$ then (2.7) becomes (2.5), which is equivalent to (2.3) with center $E=-\dfrac{\overline{B}}{A}$ and radius $R=\dfrac{\sqrt{|B|^2-AC}}{A}$.

We have thus shown that any circle or line in the plane is given by Equation (2.7), depending on whether $A>0$ or $A=0$.

Similarly, half planes in $\mathbb{C}$ are given by equations of the form

$$\Re(B\,z)>C\ \text{ or, equivalently, } \Im(B\,z)>C,$$

with B in $\mathbb{C}_{\neq 0}$ and C real.

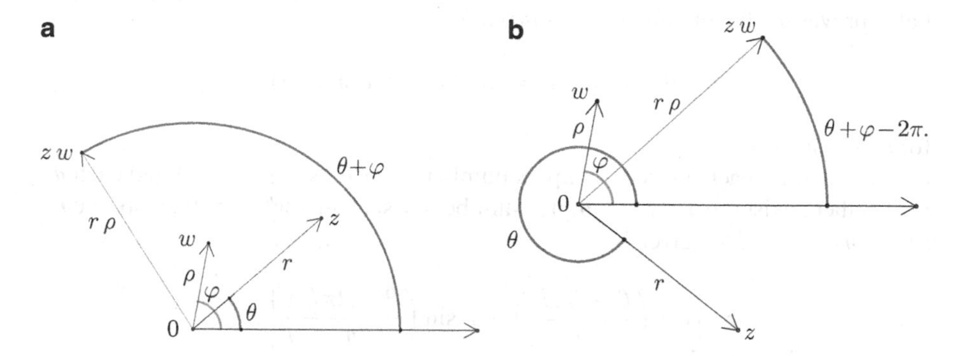

Fig. 2.3 Vector multiplication. (**a**) Sum of arguments smaller than 2π. (**b**) Sum of arguments larger than 2π

2.1.6 Polar Coordinates

A nonzero vector in $\mathbb{C}$ can be described by polar coordinates (r, θ) as well as by the rectangular coordinates (x, y) we have been using. If $z \in \mathbb{C}$ and $z \neq 0$, then we can write

$$z = x + \imath\, y = r\,(\cos\theta + \imath\, \sin\theta)\,,$$

where $r = |z|$ and $\theta = \arg z$ (an *argument* of z) $= \arcsin\dfrac{y}{r} = \arccos\dfrac{x}{r}$.

Note that the last two identities are needed to define the argument and that $\arg z$ is defined up to addition of an integral multiple of 2π. This is why we labeled θ *an* argument of z as opposed to *the* argument.[4]

If $w = \rho[\cos\varphi + \imath\, \sin\varphi]$ is another nonzero complex number, then, using the usual addition formulas for the sine and cosine functions, we have

$$zw = (r\rho)[\cos(\theta + \varphi) + \imath\, \sin(\theta + \varphi)].$$

This polar form of the multiplication formula shows that the complex multiplication of two (nonzero) complex numbers is equivalent to the real multiplication of their moduli and the addition of their arguments, giving a geometric interpretation of how the operation of multiplication acts on vectors represented in polar coordinates; see Fig. 2.3. It also shows (again) that $|z\,w| = |z|\,|w|$. Polar coordinates also provide another way to view Fig. 2.2.

In particular, it follows that if $n \in \mathbb{Z}$ and $z = r\,(\cos\theta + \imath\, \sin\theta)$ is a nonzero complex number, then

$$z^n = r^n[\cos n\theta + \imath\, \sin n\theta];$$

---

[4]The number π will be defined rigorously in Definition 3.34. Trigonometric functions will be introduced in the next chapter where some of their properties, including addition formulae, will be developed. For the moment, polar coordinates should not be used in proofs.

it also proves the famous *de Moivre's formula*:

$$(\cos\theta + \imath\ \sin\theta)^n = \cos(n\,\theta) + \imath\ \sin(n\,\theta)$$

for $n \in \mathbb{Z}$ and $\theta \in \mathbb{R}$.

Therefore, for each nonzero complex number $z = r\,(\cos\theta + \imath\ \sin\theta)$ and each n in $\mathbb{Z}_{>0}$, there exist precisely n complex numbers w such that $w^n = z$; they are the n n-th *roots* of z, and are given by

$$r^{\frac{1}{n}}\left[\cos\left(\frac{\theta + 2\pi k}{n}\right) + \imath\ \sin\left(\frac{\theta + 2\pi k}{n}\right)\right],$$

with $k = 0, 1, \ldots, n-1$.

Note that these n complex numbers are the vertices of a regular n-gon in the plane.

2.1.7 Coordinates on $\mathbb{C}$

We have already seen that we can use three sets of coordinates on $\mathbb{C}$, as follows.

1. Rectangular (x, y): Each equation x = constant (respectively y = constant) yields a line parallel to the imaginary axis (respectively real axis), while the equation $x = y$ yields the line through the origin with slope equal to 1.
2. Complex $(z, \bar{z})$: Only one of these coordinates is needed to describe a point by the equation z = constant (or $\bar{z}$ = constant), while the equation $z = \bar{z}$ yields the real axis.
3. Polar (r, θ): The equation $r = a$ with $a > 0$ is a circle of radius a centered at the origin, whereas the equation θ = constant is a ray emanating from (but not including) the origin. The equation $r = \theta$ denotes a type of spiral ending at (but not passing through) the origin.

The choice of the appropriate one among the various possible coordinates on $\mathbb{C}$ may simplify a problem. As an example we solve the following one.

Let n be a positive integer, and suppose we want to find the set of points z in $\mathbb{C}$ that satisfy the equation

$$z^n = \bar{z}^n. \tag{2.8}$$

Using rectangular coordinates would lead us to solve

$$(x + \imath\ y)^n = (x - \imath\ y)^n,$$

which is doable but far from pleasant.

Instead, we first note that certainly $z = 0$ satisfies (2.8). For $z \neq 0$, we may use the polar coordinates: the equation we are trying to solve is then equivalent to

$$r^n[\cos n\theta + \imath \sin n\theta] = r^n[\cos n\theta - \imath \sin n\theta],$$

which implies that $n\,\theta = k\,\pi$ for some integer k. Thus we immediately see that the complete solution to (2.8) is the set of $2n$ rays $\theta = \theta_o$ from the origin (including the origin) with

$$\theta_o \in \left\{0\,, \frac{\pi}{n}\,, \frac{2\pi}{n}\,, \ldots, \frac{(2n-1)\,\pi}{n}\right\}.$$

2.2 More Preliminaries that Rely on Topology, Metrics, and Sequences

We collect some facts on sets of complex numbers and functions defined on them, that mostly follow from translating to the complex system the analogous results from real analysis.

The formula $d(z, w) = |z - w|$, for z and $w \in \mathbb{C}$, defines a **metric on** $\mathbb{C}$. Thus $(\mathbb{C}, d)$ is a metric space, with a metric that agrees with the Euclidean metric on $\mathbb{R}^2$ (under the linear representation of the complex plane described earlier).

Definition 2.1. We say that a sequence (indexed by $n \in \mathbb{Z}_{>0}$) $\{z_n\}$ of complex numbers *converges* to $\alpha \in \mathbb{C}$ if given $\epsilon > 0$, there exists an $N \in \mathbb{Z}_{>0}$ such that $|z_n - \alpha| < \epsilon$ for all $n > N$; in this case we write

$$\lim_{n\to\infty} z_n = \alpha.$$

A sequence $\{z_n\}$ of complex numbers is called *Cauchy* if given $\epsilon > 0$, there exists an $N \in \mathbb{Z}_{>0}$ such that $|z_n - z_m| < \epsilon$ for all $n, m > N$.

Theorem 2.2. *If $\{z_n\}$ and $\{w_n\}$ are Cauchy sequences of complex numbers, then*

(a) *$\{z_n + \alpha\, w_n\}$ is Cauchy for all $\alpha \in \mathbb{C}$.*
(b) *$\{\overline{z}_n\}$ is Cauchy.*
(c) *$\{\,|z_n|\,\} \subset \mathbb{R}_{\geq 0}$ is Cauchy.*

Proof. (a) It suffices to assume that $\alpha \neq 0$. Given $\epsilon > 0$, choose N_1 such that $|z_n - z_m| < \frac{\epsilon}{2}$ for all $n, m > N_1$ and choose N_2 such that $|w_n - w_m| < \frac{\epsilon}{2|\alpha|}$ for all $n, m > N_2$. Choose $N = \max\{N_1, N_2\}$. Then, for all n and $m > N$, we have

$$|(z_n + \alpha\, w_n) - (z_m + \alpha\, w_m)| \leq |z_n - z_m| + |\alpha|\,|w_n - w_m| < \epsilon.$$

(b) It follows directly from $|\overline{z}_n - \overline{z}_m| = |\overline{z_n - z_m}| = |z_n - z_m|$.
(c) We know that for all z and w in $\mathbb{C}$ we have

$$||z| - |w|| \leq |z - w|.$$

Applying this inequality to z_n and z_m in the sequence, we obtain

$$||z_n| - |z_m|| \leq |z_n - z_m|,$$

and the result follows.

□

Remark 2.3. The above arguments mimic arguments in real analysis needed to establish the corresponding results for real sequences. We will, in the sequel, leave such routine arguments as exercises for the reader.

Corollary 2.4. *$\{z_n\}$ is a Cauchy sequence of complex numbers if and only if $\{\Re z_n\}$ and $\{\Im z_n\}$ are Cauchy sequences of real numbers.*

Corollary 2.5. *$(\mathbb{C}, d)$ is a complete metric space; that is, every Cauchy sequence of complex numbers converges to a complex number.*

Proof. Observe that the metric on $\mathbb{C}$ restricts to the Euclidean metric on $\mathbb{R}$, which is complete, and applies the previous corollary. □

Definition 2.6. Let $A \subseteq \mathbb{C}$. We say that A is *bounded* if the set of nonnegative real numbers $\{\,|z|\,;\ z \in A\}$ is; that is, if there exists a positive real number M such that $|z| < M$ for all z in A.

Definition 2.7. Let $c \in \mathbb{C}$ and $\epsilon > 0$. The *ϵ-ball about c*, or the *open disc* with center c and radius ϵ, is the set

$$U_c(\epsilon) = U(c, \epsilon) = \{z \in \mathbb{C}; |z - c| < \epsilon\},$$

that is, the interior of the circle with center c and radius ϵ.

Proposition 2.8. *A subset A of $\mathbb{C}$ is bounded if and only if there exist a complex number c and a positive number R such that*

$$A \subset U(c, R).$$

Remark 2.9. A proof is omitted for one of three reasons (in addition to the reason described in Remark 2.3): either it is trivial or it follows directly from results in real analysis or it appears as an exercise at the end of the corresponding chapter.[5] The third possibility is always labeled as such; when standard results in real analysis are needed, there is some indication of what they are or where to find them. For example, the next two theorems are translations to $\mathbb{C}$ of standard metric results for $\mathbb{R}^2$. It

[5]Exercises can be found at the end of each chapter and are numbered by chapter, so that Exercise 2.7 is to be found at the end of Chap. 2.

should be clear from the context when the first possibility occurs. It is recommended that the reader ensures that he/she is able to supply an appropriate proof when none is given.

Theorem 2.10 (Bolzano–Weierstrass). *Every bounded infinite set S in $\mathbb{C}$ has at least one limit point; that is, there exists at least one $c \in \mathbb{C}$ such that, for each $\epsilon > 0$, the ball $U(c, \epsilon)$ contains a point $z \in S$ with $z \neq c$.*

Theorem 2.11. *A set $K \subset \mathbb{C}$ is compact if and only if it is closed and bounded.*

We will certainly be using a number of consequences of compactness not discussed in this chapter (e.g., in a compact metric space, every sequence has a convergent subsequence) and also of connectedness, which we will not define here.

Definition 2.12. Let f be a function defined on a set S in $\mathbb{C}$. We assume that f is complex-valued, unless otherwise stated. Thus f may be viewed as either a map from S into $\mathbb{R}^2$ or into $\mathbb{C}$ and also as two real-valued functions defined on the set S.

Let c be a limit point of S and let α be a complex number. We say that the *limit of f at c* is α, and we write

$$\lim_{z \to c} f(z) = \alpha$$

if for each $\epsilon > 0$ there exists a $\delta > 0$ such that

$$|f(z) - \alpha| < \epsilon \text{ whenever } z \in S \text{ and } 0 < |z - c| < \delta.$$

Remark 2.13. The condition that c is a limit point of S ensures that there are points z in S arbitrarily close to (but different from) c so that $f(z)$ is defined there. Note that it is not required that $f(c)$ be defined.

The above definition is again a translation of language from $\mathbb{R}^2$ to $\mathbb{C}$. Thus we will be able to adopt many results (the next three theorems, in particular) from real analysis. In addition to the usual algebraic operations on pairs of functions $f : S \to \mathbb{C}$ and $g : S \to \mathbb{C}$ familiar from real analysis, such as $f + cg$ with $c \in \mathbb{C}$, fg, and $\frac{f}{g}$ (provided g does not *vanish* on S; that is, if $g(z) \neq 0$ for any $z \in S$ or, equivalently, if no $z \in S$ is a *zero* of g), we will consider other functions constructed from a single function f, that are usually not emphasized in real analysis. Among them are the following:

$$(\Re f)(z) = \Re f(z), \ (\Im f)(z) = \Im f(z), \ \overline{f}(z) = \overline{f(z)}, \ |f|(z) = |f(z)|,$$

also defined on S.

For instance, if $f(z) = z^2 = x^2 - y^2 + 2\imath\, x\, y$ for $z \in \mathbb{C}$, we have $(\Re f)(z) = x^2 - y^2$, $(\Im f)(z) = 2\, x\, y$, $\overline{f}(z) = \bar{z}^2 = x^2 - y^2 - 2\imath\, x\, y$, and $|f|(z) = |z|^2 = x^2 + y^2$ for $z \in \mathbb{C}$.

Theorem 2.14. *Let S be a subset of $\mathbb{C}$ and let f and g be functions defined on S. If c is a limit point of S, then:*

(a) $\lim_{z \to c} (f + a\, g)(z) = \lim_{z \to c} f(z) + a \lim_{z \to c} g(z)$ *for all* $a \in \mathbb{C}$
(b) $\lim_{z \to c} (fg)(z) = \lim_{z \to c} f(z) \lim_{z \to c} g(z)$
(c) $\lim_{z \to c} |f|(z) = \left| \lim_{z \to c} f(z) \right|$
(d) $\lim_{z \to c} \overline{f}(z) = \overline{\lim_{z \to c} f(z)}$

Remark 2.15. The usual interpretation of the above formulae is used here and in the rest of the book: the LHS[6] exists whenever the RHS exists, and then we have the stated equality.

Corollary 2.16. *Let S be a subset of $\mathbb{C}$, let f be a function defined on S, and $\alpha \in \mathbb{C}$. Set $u = \Re f$ and $v = \Im f$ (so that $f(z) = u(z) + \imath\, v(z)$). If c is a limit point of S, then*

$$\lim_{z \to c} f(z) = \alpha$$

if and only if

$$\lim_{z \to c} u(z) = \Re\alpha \quad \text{and} \quad \lim_{z \to c} v(z) = \Im\alpha.$$

Definition 2.17. Let S be a subset of $\mathbb{C}$, $f : S \to \mathbb{C}$ be a function defined on S, and $c \in S$ be a point in S. We say that:

(a) f is *continuous at c* if $\lim_{z \to c} f(z) = f(c)$.
(b) f is *continuous on S* if it is continuous at each c in S.
(c) f is *uniformly continuous on S* if for all $\epsilon > 0$, there is a $\delta > 0$ such that

$$|f(z) - f(w)| < \epsilon \text{ for all } z \text{ and } w \text{ in } S \text{ with } |z - w| < \delta.$$

Remark 2.18. A function f is (uniformly) continuous on S if and only if both $\Re f$ and $\Im f$ are.

Uniform continuity implies continuity, but the converse is not true in general.

Theorem 2.19. *Let f and g be functions defined in appropriate sets, that is, sets where the composition $g \circ f$ of these functions makes sense. Then the following properties hold:*

(a) *If f is continuous at c and $f(c) \neq 0$, then $\frac{1}{f}$ is defined in a neighborhood of c and is continuous at c.*
(b) *If f is continuous at c and g is continuous at $f(c)$, then $g \circ f$ is continuous at c.*

Theorem 2.20. *Let $K \subset \mathbb{C}$ be a compact set and $f : K \to \mathbb{C}$ be a continuous function on K. Then f is uniformly continuous on K.*

---

[6]LHS (RHS) are standard abbreviations for left (right) hand side and will be used throughout this book.

Proof. A continuous mapping from a compact metric space to a metric space is uniformly continuous. □

Definition 2.21. Given a sequence of functions $\{f_n\}$, all defined on the same set S in $\mathbb{C}$, we say that $\{f_n\}$ *converges uniformly* to a function f on S if for all $\epsilon > 0$ there exists an $N \in \mathbb{Z}_{>0}$ such that

$$|f(z) - f_n(z)| < \epsilon \text{ for all } z \in S \text{ and all } n > N.$$

Remark 2.22. $\{f_n\}$ converges uniformly on S (to some function f) if and only if for all $\epsilon > 0$ there exists an $N \in \mathbb{Z}_{>0}$ such that

$$|f_n(z) - f_m(z)| < \epsilon \text{ for all } z \in S \text{ and all } n \text{ and } m > N.$$

Note that in this case the limit function f is uniquely determined; it is the pointwise limit $f(z) = \lim_{n\to\infty} f_n(z)$, for all $z \in S$.

Theorem 2.23. *Let $\{f_n\}$ be a sequence of functions defined on $S \subseteq \mathbb{C}$. If:*

(1) *$\{f_n\}$ converges uniformly on S.*
(2) *Each f_n is continuous on S.*

Then the function f defined by

$$f(z) = \lim_{n\to\infty} f_n(z), \; z \in S$$

is continuous on S.

Proof. Start with two points z and c in S. Then for each natural number n we have

$$|f(z) - f(c)| \le |f(z) - f_n(z)| + |f_n(z) - f_n(c)| + |f_n(c) - f(c)|.$$

Now fix $\epsilon > 0$. By (1), the first and third term on the right-hand side are less than $\frac{\epsilon}{3}$ for n large. If we now fix c and n, it follows from (2) that the second term is less than $\frac{\epsilon}{3}$ as soon as z is close enough to c. Thus f is continuous at c. □

Definition 2.24. A *domain* or *region* in $\mathbb{C}$ is a subset of $\mathbb{C}$ which is open and connected.

Remark 2.25. Note that a domain in $\mathbb{C}$ could also be defined as an open arcwise connected subset of $\mathbb{C}$. (See also Exercise 2.20.) Also note that each point in a domain D is a limit point of D, and therefore it makes sense to ask, at each point in D, about the limit of any function defined on D.

2.3 Differentiability and Holomorphic Mappings

Up to now, the complex numbers were used mainly to supply us with a convenient alternative notation. This is about to change. The definition of the derivative of a complex-valued function of a complex variable mimics that for the derivative of a real-valued function of a real variable. However, we shall see shortly that the properties of the two classes of functions are quite different.

Definition 2.26. Let f be a function defined in some disc about $c \in \mathbb{C}$. We say that f is *(complex) differentiable at* c provided

$$\lim_{h \to 0} \frac{f(c+h) - f(c)}{h} \tag{2.9}$$

exists. In this case the limit is denoted by

$$f'(c), \ \frac{\mathrm{d}f}{\mathrm{d}z}(c), \ \left.\frac{\mathrm{d}f}{\mathrm{d}z}\right|_{z=c}, \text{ or } (Df)(c),$$

and is called the *derivative of* f *at* c.

Remark 2.27. (1) It is important that h be an arbitrary *complex* number (of small nonzero modulus) in the above definition.

(2) Note that

$$\lim_{h \to 0} \frac{f(c+h) - f(c)}{h} = \lim_{z \to c} \frac{f(z) - f(c)}{z - c}.$$

(3) If f is differentiable at c, then f is continuous at c. The converse is not true in general; see Example 2.32.4.

(4) We consider two identities for a function f defined in a neighborhood of $c \in \mathbb{C}$:

$$f(c+h) = a_0 + \epsilon(h) \text{ with } \lim_{h \to 0} \epsilon(h) = 0,$$

and

$$f(c+h) = a_0 + a_1 h + h\epsilon(h) \text{ with } \lim_{h \to 0} \epsilon(h) = 0.$$

As in real analysis, the first of these says that f is continuous at c if and only if $f(c) = a_0$; the second says that f is differentiable at c if and only if $f(c) = a_0$ and $f'(c) = a_1$. Whereas in the real case the second statement is sharp with regard to smoothness, we shall see that in the complex case, under appropriate conditions, it can be improved significantly.

Notation 2.28. If the function f is differentiable on a domain D (i.e., at each point of D), then it defines a function $f' : D \to \mathbb{C}$.

Thus for every $n \in \mathbb{Z}_{\geq 0}$ we can define inductively $f^{(n)}$, the n-th derivative of f, as follows:

$f^{(0)} = f$, and if $f^{(n)}$ is defined for $n \geq 0$, then we set $f^{(n+1)} = \left(f^{(n)}\right)'$ whenever the appropriate limits exist.

It is customary to abbreviate $f^{(2)}$ and $f^{(3)}$ by f'' and f''', respectively. Of course, $f^{(1)} = f'$.

Definition 2.29. Let f be a function defined in a neighborhood of $c \in \mathbb{C}$. Then f is *holomorphic* or *analytic at* c if it is differentiable in a neighborhood (perhaps smaller) of c. A function defined on an open set U is *holomorphic* or *analytic on* U if it is holomorphic (equivalently, differentiable) at each point of U. It should be emphasized that holomorphicity is always defined on open sets.

A function f is called *anti-holomorphic* if $\bar{f}$ is holomorphic.

The usual rules of differentiation hold. Let f and g be functions defined in a neighborhood of $c \in \mathbb{C}$, let F be a function defined in a neighborhood of $f(c)$, and let $a \in \mathbb{C}$. Then (recall Remark 2.15):

(a) $(f + ag)'(c) = f'(c) + ag'(c)$
(b) $(fg)'(c) = f(c)g'(c) + f'(c)g(c)$
(c) $(F \circ f)'(c) = F'(f(c))f'(c)$ (the chain rule)
(d) $\left(\frac{1}{f}\right)'(c) = -\frac{f'(c)}{f(c)^2}$ provided $f(c) \neq 0$
(e) if $f(z) = z^n$ with $n \in \mathbb{Z}$ (and $z \in \mathbb{C}_{\neq 0}$ if $n \leq 0$), then $f'(z) = n\, z^{n-1}$

Remark 2.30. About the chain rule (c): If $f(z) = w$ is a differentiable function of z and if $F(w) = \zeta$ is a differentiable function of w, then we often write the chain rule as

$$\frac{d\zeta}{dz} = \frac{d\zeta}{dw}\frac{dw}{dz}.$$

A "proof" follows. Let z_0 be arbitrary in the domain of f, and set $w_0 = f(z_0)$ and $\zeta_0 = F(w_0)$. Note that $w = f(z) \to w_0$ as $z \to z_0$. Now

$$\begin{aligned}
(F \circ f)'(z_0) &= \frac{d\zeta}{dz}(z_0) = \lim_{z \to z_0} \frac{\zeta - \zeta_0}{z - z_0} \\
&= \lim_{z \to z_0} \frac{(\zeta - \zeta_0)(w - w_0)}{(w - w_0)(z - z_0)} = \lim_{w \to w_0} \frac{\zeta - \zeta_0}{w - w_0} \lim_{z \to z_0} \frac{w - w_0}{z - z_0} \\
&= \frac{d\zeta}{dw}(w_0)\frac{dw}{dz}(z_0) = F'(w_0)\, f'(z_0).
\end{aligned}$$

This "proof" has an error in it, what is it?

Definition 2.31. A function defined on the complex plane is called *entire* if it is holomorphic on $\mathbb{C}$, that is, if its derivative exists at each point of $\mathbb{C}$.

Example 2.32. We illustrate some of the concepts introduced with more or less familiar examples.

1. Every polynomial (in one complex variable) is entire. These (apparently) simple objects have fairly complicated behavior, that is studied, for example, as part of complex dynamics.
2. A *rational* function is a function of the form $R = \frac{P}{Q}$, where P and Q are polynomials (in one complex variable), with Q not the zero polynomial. Note that the polynomial Q has only finitely many zeros (the number of zeros, properly counted, equals the *degree* of Q; see Exercise 3.19). The rational function R is holomorphic on $\mathbb{C} - \{\text{zeros of } Q\}$.
3. A special case of Example 2.32.2 is $R(z) = \frac{az+b}{cz+d}$ with a, b, c, and d fixed complex numbers satisfying $ad - bc \neq 0$. These rational functions are called *fractional linear transformations* or *Möbius transformations* and will be studied in detail in Sect. 8.1. They are the building blocks for much that will follow in this book—automorphisms of domains in the Riemann sphere, and Blaschke products, and as important ingredients for much current research in areas of complex analysis: Riemann surfaces, Fuchsian, and (the more general case of) Kleinian groups.
4. In real analysis it takes work to construct a continuous function on $\mathbb{R}$ that is nowhere differentiable. The situation with respect to complex differentiability is much simpler. The functions $z \mapsto \bar{z}$ and $z \mapsto |z|$ are both continuous on $\mathbb{C}$, but they are nowhere (complex) differentiable, since the corresponding limits (2.9) do not exist at any c in $\mathbb{C}$.

2.3.1 Convention

Whenever we write $z = x + \imath y$ for variables and $f = u + \imath v$ for functions, then we automatically mean that $x = \Re z$, $y = \Im z$, $u = \Re f$, and $v = \Im f$. We also write $u = u(x, y)$ and $v = v(x, y)$, as well as $u = u(z)$ and $v = v(z)$. We naturally use subscripts to denote partial derivatives with respect a given variable, so that notation such as u_x, u_y, v_x, or v_y has the obvious meaning.

2.3.2 The Cauchy-Riemann (CR) Equations

Theorem 2.33. *If $f = u + \imath v$ is differentiable at $c = a + \imath b$, then u and v have partial derivatives with respect to x and y at c, and they satisfy the* ***Cauchy–Riemann equations:***

$$u_x(a, b) = v_y(a, b), \quad u_y(a, b) = -v_x(a, b). \tag{CR}$$

Furthermore,

$$f'(c) = u_x(a,b) + \imath\, v_x(a,b) = -\imath\, u_y(a,b) + v_y(a,b).$$

Proof. First take $h = \alpha$, with α real, in the limit (2.9) appearing in the definition of differentiability and compute

$$f'(c) = u_x(a,b) + \imath\, v_x(a,b).$$

Then take $h = \imath\beta$, with β real, and compute

$$f'(c) = -\imath\, u_y(a,b) + v_y(a,b).$$

Comparing the two expressions we obtain the desired result. □

Remark 2.34. Let $f = u + \imath\, v$ be a function defined in a neighborhood of c, such that the partial derivatives u_x, v_x, u_y, and v_y exist at c. Then we will use the obvious notation:

$$f_x = u_x + \imath\, v_x \quad \text{and} \quad f_y = u_y + \imath\, v_y.$$

In this language the CR equations (CR) for the function f are written as follows:

$$f_x(c) = -\imath\, f_y(c), \tag{2.10}$$

and Theorem 2.33 may be stated as follows.
If f is differentiable at c, then f has partial derivatives with respect to x and y at c, and they satisfy the Cauchy–Riemann equation (2.10). *Furthermore,*

$$f'(c) = f_x(c) = -\imath\, f_y(c). \tag{2.11}$$

Remark 2.35. The CR equations are not sufficient for differentiability. To see this, define

$$f(z) = \begin{cases} z^5\,|z|^{-4} & \text{for } z \neq 0, \\ 0 & \text{for } z = 0. \end{cases}$$

It is easy to verify that the function f is continuous on $\mathbb{C}$. Furthermore, for α real and nonzero we have $\dfrac{f(\alpha)}{\alpha} = 1$, and for β real and nonzero we have $\dfrac{f(\imath\beta)}{\imath\beta} = 1$. Therefore $f_x(0) = 1$ and $f_y(0) = \imath$, and f satisfies the Cauchy–Riemann equation (2.10) at $z = 0$. However, f is not differentiable at $z = 0$. Indeed, if it were, we would conclude from (2.11) that $f'(0) = 1$. Now take $h = (1 + \imath)\gamma$ with γ real and nonzero and observe that $\dfrac{f(h)}{h} = -1$ so that $f'(0)$ would be equal to -1.

Remark 2.36. We may use the CR equations to try to manufacture an entire function with a given real (or imaginary) part. Let us start with the real-valued function $u(x, y) = x^2 + y^2$. If this were to be the real part of some entire function $f = u + \imath\, v$, then the CR equations would help us to determine v. Since $u_x = 2x = v_y$ and $u_y = 2y = -v_x$ must be satisfied, by integrating $2x = v_y$ with respect to y we obtain that $v(x, y) = 2\, x\, y + h(x)$, for some function h of x alone; by integrating $2y = -v_x$ with respect to x we obtain that $v(x, y) = -2\, y\, x + g(y)$, for some function g of y. It is quite obvious that these two expressions for v are incompatible, and hence there is no such function f.

The situation changes dramatically for $u(x, y) = x^2 - y^2$, where similar calculations lead to $v(x, y) = 2\, x\, y + h(x) = 2\, y\, x + g(y)$, and we may choose $h(x) = g(y) = a$, for any real value of a. We have thus obtained a family of entire functions f with prescribed real part u; these are given by

$$f(z) = u(x, y) + \imath\, v(x, y) = x^2 - y^2 + \imath\,(2xy + a) = z^2 + \imath\, a,$$

with a any real number.

We will determine later the class of real-valued functions u for which the construction outlined above leads us to an entire function f.

Definition 2.37. For a complex-valued function f defined on a region in the complex plane, such that both f_x and f_y exist in this region, set

$$f_z = \frac{1}{2}\left(f_x - \imath\, f_y\right)$$

and

$$f_{\bar{z}} = \frac{1}{2}\left(f_x + \imath\, f_y\right).$$

Remark 2.38. The partial derivatives just defined are computed as if z and $\bar{z}$ were independent variables. For instance, if $f(z) = z^2 + 5z\bar{z}^3$, then it is easy to verify that $f_z = 2z + 5\bar{z}^3$ and $f_{\bar{z}} = 15z\bar{z}^2$.

These partials not only simplify the notation: for example, the two Cauchy–Riemann equations (CR) are written as the single equation

$$f_{\bar{z}} = 0, \qquad \text{(CR complex)}$$

or, if f is differentiable at c, then

$$f'(c) = f_z(c),$$

but they also allow us to produce more concise arguments (and, as we shall see later, prettier formulae), as illustrated in the proof of the lemma below.

We use the notation $\frac{\partial f}{\partial \bar{z}}$ (respectively $\frac{\partial f}{\partial z}$) interchangeably with $f_{\bar{z}}$ (respectively f_z).

Lemma 2.39. *If f is a $\mathbf{C}^1$-complex-valued function defined in a neighborhood of $c \in \mathbb{C}$, then*

$$f(z) - f(c) = (z-c)\, f_z(c) + \overline{(z-c)}\, f_{\bar{z}}(c) + |z-c|\; \varepsilon(z,c), \tag{2.12}$$

for all $z \in \mathbb{C}$ with $|z-c|$ small, where $\varepsilon(z,c)$ is a complex-valued function of z and c such that

$$\lim_{z \to c} \varepsilon(z,c) = 0.$$

Proof. As usual we write $z = x + \imath\, y$, $c = a + \imath\, b$, and $f = u + \imath\, v$ and abbreviate $\Delta u = u(z) - u(c)$, $\Delta x = x - a$, $\Delta y = y - b$, and $\Delta z = z - c = \Delta x + \imath\, \Delta y$. By hypothesis, the real-valued function u has continuous first partial derivatives defined in a neighborhood of c, and we can define ε_1 by

$$\varepsilon_1(z,c) = \frac{\Delta u - u_x(c)\Delta x - u_y(c)\Delta y}{|\Delta z|}$$

for $z \neq c$, and $\varepsilon_1(c,c) = 0$. Then it is clear that

$$u(z) - u(c) = (x-a)u_x(a,b) + (y-b)u_y(a,b) + |z-c|\,\varepsilon_1(z,c).$$

We now show that

$$\lim_{z \to c} \varepsilon_1(z,c) = 0. \tag{2.13}$$

If we rewrite Δu as

$$\Delta u = [u(x,y) - u(x,b)] + [u(x,b) - u(a,b)]\,,$$

it follows from the (real) mean value theorem applied to the two summands on the RHS that

$$\Delta u = u_y(x, y_0)\Delta y + u_x(x_0, b)\Delta x,$$

where y_0 is between y and b and x_0 is between x and a. Thus

$$\varepsilon_1(z,c) = \frac{[u_y(x,y_0) - u_y(a,b)]\Delta y + [u_x(x_0,b) - u_x(a,b)]\Delta x}{|\Delta z|},$$

for $z \neq c$. Hence we see that

$$|\varepsilon_1(z,c)| \leq \left|u_y(x,y_0) - u_y(a,b)\right| + |u_x(x_0,b) - u_x(a,b)|\,,$$

and the claim (2.13) follows.

Similarly,

$$v(z) - v(c) = (x-a)v_x(a,b) + (y-b)v_y(a,b) + |z-c|\,\varepsilon_2(z,c),$$

with

$$\lim_{z \to c} \varepsilon_2(z, c) = 0. \tag{2.14}$$

With obvious notational conventions, we compute that

$$\begin{aligned}
\Delta f &= \Delta u + \imath\, \Delta v \\
&= [u_x(a,b) + \imath\, v_x(a,b)]\, \Delta x + \left[u_y(a,b) + \imath\, v_y(a,b)\right] \Delta y + |\Delta z|\, \varepsilon(z,c) \\
&= \frac{\Delta z + \overline{\Delta z}}{2} f_x(c) + \imath\, \frac{\overline{\Delta z} - \Delta z}{2} f_y(c) + |\Delta z|\, \varepsilon(z,c) \\
&= \Delta z f_z(c) + \overline{\Delta z} f_{\bar{z}}(c) + |\Delta z|\, \varepsilon(z,c),
\end{aligned}$$

with $\varepsilon(z, c) = \varepsilon_1(z, c) + \imath\, \varepsilon_2(z, c)$. Now equalities (2.13) and (2.14) imply that

$$\lim_{z \to c} \varepsilon(z, c) = 0. \qquad \square$$

Theorem 2.40. *If the function f has continuous first partial derivatives in a neighborhood of c that satisfy the* CR *equations at c, then f is (complex) differentiable at c.*

Proof. The theorem is an immediate consequence of (2.12), since in this case $f_{\bar{z}}(c) = 0$ and hence $f'(c) = f_z(c)$. □

Corollary 2.41. *If the function f has continuous first partial derivatives in an open neighborhood U of $c \in \mathbb{C}$ and the* CR *equations hold at each point of U, then f is holomorphic at c (in fact on U).*

Remark 2.42. The converse to this corollary is also true. It will take us some time to prove it.

Theorem 2.43. *If f is holomorphic and real-valued on a domain D, then f is constant.*

Proof. As usual we write $f = u + \imath\, v$; in this case $v = 0$. The CR equations say $u_x = v_y = 0$ and $u_y = -v_x = 0$. Thus u is constant, since D is connected. □

Theorem 2.44. *If f is holomorphic and $f' = 0$ on a domain D, then f is constant.*

Proof. As above $f = u + \imath\, v$ and $f' = u_x + \imath\, v_x = 0$. The last equation together with the CR equations say $0 = u_x = v_y$ and ; $0 = v_x = -u_y$. Thus both u and v are constant, since D is connected. □

Exercises

2.1. (a) Let $\{z_n\}$ be a sequence of complex numbers and assume

$$|z_n - z_m| < \frac{1}{1 + |n - m|}, \quad \text{for all } n \text{ and } m.$$

Show that the sequence converges.
Do you have enough information to evaluate $\lim_{n\to\infty} z_n$?
What else can you say about this sequence?

(b) Let $\{z_n\}$ be a sequence with $\lim_{n\to\infty} z_n = 0$ and let $\{w_n\}$ be a bounded sequence. Show that

$$\lim_{n\to\infty} w_n z_n = 0.$$

2.2. (a) Let z and c denote two complex numbers. Show that

$$|\bar{c}\, z - 1|^2 - |z - c|^2 = (1 - |z|^2)\,(1 - |c|^2).$$

(b) Use (a) to conclude that if c is any fixed complex number with $|c| < 1$, then

$$\{z \in \mathbb{C};\, |z - c| < |\bar{c}\, z - 1|\} = \{z \in \mathbb{C};\, |z| < 1\},$$

$$\{z \in \mathbb{C};\, |z - c| = |\bar{c}\, z - 1|\} = \{z \in \mathbb{C};\, |z| = 1\} \text{ and}$$

$$\{z \in \mathbb{C};\, |z - c| > |\bar{c}\, z - 1|\} = \{z \in \mathbb{C};\, |z| > 1\}.$$

2.3. Let a, b, and c be three distinct points on a straight line with b between a and c. Show that

$$\frac{a - b}{c - b} \in \mathbb{R}_{<0}.$$

2.4. (a) Given two points z_1, z_2 such that $|z_1| < 1$ and $|z_2| < 1$, show that for every point $z \neq 1$ in the closed triangle with vertices z_1, z_2, and 1,

$$\frac{|1 - z|}{1 - |z|} \leq K,$$

where K is a constant that depends only on z_1 and z_2.

(b) Determine the smallest value of K for $z_1 = \dfrac{1 + \imath}{2}$ and $z_2 = \dfrac{1 - \imath}{2}$.

2.5. Verify the Cauchy–Riemann equations for the function $f(z) = z^3$ by splitting f into its real and imaginary parts.

2.6. Suppose $z = x + \imath\, y$. Define

$$f(z) = \frac{xy^2\,(x + \imath\, y)}{x^2 + y^4},$$

for $z \neq 0$ and $f(0) = 0$. Show that

$$\lim \frac{f(z) - f(0)}{z} = 0$$

as $z \to 0$ along any straight line. Show that as $z \to 0$ along the curve $x = y^2$, the limit of the difference quotient is $\frac{1}{2}$, thus showing that $f'(0)$ does not exist.

2.7. Let $x = r\cos\theta$ and $y = r\sin\theta$. Show that the Cauchy–Riemann equations in polar coordinates for $F = U + \imath V$, where $U = U(r,\theta)$ and $V = V(r,\theta)$, are

$$r\,\frac{\partial U}{\partial r} = \frac{\partial V}{\partial \theta} \quad \text{and} \quad r\,\frac{\partial V}{\partial r} = -\frac{\partial U}{\partial \theta},$$

or, in alternate notation,

$$rU_r = V_\theta \quad \text{and} \quad rV_r = -U_\theta.$$

2.8. Let f be a complex-valued function defined on a region in the complex plane, and assume that both f_x and f_y exist in this region. Using the definitions of f_z and $f_{\bar{z}}$, show that for $\mathbf{C}^1$-functions f,

$$f \text{ is holomorphic if and only if } f_{\bar{z}} = 0$$

and that in this case $f_z = f'$.

2.9. Let R and Φ be two real-valued $\mathbf{C}^1$-functions of a complex variable z. Show that $f = Re^{\imath\Phi}$ is holomorphic if and only if

$$R_{\bar{z}} + \imath R\Phi_{\bar{z}} = 0.$$

2.10. Show that if f and g are $\mathbf{C}^1$-functions, then the (complex) chain rule is expressed as follows (here $w = f(z)$ and g is viewed as a function of w).

$$(g \circ f)_z = g_w\, f_z + g_{\overline{w}}\, \overline{f}_z$$

and

$$(g \circ f)_{\bar{z}} = g_w\, f_{\bar{z}} + g_{\overline{w}}\, \overline{f}_{\bar{z}}.$$

2.11. Let p be a complex-valued polynomial of two real variables:

$$p(z) = \sum a_{ij} x^i y^j.$$

Write

$$p(z) = \sum_{j \geq 0} P_j(z)\bar{z}^j,$$

where each P_j is of the form $P_j(z) = \sum b_{ij} z^i$ (a polynomial in z). Prove that p is an entire function if and only if

$$0 \equiv P_1 \equiv P_2 \equiv \ldots.$$

What can you conclude in this case for the matrix $[a_{ij}]$?

2.12. Deduce the analogues of the CR equations for anti-holomorphic functions, in rectangular, polar, and complex coordinates.

2.13. Let $f : \mathbb{C} \to \mathbb{C}$ be a holomorphic function, and set $g(z) = \overline{f}(\overline{z})$ and $h(z) = f(\overline{z})$, for z in $\mathbb{C}$. Show that g is holomorphic and h is anti-holomorphic on $\mathbb{C}$. Furthermore, h is holomorphic on $\mathbb{C}$ if and only if f is a constant function.

2.14. Let D be an arbitrary (nonempty) open connected set in $\mathbb{C}$. Describe the class of complex-valued functions on D that are both holomorphic and anti-holomorphic.

2.15. Does there exist a holomorphic function f on $\mathbb{C}$ whose real part is:

(a) $u(x, y) = \mathrm{e}^x$? Or
(b) $u(x, y) = \mathrm{e}^x(x \cos y - y \sin y)$?

Justify your answer; that is, if yes, exhibit the holomorphic function(s) and if not, prove it.

2.16. Prove the *fundamental theorem of algebra*: If $a_0, \ldots, a_{n-1}$ are complex numbers ($n \geq 1$) and $p(z) = z^n + a_{n-1}z^{n-1} + \cdots + a_0$, then there exists a number $z_0 \in \mathbb{C}$ such that $p(z_0) = 0$.

Hint: A standard method of attack:

(a) Show that there are an $M > 0$ and an $R > 0$ such that for all $|z| \geq R$, $|p(z)| \geq M$ holds.
(b) Show next that there is a $z_0 \in \mathbb{C}$ such that

$$|p(z_0)| = \min\{|p(z)|; z \in \mathbb{C}\}.$$

(c) By the change of variable $p(z + z_0) = g(z)$, it suffices to show that $g(0) = 0$.
(d) Write $g(z) = \alpha + z^m(\beta + c_1 z + \cdots + c_{n-m}z^{n-m})$ with $\beta \neq 0$. Choose γ such that

$$\gamma^m = -\frac{\alpha}{\beta}.$$

If $\alpha \neq 0$, obtain the contradiction $|g(\gamma z)| < |\alpha|$ for some z.

Note. We will later have several simpler proofs of this theorem using results from complex analysis, for instance, in Theorem 5.16 and Exercise 6.1. See also the April 2006 issue of *The American Mathematical Monthly* for yet other proofs of this fundamental result.

2.17. Conclude from the fundamental theorem of algebra that a nonconstant complex polynomial of degree n has n complex roots, counted with multiplicities.

Use this result to show that a nonconstant real polynomial that cannot be factored as a product of two nonconstant real polynomials of lower degree (i.e., a real irreducible nonconstant polynomial) has degree one or two.

2.18. Using the fundamental theorem of algebra stated in Exercise 2.16, prove the *Frobenius theorem*: If F is a field containing the reals whose dimension as a real vector space is finite, then either F is the reals or F is (isomorphic to) $\mathbb{C}$.

Hint: An outline of possible steps follows.

(a) Assume $\dim_{\mathbb{R}} F = n > 1$. Show that for θ in $F - \mathbb{R}$ there exists a nonzero *real* polynomial p with leading coefficient 1 and such that $p(\theta) = 0$.
(b) Show that there exist real numbers β and γ such that

$$\theta^2 - 2\beta\theta + \gamma = 0.$$

(c) Show that there exists a positive real number δ such that $(\theta - \beta)^2 = -\delta^2$, and therefore

$$\sigma = \frac{\theta - \beta}{\delta}$$

is an element of F satisfying $\sigma^2 = -1$.
(d) The field

$$G = \mathbb{R}(\sigma) = \{x + y\sigma : x, y \in \mathbb{R}\} \subseteq F$$

is isomorphic to $\mathbb{C}$, so without loss of generality assume $\sigma = \imath$ and $G = \mathbb{C}$.

Conclude by showing that any element of F is the root of a *complex* polynomial with leading coefficient 1 and is therefore a complex number.

2.19. Prove the following statements, where automorphism is a bijection preserving sums and products.

(a) Every automorphism of the real field is the identity.
(b) Every automorphism of the complex field fixing the reals is either the identity or conjugation.
(c) Every continuous automorphism of the complex field is either the identity or conjugation.

2.20. A domain is defined to be an open connected set. It was remarked that it could also be defined to be an open arcwise connected set. Can it be defined as an open path connected set? Justify your answer.

Chapter 3
Power Series

This chapter is devoted to an important tool for constructing holomorphic functions: convergent power series. It is the basis for the introduction of new non-algebraic holomorphic functions, the elementary transcendental functions. It turns out that power series play an even more central role in the theory of holomorphic functions, a role beyond enabling the construction of complex transcendental functions that are the extension of the real transcendental functions. A much stronger result holds. All holomorphic functions are (at least locally) convergent power series. This will be proven in the next chapter.

The first section of this chapter is devoted to a discussion of elementary properties of complex power series. Some material from real analysis, not usually treated in books or courses on that subject, is studied. The concept of a convergent power series is extended from series with real coefficients to complex power series, and tests for convergence are established. In the second section, we show that convergent power series define holomorphic functions. Section 3.3 introduces important complex-valued functions of a complex variable including the exponential function, the trigonometric functions, and the logarithm. This is followed by Sects. 3.4 and 3.5, which describe an identity principle and introduce the new class of *meromorphic* functions; these functions are holomorphic on a domain except that they "assume the value ∞" (in a controlled way) at certain isolated points, known as the *poles* of the function. Meromorphic functions are defined locally as ratios of functions having power series expansions. It will hence follow subsequently that these are locally ratios of holomorphic functions. After some more work we will be able to replace "locally" by "globally." We develop the fundamental *identity principle* and its corollary, the *principle of analytic continuation*, for functions that locally have convergent power series expansions, and discuss the zeros and poles of a meromorphic function. The principle of analytic continuation is one of the most powerful results in complex function theory. Once we show that every holomorphic function is locally defined by a power series, we will see that the principle of analytic continuation says that a holomorphic function defined on an open connected set is

R.E. Rodríguez et al., *Complex Analysis: In the Spirit of Lipman Bers*,
Graduate Texts in Mathematics 245, DOI 10.1007/978-1-4419-7323-8_3,

remarkably rigid: its behavior at a single point in the set determines its behavior at all other points of the set. Holomorphicity at a point is an extremely strong concept.

3.1 Complex Power Series

Let $A \subseteq \mathbb{C}$ and let $\{f_n\} = \{f_n\}_{n=0}^{\infty}$ be a sequence of functions defined on A (in the previous chapter sequences were indexed by $\mathbb{Z}_{>0}$; for convenience, in this chapter, they will be indexed by $\mathbb{Z}_{\geq 0}$). We form the new sequence (known as a *series*) $\{S_N\}$, where

$$S_N(z) = \sum_{n=0}^{N} f_n(z), \; z \in A,$$

and the formal *infinite series* $\sum_{n=0}^{\infty} f_n(z)$.

The sequence of complex numbers $\{S_N(z)\}_N$ is also known as the *sequence of partial sums* associated to the infinite series $\sum_{n=0}^{\infty} f_n(z)$ at the point $z \in A$. When the indices of summation are clear from the context, we often omit them. For example, $\sum f_n(\zeta)$ usually means the infinite sum $\sum_{n=0}^{\infty} f_n(\zeta)$. Other similar abbreviations and conventions are used.

Definition 3.1. We distinguish several types of convergence for a series. We say that the infinite series $\sum_{n=0}^{\infty} f_n$:

(i) *Converges* at a point $c \in A$ if $\{S_N(c)\}$ converges. In this case we write $\sum_{n=0}^{\infty} f_n(c) = \lim_{N \to \infty} S_N(c)$.
(ii) *Converges pointwise* in A if $\{S_N(z)\}$ converges for every $z \in A$.
(iii) *Converges absolutely* at a point $c \in A$ if the infinite series $\sum_{n=0}^{\infty} |f_n(c)|$ converges.
(iv) *Converges uniformly* in A if the sequence of partial sums $\{S_N\}$ converges uniformly in A.
(v) *Converges normally* on a set $B \subseteq A$ if there exists a sequence of positive constants $\{M_n\}$ such that:

    (1) $|f_n(z)| \leq M_n$ for all $z \in B$ and all n.
    (2) $\sum M_n < \infty$; that is, the series $\sum M_n$ converges.

(vi) *Diverges* at a point of A if it does not converge at that point.

We speak of the *pointwise, uniform, absolute, or normal convergence* of a series as well as the *divergence* of a series.

Remark 3.2. The fact that the sequence $\{S_N(c)\}$ converges if and only if it is a Cauchy sequence allows us to rephrase conditions (i) and (ii) as if and only if statements are useful:

(i) $\sum f_n$ *converges* at a point $c \in A$ if and only if $\{S_N(c)\}$ is Cauchy.
(ii) $\sum f_n$ *converges pointwise* in A if and only if $\{S_N(z)\}$ is Cauchy for every $z \in A$.

Remark 3.3. Many questions on convergence of complex sequences are reduced to the real case by the *trivial but important observation* that absolute convergence at a point implies convergence at that point.

Remark 3.4. Two other such observations, both *trivial but important*, are that if an infinite series $\sum_{n=0}^{\infty} f_n$ converges at a point c, then $\lim_{n\to\infty} f_n(c) = 0$ and $\lim_{n\to\infty} \sum_{k=n}^{\infty} f_k(c) = 0$.

Some of the relationships between some of the different types of convergence of a series are given in the following result.

Theorem 3.5 (Weierstrass M-test). *Normal convergence implies uniform and absolute convergence.*

Proof. With notation as in the definition of normal convergence, if $N_1 < N$ are positive integers, then

$$|S_N(z) - S_{N_1}(z)| \le \sum_{n=N_1+1}^{N} M_n \text{ for all } z \in B$$

(needed for the uniform convergence argument), and

$$\big|\, |S_N(z)| - |S_{N_1}(z)| \,\big| \le \sum_{n=N_1+1}^{N} M_n \text{ for all } z \in B$$

(needed for the absolute convergence argument).

Since the series $\sum M_n$ converges, for any given $\epsilon > 0$, we can find a positive integer N_0 such that $N > N_1 > N_0$ implies $\sum_{n=N_1+1}^{N} M_n < \epsilon$, thus concluding the proof. □

We shall be mostly interested in series of the form $\sum_{n=0}^{\infty} a_n z^n$ with $a_n \in \mathbb{C}$ (these are known as *power series*) and the *associated* real-valued series $\sum_{n=0}^{\infty} |a_n| r^n$ where $r = |z|$. We define, for each $N \in \mathbb{Z}_{\geq 0}$,

$$S_N^*(r) = \sum_{n=0}^{N} |a_n| r^n \quad \text{for } r \in \mathbb{R}_{\geq 0}$$

and observe that

$$S_{N+1}^*(r) \geq S_N^*(r) \text{ for all } N \in \mathbb{Z}_{\geq 0} \text{ and for all } r \in \mathbb{R}_{\geq 0}.$$

We refer to $S_N^*(r)$ as the *real partial sum at* r.

This nondecreasing sequence of real partial sums $\{S_N^*(r)\}$ is important because if it converges at some fixed positive r, then the series $\sum a_n z^n$ converges normally on the set $\{|z| \leq r\}$ (with $M_n = a_n\, r^n$ in the definition of normal convergence).

An *elementary but most important example* is provided by the *geometric series*:

$$1 + r + r^2 + \cdots = \sum_{n=0}^{\infty} r^n. \tag{3.1}$$

In this case $S_N^*(r) = \sum_{n=0}^{N} r^n$ for all nonnegative r, and therefore $S_N^*(1) = N + 1$, and $S_N^*(r) = \dfrac{1 - r^{N+1}}{1 - r}$ for $0 \leq r < 1$ and for $r > 1$.

Thus $\sum_{n=0}^{\infty} r^n = \dfrac{1}{1 - r}$ if $0 \leq r < 1$ and $\sum r^n$ diverges if $r \geq 1$.

It follows that the power series $\sum z^n$ converges for all $|z| < 1$; the same argument shows that in this case it converges to $\dfrac{1}{1 - z}$. Note that this power series diverges for all z with $|z| \geq 1$, since in this case $\lim_{n \to \infty} z^n = 0$ is not satisfied, a necessary condition for the convergence of the series, as we saw in Remark 3.4.

We now introduce two special cases of divergence of a sequence of real numbers. It will be useful to regard these sequences as convergent sequences with infinite limits.

Definition 3.6. A sequence of real numbers $\{b_n\}$ *converges to* $+\infty$ if for all $M > 0$ there exists an $N \in \mathbb{Z}_{>0}$ such that $b_n > M$ for all $n > N$. In this case we shall write $\lim_{n \to \infty} b_n = +\infty$. A similar definition applies to real sequences converging to $-\infty$.

With this notation either:

(a) $\lim_{N\to\infty} S_N^*(r)$ exists and is finite: that is, $\sum_{n=0}^{\infty} |a_n| r^n$ converges. In this case we write $\sum_{n=0}^{\infty} |a_n| r^n < +\infty$,

or

(b) $\lim_{N\to\infty} S_N^*(r) = +\infty$: that is, $\sum_{n=0}^{\infty} |a_n|\, r^n$ diverges (in the previous sense). In this case we write $\sum_{n=0}^{\infty} |a_n|\, r^n = +\infty$.

For real series, we have the **comparison test**. Assume $0 \le a_k \le b_k$ for all large values of k; that is, assume there exists $N \in \mathbb{N}$ such that $0 \le a_k \le b_k$ for all $k \ge N$. Then the following results are easy to prove:

(a) If $\sum a_n = +\infty$, then $\sum b_n = +\infty$.
(b) If $\sum b_n < +\infty$, then $\sum a_n < +\infty$.

As an application of the comparison test we prove the following very useful result.

Lemma 3.7 (Abel's Lemma). *Assume that $0 < r < r_0$. If there exists a positive number M such that*

$$|a_n r_0^n| \le M \text{ for all } n \in \mathbb{Z}_{>0},$$

then the series $\sum a_n z^n$ converges normally for all z with $|z| \le r$.

In particular, it converges absolutely and uniformly for all z with $|z| \le r$.

Proof. For all $|z| \le r$ we have

$$|a_n z^n| = |a_n|\, |z|^n \le |a_n|\, |r|^n = |a_n| \left(\frac{r}{r_0}\right)^n r_0^n \le M \left(\frac{r}{r_0}\right)^n .$$

Since $0 < \frac{r}{r_0} < 1$, the normal convergence of $\sum a_n z^n$ follows by comparison with the geometric series or by an application of the Weierstrass M-test with $M_n = M\left(\frac{r}{r_0}\right)^n$. □

If S is any nonempty set of real numbers, then the least upper bound or *supremum* of S is denoted by $\sup S$, and the greatest lower bound or *infimum* of S is denoted by $\inf S$. The possibilities that $\sup S = +\infty$ or $\inf S = -\infty$ are allowed.

Definition 3.8. The *radius of convergence* ρ of the power series $\sum a_n z^n$ is given by

$$\rho = \sup\{r \in \mathbb{R}_{\geq 0};\ \sum |a_n|\, r^n < +\infty\}.$$

Note that $0 \leq \rho \leq +\infty$. As a result of the next theorem it makes sense to define the set $\{z \in \mathbb{C};\ |z| < \rho\}$ as the *disc of convergence* of the power series $\sum a_n z^n$.

Theorem 3.9. *Let $\sum a_n z^n$ be a power series with radius of convergence $\rho > 0$. Then*

(a) *For any $0 < r < \rho$, the series $\sum a_n z^n$ converges normally, absolutely, and uniformly for $|z| \leq r$.*
(b) *The series $\sum a_n z^n$ diverges for $|z| > \rho$.*

Proof. (a) For any r_0 satisfying $r < r_0 < \rho$, the series $\sum |a_n|\, r_0^n$ converges. Then $\lim_{n\to\infty} |a_n|\, r_0^n = 0$, and thus there exists an $M > 0$ with $|a_n|\, r_0^n \leq M$ for all n in $\mathbb{Z}_{>0}$. Now apply Abel's lemma.
(b) We claim that for $|z| > \rho$, the sequence $\{|a_n|\, |z|^n\}$ is not even bounded. Otherwise Abel's lemma (with $r_0 = |z|$) would guarantee the existence of an r with $\rho < r < |z|$ and $\sum |a_n|\, r^n < +\infty$. This contradicts the definition of ρ. □

The theorem makes no claim about the convergence or divergence of a power series on the boundary of its disc of convergence. There are power series with finite positive radii of convergence that diverge at every boundary point, as well as others that converge at all or some boundary points (see Exercise 3.1). We will explore this situation a little further in Theorem 3.22.

Corollary 3.10. *Let $\sum a_n z^n$ be a power series with radius of convergence $\rho > 0$. Then the function defined by $S(z) = \sum a_n z^n$ is continuous for $|z| < \rho$.*

Proof. It follows immediately from Theorems 2.23 and 3.9. □

We now turn to the obvious and important question: *how do we compute ρ?*

To answer this question we introduce the concepts of limit inferior and limit superior of real sequences; these concepts appropriately belong to real analysis and are extremely useful in the current context.

Definition 3.11. Let $\{u_n\}$ be a real sequence. Here we use $\equiv$ to indicate equivalent names[1] and define

$$\overline{\lim_n}\ u_n \equiv \limsup_n u_n \equiv \text{upper limit of } \{u_n\} \equiv \text{limit superior of } \{u_n\}$$

$$= \lim_{p\to\infty} \sup_{n\geq p} \{u_n\} = \inf_p \sup_{n\geq p} \{u_n\},$$

[1]We will also use this same notation with a different meaning in other places, such as $f \equiv 0$ or $f \equiv g$, to emphasize that these functions are (identically) equal.

and

$$\underline{\lim_n}\ u_n \equiv \liminf_n u_n \equiv \text{ lower limit of } \{u_n\} \equiv \text{ limit inferior of } \{u_n\}$$

$$= \lim_{p\to\infty} \inf_{n\geq p} \{u_n\} = \sup_p \inf_{n\geq p} \{u_n\}.$$

Note that every real sequence has a limit superior as well as a limit inferior, which are either real numbers or $+\infty$ or $-\infty$.

3.1.1 *Properties of Limits Superior and Inferior*

Let $\{u_n\}$ and $\{v_n\}$ be real sequences. Then we have:

(a) $\liminf_n u_n \leq \limsup_n u_n$.

(b) $\liminf_n(-u_n) = -\limsup_n u_n$.

(c) If $r > 0$, then

$$\liminf_n(ru_n) = r \liminf_n u_n\ , \text{ and } \limsup_n(ru_n) = r \limsup_n u_n.$$

More generally,

(d) If $\lim_n v_n$ exists, then

$$\liminf_n(u_n v_n) = (\liminf_n u_n)(\lim_n v_n), \text{ and}$$

$$\limsup_n(u_n v_n) = (\limsup_n u_n)(\lim_n v_n),$$

provided the right-hand sides are not indeterminates, that is, not of the form $0 \cdot (\pm\infty)$ nor $(\pm\infty) \cdot 0$.

(e) If $u_n \leq v_n$ for all n, then

$$\liminf_n u_n \leq \liminf_n v_n\ , \text{ and } \limsup_n u_n \leq \limsup_n v_n.$$

(f) $\lim_n u_n = L$ if and only if $\liminf_n u_n = L = \limsup_n u_n$ (L is allowed to be $\pm\infty$ in this context).

(g) $\liminf_n(u_n + v_n) \geq \liminf_n u_n + \liminf_n v_n$, and $\limsup_n(u_n + v_n) \leq \limsup_n u_n + \limsup_n v_n$.

Remark and Exercise: If f is a real-valued function of a complex variable, defined on a set S in $\mathbb{C}$ and if c, is a limit point of S, then it is possible to define $\overline{\lim_{z\to c}}\ f(z)$ and $\underline{\lim_{z\to c}}\ f(z)$.

Example 3.12. Let $u_n = \sin\left(\frac{n\pi}{2}\right), n = 0, 1, 2, \ldots$; this is the sequence: $\{0, 1, 0, -1, \ldots\}$. Hence for all $p \in \mathbb{Z}_{\geq 0}$ we have

$a_p = \sup_{n \geq p} u_n = 1$ and thus $\limsup_n u_n = \lim_{p \to \infty} a_p = 1$, and

$b_p = \inf_{n \geq p} u_n = -1$ and thus $\liminf_n u_n = \lim_{p \to \infty} b_p = -1$.

3.1.2 The Radius of Convergence

The **ratio test** will be well known to most readers (see Exercise 3.5). Suppose that $v_n > 0$ for all nonnegative integers n. Then the following properties hold:

(a) If $\lim_{n \to \infty} \frac{v_{n+1}}{v_n} = L < 1$, then $\sum v_n$ converges.

(b) If $\lim_{n \to \infty} \frac{v_{n+1}}{v_n} = L > 1$, then $\sum v_n$ diverges.

Perhaps less familiar (but extremely useful in the current setting) is the **root test.** Suppose that $v_n \geq 0$ for all nonnegative integers n:

(a) If $\overline{\lim_n}(v_n)^{\frac{1}{n}} = L < 1$, then $\sum v_n$ converges.

(b) If $\overline{\lim_n}(v_n)^{\frac{1}{n}} = L > 1$, then $\sum v_n$ diverges.

Proof. (a) Choose $\epsilon > 0$ such that $0 < L + \epsilon < 1$; then there exists a $P \in \mathbb{Z}_{>0}$ such that

$$\sup_{n \geq p} \left\{ v_n^{\frac{1}{n}} \right\} < L + \epsilon \text{ for all } p \geq P.$$

Thus

$$v_n < (L + \epsilon)^n \text{ for all } n \geq P,$$

and comparison with the geometric series yields convergence.

(b) Suppose that $\sum v_n < +\infty$. Then $\lim_n v_n = 0$ and thus there exists a P in $\mathbb{Z}_{>0}$ such that $n \geq P$ implies that $v_n < 1$ for all $n \geq P$. Hence $(v_n)^{\frac{1}{n}} < 1$ for all $n \geq P$ and therefore $\overline{\lim_n}(v_n)^{\frac{1}{n}} \leq 1$. □

Remark 3.13. The root test applies (with the same value of L) whenever the ratio test applies. However, the converse is not true. To see this take a sequence where the ratios are alternately $\frac{1}{2}$ and $\frac{1}{8}$; then the root test will apply with $L = \frac{1}{4}$, but the sequence of ratios is obviously divergent.

Example 3.14. For general series, the two tests are not strong enough to determine convergence, as this example will show. However, we will see that we can compute the radius of convergence for any power series with the root test.

Consider the series $\sum_{n\geq 1} n^{-s}$, with s a positive real number. Using standard calculus techniques, one shows that

$$\lim_{n\to\infty} \frac{(n+1)^{-s}}{n^{-s}} = 1 = \lim_{n\to\infty} (n^{-s})^{\frac{1}{n}} ;$$

thus both the ratio and the root tests are inconclusive in this case.

The integral test, not discussed in this book, easily handles this example, showing that the series converges for $s > 1$ and diverges (converges to $+\infty$) for $0 < s \leq 1$.

We now return to the study of the complex power series $\sum_{n=0}^{\infty} a_n z^n$ and the problem of computing its radius of convergence.

Theorem 3.15. *If $\sum a_n z^n$ is a power series with $a_n \neq 0$ for all n, and $\lim_n \left|\frac{a_{n+1}}{a_n}\right| = L$, with $0 \leq L \leq +\infty$, then its radius of convergence ρ is $\frac{1}{L}$; in other words,*

$$\frac{1}{\rho} = L,$$

where $\frac{1}{+\infty} = 0$.

The hypotheses required for this result to hold are strong. As pointed out in Remark 3.13, the ratio test is not as effective as the root test. The next result provides a way of computing the radius of convergence for *any* power series.

Theorem 3.16 (Cauchy–Hadamard). *The radius of convergence ρ of the power series $\sum_{n=0}^{\infty} a_n z^n$ is given by*

$$\frac{1}{\rho} = \overline{\lim_n} \, |a_n|^{\frac{1}{n}} .$$

Proof. Let $L = \overline{\lim_n} \, |a_n|^{\frac{1}{n}}$. Thus $\overline{\lim_n} \, |a_n r^n|^{\frac{1}{n}} = rL$ for all $r \geq 0$, and we conclude by the root test that the associated series $\sum |a_n| \, r^n$ converges for $0 \leq r < \frac{1}{L}$ and diverges for $r > \frac{1}{L}$. Thus $\rho = \frac{1}{L}$. □

3.2 More on Power Series

Our next result is a technical lemma, useful in proving that the product of two convergent power series is a convergent power series.

Lemma 3.17. *If $\sum u_n$ and $\sum v_n$ are absolutely convergent series and*

$$w_n = \sum_{p=0}^{n} u_p v_{n-p},$$

then $\sum w_n$ is absolutely convergent and $\sum w_n = (\sum u_n)(\sum v_n)$.

Proof. Let $\alpha_p = \sum_{n \geq p} |u_n|$ and $\beta_p = \sum_{n \geq p} |v_n|$. Then

$$\lim_p \alpha_p = 0 = \lim_p \beta_p. \tag{3.2}$$

Also

$$\sum_{n=0}^{N} |w_n| \leq \sum_{n=0}^{\infty} |u_n| \sum_{n=0}^{\infty} |v_n| = \alpha_0 \beta_0 < +\infty,$$

and therefore $\sum_{n=0}^{\infty} |w_n| < +\infty$. Thus we have proven the absolute convergence of the new series $\sum w_n$.

To show the required equality, choose m and n with $m \geq 2n$ and consider

$$\left| \sum_{k=0}^{m} w_k - \left(\sum_{k=0}^{n} u_k \right) \left(\sum_{k=0}^{n} v_k \right) \right| = \mathfrak{L}.$$

We have to show that $\mathfrak{L} \to 0$ as $n \to \infty$ (we already know each of the above series converges). We rewrite

$$\mathfrak{L} = \left| \sum_{k=0}^{m} \sum_{i=0}^{k} u_i v_{k-i} - \sum_{j=0}^{n} \sum_{i=0}^{n} u_i v_j \right|.$$

By looking at the diagrams in the (i,k) plane shown in Fig. 3.1, we see that

$$\begin{aligned} \sum_{k=0}^{m} \sum_{i=0}^{k} u_i v_{k-i} &= \sum_{i=0}^{m} \sum_{k=i}^{m} u_i v_{k-i} \\ &= \sum_{i=0}^{m} u_i \sum_{k=i}^{m} v_{k-i} = \sum_{i=0}^{m} u_i \sum_{j=0}^{m-i} v_j = \sum_{i=0}^{m} \sum_{j=0}^{m-i} u_i v_j. \end{aligned} \tag{3.3}$$

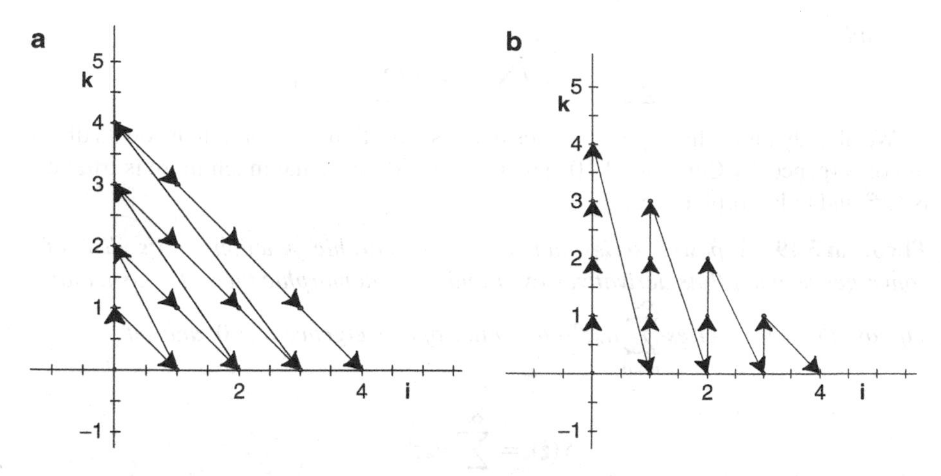

Fig. 3.1 The (i,k) plane. (**a**) The first sum in (3.3). (**b**) The second sum

Thus we can estimate

$$\begin{aligned}\mathfrak{L} &= \left|\sum_{i=0}^{n}\left[\sum_{j=0}^{m-i} u_i v_j - \sum_{j=0}^{n} u_i v_j\right] + \sum_{i=n+1}^{m}\sum_{j=0}^{m-i} u_i v_j\right| \\ &\le \left|\sum_{i=0}^{n}\sum_{j=n+1}^{m-i} u_i v_j\right| + \left|\sum_{i=n+1}^{m}\sum_{j=0}^{m-i} u_i v_j\right| \\ &\le \sum_{i=0}^{\infty}\sum_{j=n+1}^{\infty} |u_i|\,|v_j| + \sum_{i=n+1}^{\infty}\sum_{j=0}^{\infty} |u_i|\,|v_j| \\ &= \alpha_0\beta_{n+1} + \beta_0\alpha_{n+1},\end{aligned}$$

and the last expression approaches 0 as n goes to ∞, by (3.2). □

It is straightforward to prove the following result (see Exercise 3.6).

Theorem 3.18. *Suppose the power series* $\sum a_n z^n$ *and* $\sum b_n z^n$ *have radii of convergence that are both greater than or equal to a positive number* ρ*. Then*

(a) $\sum(a_n + b_n)z^n$ *and* $\sum c_n z^n$*, where* $c_n = \sum_{i=0}^{n} a_i b_{n-i}$*, have radii of convergence that are at least* ρ*.*
(b) *For* $|z| < \rho$*, we have*

$$\sum(a_n + b_n)z^n = \sum a_n z^n + \sum b_n z^n$$

and

$$\sum c_n z^n = \left(\sum a_n z^n\right)\left(\sum b_n z^n\right).$$

We already know that a power series defines a continuous function inside its disc of convergence, by Corollary 3.10. The next result shows that much more is true: it is $\mathbf{C}^\infty$ and holomorphic there.

Theorem 3.19. *A power series defines a holomorphic function in its disc of convergence, and all its derivatives exist and are holomorphic there. In particular, suppose the power series* $\sum_{n=0}^{\infty} a_n z^n$ *has radius of convergence* $\rho > 0$, *and set*

$$S(z) = \sum_{n=0}^{\infty} a_n z^n$$

for $|z| < \rho$. *Then the power series* $\sum_{n=1}^{\infty} n a_n z^{n-1}$ *also has radius of convergence* ρ, *and*

$$S'(z) = \sum_{n=1}^{\infty} n a_n z^{n-1}, \quad \textit{for all } |z| < \rho.$$

Proof. Let ρ' be the radius of convergence of $\sum_{n=1}^{\infty} n a_n z^{n-1}$. Then

$$\frac{1}{\rho'} = \overline{\lim_n}(n\,|a_n|)^{\frac{1}{n}} = \lim_n n^{\frac{1}{n}} \cdot \overline{\lim_n}\,|a_n|^{\frac{1}{n}} = 1 \cdot \overline{\lim_n}\,|a_n|^{\frac{1}{n}} = \frac{1}{\rho},$$

where the first and last equalities follow from Hadamard's theorem and the second one from Exercise 3.4. Thus we can define a continuous function T on $|z| < \rho$ by

$$T(z) = \sum_{n=1}^{\infty} n a_n z^{n-1}.$$

Note that if A and $B \in \mathbb{C}$ and $n \in \mathbb{Z}_{>1}$, then

$$A^n - B^n = (A - B)(A^{n-1} + A^{n-2}B + \cdots + B^{n-1}).$$

Let us set for $h \neq 0$ and $|h|$ sufficiently small and for fixed z with $|z| < \rho$,

$$f(h) = \left|\frac{S(z+h) - S(z)}{h} - T(z)\right|$$

$$= \left| \sum_{n=2}^{\infty} a_n \left[\frac{(z+h)^n - z^n}{h} - nz^{n-1} \right] \right|$$

$$= \left| \sum_{n=2}^{\infty} a_n \left[(z+h)^{n-1} + (z+h)^{n-2}z + \cdots + z^{n-1} - nz^{n-1} \right] \right|.$$

Now we choose r so that $|z| < r < \rho$, and then we choose h with $|z+h| < r$. Under these restrictions,

$$\begin{aligned} &\left|(z+h)^{n-1} + (z+h)^{n-2}z + \cdots + z^{n-1} - nz^{n-1}\right| \\ &\quad \le \left|(z+h)^{n-1} + z(z+h)^{n-2} + \cdots + z^{n-1}\right| + \left|nz^{n-1}\right| \\ &\quad \le 2nr^{n-1}, \end{aligned}$$

and thus

$$0 \le f(h) \le \sum_{n=2}^{\infty} 2\,|a_n|\,nr^{n-1} < +\infty.$$

Therefore given any $\epsilon > 0$ there exists a positive integer N with

$$\sum_{n=N}^{\infty} 2\,|a_n|\,nr^{n-1} < \frac{\epsilon}{2},$$

and we can write

$$0 \le f(h) \le \sum_{n=2}^{N-1} |a_n| \left| \frac{(z+h)^n - z^n}{h} - n\,z^{n-1} \right| + \sum_{n=N}^{\infty} 2\,|a_n|\,n\,r^{n-1}.$$

The first sum, being a finite sum, goes to 0 as $h \to 0$. Hence there exists a $\delta > 0$ such that $0 < |h| < \delta$ implies that the first term is at most $\frac{\epsilon}{2}$. Thus $0 \le f(h) < \epsilon$ for $0 < |h| < \delta$, and therefore $T(z) = S'(z)$. □

Remark 3.20. The theorem tells us that under certain circumstances we can interchange the order of computing limits: if the power series $\sum a_n z^n$ has positive radius of convergence and if we let

$$S_N(z) = \sum_{n=0}^{N} a_n z^n,$$

then the theorem states that

$$\lim_{h\to 0} \lim_{N\to\infty} \frac{S_N(z+h) - S_N(z)}{h} = \lim_{N\to\infty} \lim_{h\to 0} \frac{S_N(z+h) - S_N(z)}{h}.$$

Corollary 3.21. *If* $S(z) = \sum_{n=0}^{\infty} a_n z^n$ *for* $|z| < \rho$, *then for all* $n \in \mathbb{Z}_{\geq 0}$ *and all* $|z| < \rho$, *the power series for the derivatives* $S^{(n)} = \frac{d^n S}{dz^n}$ *are computed by term–term differentiation of* $\sum a_n z^n$, *and*

$$a_n = \frac{S^{(n)}(0)}{n!}.$$

Proof. Applying the theorem and induction on n shows that

$$S^{(n)}(z) = n!\, a_n + \frac{(n+1)!}{1!} a_{n+1}\, z + \cdots = \sum_{k=n}^{\infty} \frac{k!}{(n-k)!} a_k\, z^{n-k}$$

for all $|z| < \rho$. □

The results obtained so far have provided information about the behavior of a power series inside its disc of convergence. Our next result deals with a point on the boundary of this disc.

Theorem 3.22 (Abel's Limit Theorem). *Assume that the power series* $\sum a_n z^n$ *has finite radius of convergence* $\rho > 0$. *If* $\sum a_n z_0^n$ *converges for some* z_0 *with* $|z_0| = \rho$, *then* $f(z) = \sum a_n z^n$ *is defined for* $\{|z| < \rho\} \cup \{z_0\}$ *and we have*

$$\lim_{z \to z_0} f(z) = f(z_0),$$

as long as z *approaches* z_0 *from inside the circle of convergence and*

$$\frac{|z - z_0|}{\rho - |z|}$$

remains bounded.

Proof. By the change of variable $w = \frac{z}{z_0}$ we may assume that $\rho = 1 = z_0$ (replace a_n by $a_n z_0^n$). Thus $\sum a_n$ converges to $f(1)$. By changing a_0 to $a_0 - f(1)$, we may assume that $f(1) = \sum a_n = 0$. Thus we are assuming that $|z| < 1$ (with $1 - |z|$ small) and that $\frac{|1-z|}{1-|z|} \leq M$ for some fixed $M > 0$. (Recall Exercise 2.4.) Let

$$s_n = a_0 + a_1 + \cdots + a_n.$$

Then $\lim_n s_n = 0$,

$$\begin{aligned} S_n(z) &= a_0 + a_1 z + \cdots + a_n z^n \\ &= s_0 + (s_1 - s_0)z + \cdots + (s_n - s_{n-1})z^n \end{aligned}$$

$$= s_0(1-z) + s_1(z-z^2) + \cdots + s_{n-1}(z^{n-1}-z^n) + s_n z^n$$
$$= (1-z)(s_0 + s_1 z + \cdots + s_{n-1}z^{n-1}) + s_n z^n,$$

and hence

$$f(z) = \lim_{n\to\infty} S_n(z) = (1-z)\sum_{n=0}^{\infty} s_n z^n.$$

Given $\epsilon > 0$, choose $N \in \mathbb{Z}_{>0}$ such that $|s_n| < \epsilon$ for $n > N$. Then

$$\begin{aligned}
|f(z)| &\le |1-z| \left(\left| \sum_{n=0}^{N} s_n z^n \right| + \sum_{n=N+1}^{\infty} |s_n|\,|z|^n \right) \\
&\le |1-z| \left(\left| \sum_{n=0}^{N} s_n z^n \right| + \epsilon \frac{|z|^{N+1}}{1-|z|} \right) \\
&\le |1-z| \left| \sum_{n=0}^{N} s_n z^n \right| + \epsilon M,
\end{aligned}$$

and thus we conclude that $\lim_{z\to 1} f(z) = 0$. □

Remark 3.23. Observe that we have not needed nor have we used polar coordinates in our formal development thus far. After completing the next section, we will be free to do so.

3.3 The Exponential Function, the Logarithm Function, and Some Complex Trigonometric Functions

In this section we use power series to define and study several basic (and in some sense, familiar) transcendental functions.

3.3.1 The Exponential Function

We discover the exponential function by looking for functions that are solutions of the ordinary differential equation

$$f'(z) = f(z)$$

subject to the initial condition

$$f(0) = 1.$$

We try to find such a solution defined by a power series

$$f(z) = a_0 + a_1 z + \cdots + a_n z^n + \cdots$$

that converges near $z = 0$. Then (by Theorem 3.19)

$$f'(z) = a_1 + 2a_2 z + \cdots + n a_n z^{n-1} + \cdots$$

and thus we must have

$$a_0 = 1,\ a_1 = a_0 = 1,\ a_2 = \frac{1}{2} \cdot a_1 = \frac{1}{2!},\ \ldots;$$

that is,

$$a_{n+1} = \frac{1}{n+1} \cdot a_n = \frac{1}{(n+1)!} \quad \text{(by induction)}.$$

Hence

$$f(z) = 1 + z + \frac{z^2}{2!} + \cdots + \frac{z^n}{n!} + \cdots = \sum_{n=0}^{\infty} \frac{z^n}{n!}. \tag{3.4}$$

Now note that

$$\frac{1}{\rho} = \lim_n \left| \frac{a_{n+1}}{a_n} \right| = \lim_n \frac{n!}{(n+1)!} = 0;$$

thus $\rho = +\infty$, and the power series (3.4) defines an entire transcendental function.

We write e^z and $\exp z$ for $f(z)$ and call it the **exponential** function.

Recall that an *algebraic number* is a complex number that is the root of a nonconstant polynomial equation with rational (equivalently integer) coefficients, and a *transcendental number* is one that is not algebraic. A similar concept applies to functions.

Definition 3.24. Let f be a function defined on a nonempty domain $D \subseteq \mathbb{C}$. We say that f is an *algebraic function* if there exists a nonconstant complex polynomial $F = F(z, w)$ (in two variables) such that $F(z, f(z)) = 0$ for all z in D. Otherwise, f is called *transcendental*.[2]

Under the above definition, polynomials and rational functions are algebraic in their domains of definition. For holomorphic functions f on the domain D, we will be able to conclude by the principle of analytic continuation that the property of being algebraic or transcendental continues to hold if we replace D by a subdomain or superdomain (to which it extends). Exercise 3.9 shows that the exponential function is a transcendental function.

---

[2]It should be noted that this definition is appropriate for analysis; it is not so for algebra and number theory where it is appropriate to require that F be a polynomial with integer coefficients, that is, that it belongs to $\mathbb{Z}[z, w]$ rather than $\mathbb{C}[z, w]$ as in our case.

Example 3.25. At the end of the next section we will see that (well-defined, that is, single-valued) holomorphic power functions exist on appropriate subdomains of $\mathbb{C}$. It will follow that there exist precisely two square root functions on any simply connected domain D not containing the origin. If f is one of these, it satisfies the polynomial equation $F(z, f(z)) = 0$ where $F(z, w) = w^2 - z$.

Proposition 3.26. *Let $c \in \mathbb{C}$. The function $f(z) = c\,\mathrm{e}^z$ is the unique power series, and also the unique entire function, satisfying*

$$f'(z) = f(z) \text{ and } f(0) = c. \tag{3.5}$$

Proof. It is trivial that $z \mapsto c\mathrm{e}^z$ satisfies (3.5) and is the unique power series to do so; we already know that this is an entire function. We postpone the proof that this is the unique entire function that satisfies (3.5) until after we establish the next two propositions. □

Proposition 3.27. *For all $z \in \mathbb{C}$,*

$$\mathrm{e}^z\mathrm{e}^{-z} = 1.$$

Thus

$$\mathrm{e}^z \neq 0 \text{ for all } z \in \mathbb{C}.$$

Proof. Set $h(z) = \mathrm{e}^z\mathrm{e}^{-z}$ for all $z \in \mathbb{C}$. Then h is an entire function, and the rules for differentiation tell us that

$$h'(z) = \mathrm{e}^z\mathrm{e}^{-z} - \mathrm{e}^z\mathrm{e}^{-z} = 0$$

for all z. Therefore h must be constant, by Theorem 2.44; the result follows since $h(0) = 1$. □

Proposition 3.28. $\mathrm{e}^{z+c} = \mathrm{e}^z\mathrm{e}^c$ *for all z and c in $\mathbb{C}$.*

Proof. Define $h(z) = \dfrac{\mathrm{e}^{z+c}}{\mathrm{e}^c}$, with c fixed in $\mathbb{C}$. The function h has a power series expansion that converges for all $z \in \mathbb{C}$, $h'(z) = h(z)$ for all $z \in \mathbb{C}$, and $h(0) = 1$. Thus $h(z) = \mathrm{e}^z$, as needed. □

3.3.1.1 Conclusion of Proof of Proposition 3.26

If $g(z)$ is any entire function satisfying (3.5), consider the function $h(z) = \frac{g(z)}{\mathrm{e}^z}$. Then h is also entire, and the rules for differentiation tell us that

$$h'(z) = \frac{g'(z)\mathrm{e}^z - g(z)\mathrm{e}^z}{\mathrm{e}^{2z}} = 0 \text{ for all } z \in \mathbb{C}.$$

Thus h is constant, by Theorem 2.44.

Proposition 3.29. $\overline{e^z} = e^{\bar{z}}$ *for all* $z \in \mathbb{C}$.

Proof. This follows directly from the definition (3.4) of the exponential function. □

As an immediate consequence we have

Proposition 3.30. *For any* z *in* $\mathbb{C}$, *write* $z = x + \imath\, y$, *with* x *and* y *in* $\mathbb{R}$. *Then*

$$|e^{\imath y}|^2 = e^{\imath y} e^{-\imath y} = 1,$$

and thus

$$|e^z| = e^x.$$

The exponential function leads us immediately to our next section, the complex trigonometric functions.

3.3.2 The Complex Trigonometric Functions

We define two entire functions, the *cosine* and the *sine* functions, by

$$\cos z = \frac{e^{\imath z} + e^{-\imath z}}{2} = 1 - \frac{z^2}{2!} + \frac{z^4}{4!} - \cdots \tag{3.6}$$

and

$$\sin z = \frac{e^{\imath z} - e^{-\imath z}}{2\imath} = z - \frac{z^3}{3!} + \frac{z^5}{5!} - \cdots . \tag{3.7}$$

It is then easy to verify the following familiar properties:

$$\cos z + \imath\, \sin z = e^{\imath z}, \quad \cos^2 z + \sin^2 z = 1, \tag{3.8}$$

$$\cos(-z) = \cos z, \quad \sin(-z) = -\sin z,$$

and

$$\cos' z = -\sin z, \quad \sin' z = \cos z$$

for all z in $\mathbb{C}$.

Remark 3.31. Observe that the equality $e^{\imath z} = \cos z + \imath\, \sin z$ given in (3.8), called *Euler's formula*, is *not* (in general) the decomposition of the complex number $e^{\imath z}$ into its real and imaginary parts, since $\cos z$ and $\sin z$ are not necessarily real numbers.

For the same reason, the equality $\cos^2 z + \sin^2 z = 1$ *does not* imply that the cosine and the sine functions are bounded on $\mathbb{C}$, as they are on $\mathbb{R}$. In fact, they are not bounded on $\mathbb{C}$; see Exercise 3.13.

However, if $z = x + \imath y$, then we can compute the real and imaginary part of e^z as

$$e^z = e^x e^{\imath y} = e^x \cos y + \imath e^x \sin y. \tag{3.9}$$

In the next section we will formally define π, and then, after some calculations, we will be able to obtain from (3.8) the beautiful *Euler identity* (connecting perhaps the four most interesting numbers in mathematics):

$$e^{\pi \imath} + 1 = 0.$$

Remark 3.32. It should be observed that the functions sine and cosine defined above agree, for real values of the independent variable z, with the familiar real-valued functions with the same names. The easiest way to conclude this is from the power series expansions of these functions at $z = 0$. Also note that $\sin z$ and $\cos z$ form a basis for the power series solutions to the ordinary differential equation

$$f''(z) + f(z) = 0;$$

that is, every solution f of this equation is of the form

$$f(z) = a \cos z + b \sin z$$

for some complex constants a and b and for all z in $\mathbb{C}$.

Similarly, using either this last characterization of the sine and cosine functions, or the formula for the exponential of a sum, one establishes that for all z and $c \in \mathbb{C}$,

$$\cos(z + c) = \cos z \cos c - \sin z \sin c$$

and

$$\sin(z + c) = \sin z \cos c + \cos z \sin c.$$

3.3.3 The Definition of π and the Logarithm Function

In this section we show that the exponential function is periodic and hence not injective on $\mathbb{C}$. However, since it is locally injective, it will have local inverses.

In order to define a local inverse to the exponential function, our first task is then to establish the **periodicity of** e^z. We will need to use some elementary results from calculus. For $x \in \mathbb{R}$, $\sin x$ and $\cos x$ (as defined by (3.7) and (3.6), respectively) are real numbers. From $\sin^2 x + \cos^2 x = 1$, we conclude that

$$-1 \leq \cos x \leq 1.$$

Integrating for $x \geq 0$ we have $\int_0^x \cos t \, dt \leq \int_0^x dt$, or $\sin x \leq x$. Also, for $x > 0$, we must have $\sin x < x$.[3] Equivalently, $-\sin x > -x$. Thus, $\int_0^x (-\sin t)\, dt > \int_0^x (-t)\, dt$. We conclude that $\cos x - 1 > -\frac{x^2}{2}$ or, equivalently, $\cos x > 1 - \frac{x^2}{2}$. Repeating the argument, we obtain the following two inequalities for $x \neq 0$:

$$\sin x > x - \frac{x^3}{6} \quad \text{and} \quad \cos x < 1 - \frac{x^2}{2} + \frac{x^4}{24}. \tag{3.10}$$

Definition 3.33. Let f be a complex-valued function defined on $\mathbb{C}$ and let $c \in \mathbb{C}$, $c \neq 0$. We say that f has *period* c (and call f *periodic*) if and only if $f(z+c) = f(z)$ for all $z \in \mathbb{C}$.

We show next that the exponential function is periodic (i.e., it has a period). Note that $e^{z+c} = e^z$ for all z if and only if $e^c = 1$. It follows from (3.9) that $c = \imath\,\omega$, with $\omega \in \mathbb{R}$. Traditionally ω (and not $\imath\,\omega$) is called a period of the exponential function. We want to determine the smallest such positive ω.

First note that $\cos 0 = 1$ (obvious). It follows from the inequalities (3.10) that

$$\cos\sqrt{3} < 1 - \frac{3}{2} + \frac{9}{24} = -\frac{1}{8} < 0,$$

and then continuity implies that there exists a $y_0 \in (0, \sqrt{3})$ such that $\cos y_0 = 0$.

But then from $\cos^2 y_0 + \sin^2 y_0 = 1$ we obtain that $\sin y_0 = \pm 1$, and thus $e^{\imath y_0} = \cos y_0 + \imath \sin y_0 = \pm\imath$ and $e^{4\imath y_0} = 1$. We conclude that $4y_0$ is a period of the exponential function. We claim that this is the smallest positive period and that any other period is an integral multiple of this one.

Proof. If $0 < y < y_0 < \sqrt{3}$, then $y^2 < 3$ and $1 - \frac{y^2}{6} > \frac{1}{2}$; thus $\sin y > y\,(1 - \frac{y^2}{6}) > \frac{y}{2} > 0$, and we conclude that $\cos y$ is strictly decreasing on $[0, y_0]$.

Since $\cos^2 y + \sin^2 y = 1$ and $\sin y > 0$ on $(0, y_0)$, we conclude that $\sin y$ is strictly increasing here. Thus $0 < \sin y < \sin y_0 = 1$.

Hence $e^{\imath y} \neq \pm 1, \pm\imath$, and therefore $e^{4\imath y} \neq 1$ for $0 < y < y_0$. Thus $\omega_0 = 4y_0$ is the smallest positive period.

If ω is an arbitrary period of exp (recall that $\omega \in \mathbb{R}$), then so is $|\omega|$ and there is an $n \in \mathbb{Z}_{>0}$ such that $n\omega_0 \leq |\omega| < (n+1)\omega_0$. If $n\omega_0 \neq |\omega|$, then $0 < (n+1)\omega_0 - |\omega|$ is a positive period less than ω_0. Since this is impossible by definition of ω_0, we must have $|\omega| = n\,\omega_0$. □

Definition 3.34 (Definition of π). We define the real number π by $4y_0 = 2\pi$. Thus $e^z = 1$ if and only if $z = 2\pi\imath n$ with $n \in \mathbb{Z}$.

---

[3]The function $x \mapsto x - \sin x$ is certainly nondecreasing on $[0, +\infty)$ since its derivative is the function $x \mapsto 1 - \cos x \geq 0$. The inequality $\sin x < x$ certainly holds for $x > 1$. If for some x_0 in $(0, 1]$ we would have $\sin x_0 = x_0$, then we would conclude from the mean value theorem that for some $\tilde{x} \in (0, x_0)$, $\cos\tilde{x} = 1$ which leads to the contradiction $\sin\tilde{x} = 0$.

Traditionally 2π is defined as the smallest positive period of the (complex) exponential function. After some more work, it will become clear that our definition of the constant π agrees with the standard approach used in real analysis.

As in real analysis, the inverse to the exponential function should be a **logarithm**. We now turn to its definition. Since $\frac{d}{dx}(e^x) = e^x > 0$ for all $x \in \mathbb{R}$, e^x is strictly increasing on $\mathbb{R}$. Hence there exists an inverse function denoted by log (sometimes written as ln):

$$\exp : \mathbb{R} \to (0, +\infty),$$
$$\log : (0, +\infty) \to \mathbb{R},$$

and we have the well-known properties

$$\log e^x = x \text{ for all } x \in \mathbb{R} \text{ and } e^{\log x} = x \text{ for all } x \in \mathbb{R}_{>0}.$$

We know that $e^z \neq 0$ for all $z \in \mathbb{C}$; we thus can expect to define a complex logarithm. The problem is that the exponential function

$$\exp : \mathbb{C} \to \mathbb{C}_{\neq 0}$$

is not one-to-one. Let us write $z \neq 0$ in polar coordinates[4] $z = re^{\imath\theta}$ (this agrees with our previous way of writing polar coordinates). Here $r = |z|$ and $\theta = \arg z$. The argument of z is defined up to addition of $2\pi n$ with $n \in \mathbb{Z}$. We also define

$$\log z = \log |z| + \imath \arg z;$$

it is a multivalued function[5] on $\mathbb{C}_{\neq 0}$.

The *principal branch of* $\arg z$, $\operatorname{Arg} z$, is restricted to lie in $(-\pi, \pi]$. It is a function on $\mathbb{C} - \{0\}$ that has a jump discontinuity on the negative real axis. We define the *principal branch of the logarithm* by the formula

$$\operatorname{Log} z = \log |z| + \imath \operatorname{Arg} z.$$

It is easily seen to be a continuous function on $\mathbb{C} - (-\infty, 0]$; moreover, it is $\mathbf{C}^1$ on this set. We note the following.

Properties of the Complex Logarithm.

1. $e^{\log z} = e^{\log|z| + \imath \arg z} = |z|\, e^{\imath \arg z} = z$ for all $z \in \mathbb{C}_{\neq 0}$.
2. $\operatorname{Log} z$ is holomorphic on $\mathbb{C} - (-\infty, 0]$, with $\dfrac{d}{dz}\operatorname{Log} z = \dfrac{1}{z}$ there.

Proof. Write $z = re^{\imath\theta}$ with $-\pi < \theta < \pi$. Thus

$$\operatorname{Log} z = \log r + \imath\theta = u + \imath v.$$

---

[4]Having defined π, polar coordinates now rest on a solid foundation and can be used in proofs.

[5]Thus not a function.

Calculate $u_r = \frac{1}{r}$, $v_\theta = 1$, $u_\theta = 0$, and $v_r = 0$, and thus $ru_r = v_\theta$ and $rv_r = -u_\theta$. Hence (as observed earlier) Log is $\mathbf{C}^1$ and satisfies CR on $\mathbb{C} - (-\infty, 0]$ (see Exercise 2.7); thus it is holomorphic there. We can now compute formally using the chain rule:

$$e^{\text{Log}\, z} = z.$$

Thus

$$e^{\text{Log}\, z} \frac{d}{dz} \text{Log}\, z = 1,$$

and we conclude that

$$\frac{d}{dz} \text{Log}\, z = e^{-\text{Log}\, z} = \frac{1}{z}.$$

3. Log is injective.

4.

$$\text{Log}\, z_1 z_2 = \text{Log}\, z_1 + \text{Log}\, z_2 + \delta,$$

where

$$\delta = \begin{cases} 2\imath\pi, & \text{if } -2\pi < \text{Arg}z_1 + \text{Arg}z_2 \le -\pi; \\ 0, & \text{if } -\pi < \text{Arg}z_1 + \text{Arg}z_2 \le \pi; \\ -2\imath\pi, & \text{if } \pi < \text{Arg}z_1 + \text{Arg}z_2 \le 2\pi. \end{cases}$$

□

Definition 3.35. A continuous function f on a domain $D \subset \mathbb{C}$ not containing the origin is called a *branch of the logarithm on* D if for all $z \in D$, we have $e^{f(z)} = z$.

Later we will establish appropriate conditions on the domain D for a branch of the logarithm to exist on it.

We have seen that Log is a branch of the logarithm on $D = \mathbb{C} - (-\infty, 0]$ (or in any subdomain of D); other branches of the logarithm in D are given by changing the choice we made for arg. Our next result shows that this is a general property.

Theorem 3.36. *Let D be a domain in $\mathbb{C}$ with $0 \notin D$. If f is a branch of the logarithm on D, then g is also a branch of the logarithm in D if and only if there is an $n \in \mathbb{Z}$ such that $g(z) = f(z) + 2\pi\imath n$ for all z in D.*

Proof. If $g = f + 2\pi\imath n$ with $n \in \mathbb{Z}$, then $e^{g(z)} = e^{f(z)} e^{2\pi\imath n} = z$ for all z in D.

For a proof of the converse, define

$$h(z) = \frac{f(z) - g(z)}{2\pi\imath}, \quad z \in D.$$

Then $e^{2\pi\imath h(z)} = e^{f(z)} e^{-g(z)} = z\frac{1}{z} = 1$. Thus, for each $z \in D$, there is an $n \in \mathbb{Z}$ such that $h(z) = n$. Hence $h(D) \subseteq \mathbb{Z}$. Since h is continuous and D is connected, $h(D) = \{n\}$ for some fixed $n \in \mathbb{Z}$. □

Corollary 3.37. *Every branch of the logarithm on a domain D (with $0 \notin D$) is holomorphic on D.*

Proof. Holomorphicity is a local property, and there exist holomorphic branches of the logarithm in every sufficiently small disc that does not contain the origin. □

Theorem 3.38. *For $z \in \mathbb{C}$ with $|z| < 1$,*

$$\mathrm{Log}(1+z) = \sum_{n=1}^{\infty} (-1)^{n-1} \frac{z^n}{n} = z - \frac{z^2}{2} + \frac{z^3}{3} - \cdots .$$

Proof. We first compute the radius of convergence of the given series using the ratio test: $\frac{1}{\rho} = \lim_n \left| \frac{n}{n+1} \right| = 1$. Thus the function defined by

$$f(z) = \sum_{n=1}^{\infty} (-1)^{n-1} \frac{z^n}{n} \tag{3.11}$$

is holomorphic in $|z| < 1$. We calculate

$$f'(z) = 1 - z + z^2 - \cdots = \frac{1}{1+z} \quad \text{for } |z| < 1.$$

Let $g(z) = \mathrm{e}^{f(z)}$; then

$$g'(z) = \mathrm{e}^{f(z)} f'(z) = \frac{\mathrm{e}^{f(z)}}{1+z},$$

and

$$g''(z) = \frac{(1+z)\mathrm{e}^{f(z)} f'(z) - \mathrm{e}^{f(z)}}{(1+z)^2} = 0.$$

Thus $g'(z) = \alpha$, a constant, and

$$\mathrm{e}^{f(z)} = \alpha(1+z).$$

Now $f(0) = 0$ tells us that $\alpha = 1$. Thus $f(z)$ defines a branch of $\log(1+z)$ in $|z| < 1$, and hence it differs from the principal branch of $\log(1+z)$ by an integral multiple of $2\pi\imath$. But it follows from the definition of f that for every $x \in (-1, 1)$, $f(x)$ is a real number. Thus $f(z) = \mathrm{Log}(1+z)$. □

Definition 3.39. Let D be a domain in $\mathbb{C}$ with $0 \notin D$. A branch f of the logarithm on D is the *principal branch of the logarithm* on this domain if $f(z) = \mathrm{Log}\, z$ for all $z \in D \cap (\mathbb{C} - (-\infty, 0])$. This means, in particular, that if $x > 0$ belongs to D, then $f(x) \in \mathbb{R}$.

We have established in the last theorem that the principal branch of the logarithm on the disc $U(1,1)$ is given by the power series:

$$\operatorname{Log} z = \sum_{n=1}^{\infty} (-1)^{n-1} \frac{(z-1)^n}{n}, \quad z \in U(1,1).$$

Complex exponentials are defined by $z^c = e^{c \log z}$ for $c \in \mathbb{C}$ and $z \in \mathbb{C}_{\neq 0}$ and the principal branch of z^c by $e^{c \operatorname{Log} z}$. Thus if L is a branch of the logarithm on a domain D, then for any $c \in \mathbb{C}$, $f_c : z \mapsto e^{cL(z)}$ is a holomorphic function on D with the usual expected properties. For example, if $n \in \mathbb{Z}$, $f_{\frac{1}{n}}^n(z) = z$ for all $z \in D$.

3.4 An Identity Principle

Holomorphic functions are remarkably rigid: if f is a holomorphic function defined on an open connected set D, then the knowledge of its behavior at a single point $c \in D$ (i.e., knowing the values of all the derivatives of f at the point c) or the values of f on a sequence in D with a limit point in D are sufficient to uniquely determine its properties (in particular, its value $f(d)$) at any arbitrary point $d \in D$.[6] We now start on the exciting journey to establish this and other beautiful results, at first for what seems to be only a subclass of the holomorphic functions.

Definition 3.40. A function f defined in a neighborhood of $c \in \mathbb{C}$ has *a power series expansion at c* if there exists an $r > 0$ such that

$$f(z) = \sum_{n=0}^{\infty} a_n (z-c)^n \text{ for } |z-c| < r \leq \rho,$$

where $\rho > 0$ is the radius of convergence of the power series $\sum a_n w^n$ (in the variable $w = z - c$).

Theorem 3.41. *Let f be a function defined in a neighborhood of $c \in \mathbb{C}$ that has a power series expansion at c with radius of convergence $\rho > 0$. Then*

(a) *f is holomorphic and $\mathbf{C}^\infty$ in a neighborhood of c.*
(b) *If g also has a power series expansion at c and if the product $f \cdot g$ is identically zero in a neighborhood of c, then either f or g is identically zero in some neighborhood of c.*
(c) *There exists a function h defined in a neighborhood of c that has a power series expansion at c, with the same radius of convergence ρ, such that $h' = f$. The function h is unique up to an additive constant.*

[6] We do not describe any algorithms for computing these values.

Proof. Without loss of generality we assume $c = 0$:

(a) Already verified in Theorem 3.19.
(b) For some $r > 0$, we have

$$f(z) = \sum_{n=0}^{\infty} a_n z^n \text{ and } g(z) = \sum_{n=0}^{\infty} b_n z^n \text{ for all } |z| < r.$$

Suppose that neither f nor g vanish identically in any neighborhood of $c = 0$, and choose the smallest nonnegative integers N and M such that $a_N \neq 0$ and $b_M \neq 0$. We know that

$$(f \cdot g)(z) = \sum_{n=0}^{\infty} c_n z^n \text{ for } |z| < r,$$

where

$$c_n = \sum_{p+q=n} a_p b_q.$$

Thus $c_n = 0$ for $0 \leq n \leq N + M - 1$ and $c_{N+M} = a_N b_M \neq 0$. But

$$c_{N+M} = \frac{1}{(N+M)!} \left(\frac{d^{N+M}(f \cdot g)}{dz^{N+M}} \right)(0),$$

and therefore $f \cdot g$ cannot be identically zero near 0.

(c) Define $h(z) = \sum_{n=0}^{\infty} \frac{a_n}{n+1} z^{n+1}$. Then the radius of convergence ρ' of h satisfies

$$\frac{1}{\rho'} = \limsup_n \left| \frac{a_n}{n+1} \right|^{\frac{1}{n}} = \limsup_n |a_n|^{\frac{1}{n}} = \frac{1}{\rho}.$$

It now follows from Theorem 3.19 that $h'(z) = f(z)$ for all $|z| < \rho$. If H is another function that has a power series expansion at 0 and satisfies $H'(z) = f(z)$ for all $|z| < \rho$, then the holomorphic function $H - h$ has derivative equal to zero for all $|z| < \rho$ and must be a constant. □

We will see later that (c) can be strengthened significantly: if f is holomorphic on a simply connected open set D, then h extends to all of D.

The following lemma is a useful tool with significant applications beyond the immediate one.

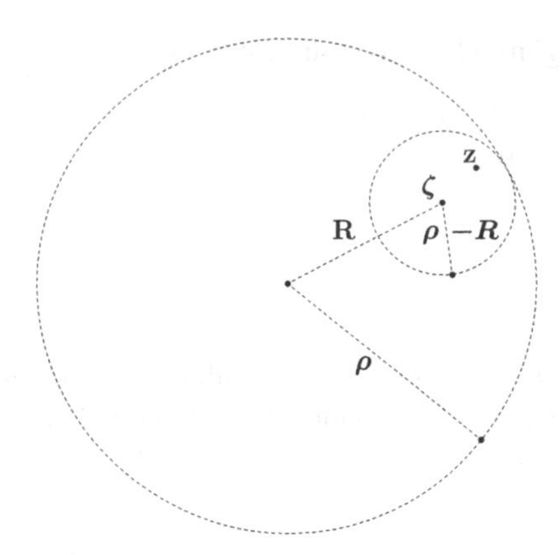

Fig. 3.2 Radii of convergence

Lemma 3.42. *If* $S(z) = \sum_{n=0}^{\infty} a_n z^n$ *has radius of convergence* $\rho > 0$*, then for any* $c \in \mathbb{C}$ *with* $|c| < \rho$*, the power series* $\sum_{n=0}^{\infty} \frac{S^{(n)}(c)}{n!} w^n$ *has radius of convergence* $\geq \rho - |c|$*, and*

$$S(z) = \sum_{n=0}^{\infty} \frac{S^{(n)}(c)}{n!}(z-c)^n \text{ for } |z-c| < \rho - |c| .$$

Proof. Set $R = |c| < \rho$ (see Fig. 3.2). The argument consists of two steps:

(I) We show first that $\sum_{p=0}^{\infty} \frac{S^{(p)}(c)}{p!} w^p$ is absolutely convergent for $|w| < \rho - R$.

We know from Corollary 3.21 that

$$S^{(p)}(c) = \sum_{n=p}^{\infty} a_n \frac{n!}{(n-p)!} c^{n-p}.$$

If we let $q = n - p$ and set $b_{p+q} = |a_n|$, then it follows that

$$\left|S^{(p)}(c)\right| \leq \sum_{q=0}^{\infty} b_{p+q} \frac{(p+q)!}{q!} R^q.$$

Now choose $r \in \mathbb{R}$ with $R < r < \rho$. Then

$$\sum_{p=0}^{\infty} \left| \frac{S^{(p)}(c)}{p!} \right| (r-R)^p \leq \sum_{p,q} b_{p+q} \frac{(p+q)!}{p!\,q!} R^q (r-R)^p.$$

Returning to the original variable $n = p+q$, we continue to estimate the above expression

$$= \sum_{n=0}^{\infty} b_n \sum_{p=0}^{n} \frac{n!}{p!(n-p)!} R^{n-p} (r-R)^p$$

$$= \sum_{n=0}^{\infty} b_n (r-R+R)^n$$

$$= \sum_{n=0}^{\infty} |a_n|\, r^n < +\infty,$$

where the last series converges because the power series $\sum_{n=0}^{\infty} a_n z^n$ converges absolutely in $|z| < \rho$.

(II) We show next that

$$S(z) = \sum_{p=0}^{\infty} \frac{S^{(p)}(c)}{p!} (z-c)^p$$

for $|z-c| < \rho - R$.

We know from (I) that the series

$$\sum_{p=0}^{\infty} \frac{S^{(p)}(c)}{p!} (z-c)^p = \sum_{p=0}^{\infty} \sum_{n=p}^{\infty} \frac{a_n\, n!}{p!(n-p)!} c^{n-p} (z-c)^p$$

converges absolutely for $|z-c| < \rho - R$. Hence we may rearrange the order of the terms and sums (see Exercise 3.16), and the argument proceeds exactly as in part (I).

□

Example 3.43. We study the holomorphic function

$$S(z) = \frac{1}{1-z},\ z \in \mathbb{C}_{\neq 1};$$

it satisfies

$$S(z) = 1 + z + z^2 + \cdots \text{ for } |z| < 1;$$

that is, S has a power series expansion at the origin with radius of convergence $\rho = 1$. Thus for the function S defined on $\mathbb{C}_{\neq 1}$, the power series representation with center at 0 is only valid for $|z| < 1$ and not for any other values of z.

Let us take $c = -\frac{1}{2}$. Then $S^{(p)}(z) = p!\,(1-z)^{-1-p}$ and thus $S^{(p)}\left(-\frac{1}{2}\right) = p!\,\left(\frac{2}{3}\right)^{1+p}$. A calculation shows that the power series $\sum_{p=0}^{\infty} \left(\frac{2}{3}\right)^{1+p} w^p$ has radius of convergence $\rho' = \frac{3}{2}$, and it follows from the lemma that

$$S(z) = \frac{1}{1-z} = \sum_{p=0}^{\infty} \left(\frac{2}{3}\right)^{1+p} \left(z+\frac{1}{2}\right)^p \quad \text{for } \left|z+\frac{1}{2}\right| < \frac{3}{2}.$$

We do not yet have (as we soon will) the machinery to conclude the last expansion without calculation.Recall that the power series expansion for S at 0 is just a consequence of the convergence of the geometric series (see (3.1)). So is the expansion for S at $-\frac{1}{2}$, as can be seen from

$$\frac{1}{1-z} = \frac{1}{\frac{3}{2} - \left(z+\frac{1}{2}\right)} = \frac{2}{3}\frac{1}{1-\frac{2}{3}\left(z+\frac{1}{2}\right)}.$$

Note that $\rho - |c| = \frac{1}{2} < \rho' = \frac{3}{2}$. What we see in this example is not an accident, as will soon become clear: the radius of convergence of each power series is the distance from its center to the nearest singularity of the function.

Corollary 3.44. *If $f(z) = \sum a_n z^n$ for $|z| < \rho$, then f has a power series expansion at each point c with $|c| < \rho$.*

The next result is called an (perhaps, "the") *identity principle*; it provides necessary and sufficient conditions for a function that has a power series expansion at each point of a connected domain of definition to vanish identically. It is usually applied in the form given by Corollary 3.46.

Theorem 3.45. *Let f be a function defined on a domain D in $\mathbb{C}$. Assume that f has a power series expansion at each point of D, and let $c \in D$. The following conditions are equivalent:*

(a) *$f^{(n)}(c) = 0$ for $n = 0, 1, 2, \ldots$.*
(b) *$f \equiv 0$ in a neighborhood of c.*
(c) *There exists a sequence $\{z_n\}$ consisting of distinct points of D with $\lim_n z_n = c$ and $f(z_n) = 0$ for each n.*
(d) *$f \equiv 0$ on D.*

Proof. First note that there are obvious implications: $(a) \Leftrightarrow (b)$ and $(d) \Rightarrow (b) \Rightarrow (c)$. To complete the proof, we will now show that $(c) \Rightarrow (a)$ and that $(a) \Rightarrow (d)$.

$(c) \Rightarrow (a)$: We know that $f(z) = \sum_{n=0}^{\infty} a_n(z-c)^n$ for all $|z-c| < \rho$

with $\rho > 0$. Furthermore, $a_k = \dfrac{f^{(k)}(0)}{k!}$ for all k, and in particular $a_0 = f(c) = \lim_n f(z_n) = 0$. Assume by induction that

$$0 = f(c) = \cdots = f^{(n)}(c)$$

for some integer $n \geq 0$. Then

$$f(z) = \sum_{p=n+1}^{\infty} a_p(z-c)^p = (z-c)^{n+1} \sum_{p=0}^{\infty} a_{n+1+p}(z-c)^p = (z-c)^{n+1} g(z).$$

Now, without loss of generality, we assume $z_n \neq c$ for all n. The function g has a power series expansion at c, and obviously $g(z_n) = 0$ if and only if $f(z_n) = 0$. Thus $g(c) = 0 = a_{n+1}$.

$(a) \Rightarrow (d)$: Let us define

$$D' = \{z \in D;\, f \equiv 0 \text{ in a neighborhood of } z\}.$$

The set D' is trivially open in D. Since we already know that (a) and (b) are equivalent, we can write

$$D' = \bigcap_{n=0}^{\infty} \{z \in D;\, f^{(n)}(z) = 0\};$$

then D' is the intersection of a countable family of closed subsets of D, and hence closed in D. But D' is not empty because $c \in D'$, and since D is connected we obtain $D' = D$. □

3.5 Zeros and Poles

The most important and the first practical consequence of the work of the last section is the next corollary. The results of Sect. 3.4 will also allow us to introduce an important class of functions; these are the meromorphic functions, taking values in the extended complex plane $\mathbb{C} \cup \{\infty\}$ rather than just in $\mathbb{C}$.

Corollary 3.46 (Principle of Analytic Continuation). *Let D be a domain in $\mathbb{C}$, and let f and g be functions defined on D having power series expansions at each point of D. If f and g agree on a sequence of distinct points in D with a limit point in D, or if they have identical power series expansions at a single point in D, then $f \equiv g$ (on D).*

Example 3.47. The exponential function e^z is the unique extension of e^x in the class of functions under study; that is, it is the unique function on $\mathbb{C}$ that has a power series expansions at each point and agrees with e^x at each point $x \in \mathbb{R}$.

Remark 3.48. One often uses *analytic continuation along a path.* To start with the simplest case, let f be a convergent power series centered at $c \in \mathbb{C}$ with radius of convergence $\rho > 0$; thus an analytic function on $\{|z-c| < \rho\}$. If c^* is a point with $|c^* - c| < \rho$, then the power series expansion f^* of f about c^* has a radius of convergence $\rho^* \geq \rho - |c^* - c|$. If $\rho^* > \rho - |c^* - c|$, then we have extended f to $\{|z-c| < \rho\} \cup \{|z-c^*| < \rho^*\}$, and we say that f^* is a *direct analytic continuation* to c^* of the power series f at c. We have seen this phenomenon in Example 3.43.

More generally, it is of interest to start with a continuous path γ in $\mathbb{C}$ from c to d. Assume that the path can be covered by discs $D_0, D_1, \ldots, D_n$ with centers $c = c_0, c_1, \ldots, c_n = d$ and radii $r_0, r_1, \ldots, r_n$, respectively, such that $D_j \cap D_{j+1}$ is not empty for $0 \leq j \leq n-1$. Assume further that there are convergent power series f_j centered at c_j with radius of convergence $\geq r_j$ and such that f_{j+1} is a direct analytic continuation to c_{j+1} of f_j at c_j; then we say that f_n is an *analytic continuation along* γ of f_0.

Example 3.49. We outline two applications of the concepts discussed for holomorphic functions. (We shall see that these are precisely the functions having convergent power series expansions.) Both can be established after the material in the next two chapters.

If D is a simply connected domain and f is a holomorphic function in a neighborhood of $c \in D$ that can be continued analytically along any path in D starting at c, then this continuation defines an analytic function on D. In more generality, let γ_1 and γ_2 be two continuous paths in a domain D with the same end points. Let f be a holomorphic function in a neighborhood of the initial point c of the two paths that can be continued analytically along all paths in D starting at c. If γ_1 and γ_2 are homotopic in D with fixed end points,[7] then the continuations of f along the two paths lead to the same holomorphic function in a neighborhood of their common end point. This result is known as the *monodromy theorem.*

Let a, b, and $c \in \mathbb{C}$ be subject to the conditions that each of the three differences $c-1$, $a-b$, and $(a+b)-c$ is not an integer. We introduce[8] the *hypergeometric functions* by solving the ordinary differential equation

[7] These concepts are defined in the next chapter.

[8] See L.V. Ahlfors, *Complex Analysis* (third edition), McGraw-Hill, 1979 for more details, including the motivation for the form of the differential equation.

$$z(1-z)w'' + [c - (1+a+b)z]w' - abw = 0.$$

This equation has a solution that is holomorphic in a neighborhood of the origin. The *hypergeometric function* $F_{a,b,c}$ is that solution normalized to assume the value 1 at 0. While it is easy to obtain the Taylor series expansion for $F_{a,b,c}$ at 0 and then compute that its radius of convergence is 1, the monodromy theorem and the theory of solutions of linear differential equations allow us to reach the same conclusion without calculations; the same facts lead us to conclude that $F_{a,b,c}$ can be analytically continued to become an analytic function on any simply connected domain in $\mathbb{C}_{\neq 1}$.

Another immediate consequence of Theorem 3.45 is

Corollary 3.50. *If K is a compact subset of a domain D and f is a nonconstant function that has a power series expansion at each point of D, then f has finitely many zeros in K.*

Definition 3.51. Let $c \in \mathbb{C}$. Assume that

$$f(z) = \sum_{n=0}^{\infty} a_n (z-c)^n \quad \text{for all} \quad |z-c| < \rho \text{ and for some } \rho > 0.$$

If f is not identically zero, it follows from Theorem 3.45 that there exists an $N \in \mathbb{Z}_{\geq 0}$ such that

$$a_N \neq 0 \text{ and } a_n = 0 \text{ for all } n \text{ such that } 0 \leq n < N.$$

Thus

$$f(z) = (z-c)^N \sum_{p=0}^{\infty} a_{N+p}(z-c)^p = (z-c)^N g(z),$$

with g having a power series expansion at c and $g(c) \neq 0$. We define

$$N = \nu_c(f) = \text{ order (of the zero) of } f \text{ at } c.$$

Note that $N \geq 0$, and $N = 0$ if and only if $f(c) \neq 0$. If $N = 1$, then we say that f has a *simple* zero at c.

Definition 3.52. (a) Let f be defined in a deleted neighborhood of $c \in \mathbb{C}$ (see the Standard Terminology summary). We say that

$$\lim_{z \to c} f(z) = \infty$$

if for all $M > 0$, there exists a $\delta > 0$ such that

$$0 < |z-c| < \delta \Rightarrow |f(z)| > M.$$

(b) Let $\alpha \in \widehat{\mathbb{C}}$, and let f be defined in $|z| > M$ for some $M > 0$ (equivalently, we say that f is defined in a *deleted neighborhood* of ∞ in $\widehat{\mathbb{C}}$). We say

$$\lim_{z \to \infty} f(z) = \alpha$$

provided

$$\lim_{z \to 0} f\left(\frac{1}{z}\right) = \alpha.$$

(c) The above defines the concept of continuous maps between sets in the Riemann sphere $\widehat{\mathbb{C}}$.

(d) A function f defined in a neighborhood of ∞ is *holomorphic (has a power series expansion)* at ∞ if and only if $g(z) = f\left(\frac{1}{z}\right)$ is holomorphic (has a power series expansion) at $z = 0$, where we define $g(0) = f(\infty)$.

Definition 3.53. Let $U \subset \mathbb{C}$ be a neighborhood of a point c. A function f that is holomorphic in $U' = U - \{c\}$, a deleted neighborhood of the point c, has a *removable singularity* at c if there is a holomorphic function in U that agrees with f on U'. Otherwise c is called a *singularity* of f. Note that all singularities are isolated points.

Let us consider two functions f and g having power series expansions at each point of a domain D in $\widehat{\mathbb{C}}$. Assume that neither function vanishes identically on D and fix $c \in D \cap \mathbb{C}$. Let

$$F(z) = \frac{f(z)}{(z-c)^{\nu_c(f)}} \quad \text{and} \quad G(z) = \frac{g(z)}{(z-c)^{\nu_c(g)}}$$

for $z \in D$. Then the functions F and G have removable singularities at c, do not vanish there, and have power series expansions at each point of D. Furthermore, we define a new function h on D by

$$h(z) = \frac{f}{g}(z) = \frac{(z-c)^{\nu_c(f)}\, F(z)}{(z-c)^{\nu_c(g)}\, G(z)} \quad \text{for all } z \in D$$

and fixed $c \in D \cap \mathbb{C}$.

There are exactly three distinct possibilities for the behavior of the function h at $z = c$, which lead to the following definitions.

Definition 3.54. (I) If $\nu_c(g) > \nu_c(f)$, then $h(c) = \infty$ (this defines $h(c)$, and the resulting function h is continuous at c). We say that h has a *pole of order* $\nu_c(g) - \nu_c(f)$ at c. If $\nu_c(g) - \nu_c(f) = 1$, we say that the pole is *simple*.

(II) If $\nu_c(g) = \nu_c(f)$, then the singularity of h at c is *removable*, and, by definition, $h(c) = \dfrac{F(c)}{G(c)} \neq 0$.

(III) If $\nu_c(g) < \nu_c(f)$, then the singularity is again removable and in this case $h(c) = 0$.

In all cases we set $\nu_c(h) = \nu_c(f) - \nu_c(g)$ and call it the *order* or *multiplicity* of h at c.

In cases (II) and (III) of the definition, h has a power series expansion at c as a consequence of the following result.

Theorem 3.55. *If a function f has a power series expansion at c and $f(c) \neq 0$, then $\frac{1}{f}$ also has a power series expansion at c.*

Proof. Without loss of generality we assume $c = 0$ and $f(0) = 1$. Thus

$$f(z) = \sum_{n=0}^{\infty} a_n z^n, \ a_0 = 1,$$

and the radius of convergence of the series is nonzero. We want to find the reciprocal power series, that is, a series g with positive radius of convergence, that we write as

$$g(z) = \sum_{n=0}^{\infty} b_n z^n$$

and satisfies

$$\left(\sum a_n z^n\right)\left(\sum b_n z^n\right) = 1.$$

The LHS and the RHS are both power series, where the RHS is a power series expansion whose coefficients are all equal to zero except for the first one. Equating the first two coefficients on both sides, we obtain

$$a_0 b_0 = 1, \quad \text{from where } b_0 = 1, \text{ and}$$
$$a_1 b_0 + a_0 b_1 = 0, \quad \text{from where } b_1 = -a_1 b_0 = -a_1.$$

Similarly, using the n-th coefficient of the power series when expanded for the LHS, for $n \geq 1$, we obtain

$$a_n b_0 + a_{n-1} b_1 + \cdots + a_0 b_n = 0.$$

Thus by induction we define

$$b_n = -\sum_{j=0}^{n-1} b_j a_{n-j}, \ n \geq 1.$$

Since $\rho > 0$, we have $\frac{1}{\rho} < +\infty$. Since $\limsup_n |a_n|^{\frac{1}{n}} = \frac{1}{\rho}$, there exists a positive number k such that $|a_n| \leq k^n$.

We show by the use of induction, once again, that $|b_n| \leq 2^{n-1} k^n$ for all $n \geq 1$. For $n = 1$, we have $b_1 = -a_1$ and hence $|b_1| = |a_1| \leq k$. Suppose the inequality holds for $1 \leq j \leq n$ for some $n \geq 1$. Then

$$\begin{aligned} |b_{n+1}| &\leq \sum_{j=0}^{n} |b_j| \, |a_{n+1-j}| = |a_{n+1}| + \sum_{j=1}^{n} |b_j| \, |a_{n+1-j}| \\ &\leq k^{n+1} + \sum_{j=1}^{n} 2^{j-1} k^j k^{n+1-j} \\ &= k^{n+1}(1 + 2^n - 1). \end{aligned}$$

Thus there is a reciprocal series, with radius of convergence σ satisfying

$$\frac{1}{\sigma} = \limsup_n |b_n|^{\frac{1}{n}} \leq \lim_n (2^{1-\frac{1}{n}})k = 2k$$

and therefore nonzero. □

Corollary 3.56. *Let D be a domain in $\widehat{\mathbb{C}}$ and f a function defined on D. If f has a power series expansion at each point of D and $f(z) \neq 0$ for all $z \in D$, then $\frac{1}{f}$ has a power series expansion at each point of D.*

Definition 3.57. For each domain $D \subseteq \widehat{\mathbb{C}}$, we define

$$\mathbf{H}(D) = \{f : D \to \mathbb{C}; f \text{ has a power series expansion at each point of } D\}.$$

We will see in Chap. 5 that $\mathbf{H}(D)$ is the set of holomorphic functions on D.

Corollary 3.58. *Assume that D is a domain in $\widehat{\mathbb{C}}$. The set $\mathbf{H}(D)$ is an integral domain and an algebra over $\mathbb{C}$. Its units are the functions that never vanish on D.*

Definition 3.59. Let D be a domain in $\widehat{\mathbb{C}}$. A function $f : D \to \widehat{\mathbb{C}}$ is *meromorphic* on D if it is locally[9] the ratio of two functions having power series expansions (with the denominator not identically zero). The set of meromorphic functions on D is denoted by $\mathbf{M}(D)$.

Recall that, by our convention, $\mathbf{M}(D)_{\neq 0}$ is the set of meromorphic functions with the constant function 0 omitted, where $0(z) = 0$ for all z in D.

---

[9]A property P is satisfied *locally* on an open set D if for each point $c \in D$, there exists a neighborhood $U \subset D$ of c such that P is satisfied in U.

Corollary 3.60. *Let D be a domain in $\widehat{\mathbb{C}}$, let c be any point in $D \cap \mathbb{C}$, and let $f \in \mathbf{M}(D)_{\neq 0}$. There exist a connected neighborhood U of c in D, an integer $n = \nu_c f$, and a unit $g \in \mathbf{H}(U)$ such that*

$$f(z) = (z-c)^n\, g(z) \ \textit{for all} \ z \in U.$$

Remark 3.61. If $\infty \in D$, there exists an appropriate version of the above Corollary for $c = \infty$ exists; see Exercise 3.7.

Corollary 3.62. *If D is a domain in $\widehat{\mathbb{C}}$, then the set $\mathbf{M}(D)$ is a field and an algebra over $\mathbb{C}$.*

Corollary 3.63. *If D be a domain and $c \in D$, then*

$$\nu_c : \mathbf{M}(D)_{\neq 0} \to \mathbb{Z}$$

is a homomorphism; that is, $\nu_c(f \cdot g) = \nu_c(f) + \nu_c(g)$ for all f and g in $\mathbf{M}(D)_{\neq 0}$.

Defining $\nu_c(0) = +\infty$, we also have

$$\nu_c(f+g) \geq \min\{\nu_c(f), \nu_c(g)\} \ \textit{for all} \ f \ \textit{and} \ g \ \textit{in} \ \mathbf{M}(D);$$

that is, ν_c is a (discrete) valuation[10] *(of rank one) on $\mathbf{M}(D)$.*

Remark 3.64. The converse statement also holds; it is nontrivial and not established in this book.

The next corollary defines the term *Laurent series*, the natural generalization of power series for functions in $\mathbf{H}(D)$ to functions in $\mathbf{M}(D)$.

Corollary 3.65. *If $f \in \mathbf{M}(D)_{\neq 0}$ and $c \in D \cap \mathbb{C}$, then f has a Laurent series expansion at c; that is, there exists a $\mu \in \mathbb{Z}$ ($\mu = \nu_c(f)$), a sequence of complex numbers $\{a_n\}_{n=\mu}^{\infty}$ with $a_\mu \neq 0$, and a deleted neighborhood U' of c such that*

$$f(z) = \sum_{n=\mu}^{\infty} a_n (z-c)^n$$

for all $z \in U'$. The corresponding power series

$$\sum_{n=\max(0,\mu)}^{\infty} a_n (z-c)^n$$

converges uniformly and absolutely on compact subsets of $U = U' \cup \{c\}$.

[10] Standard, but not universal, terminology.

Remark 3.66. If $\infty \in D$, then for all sufficiently large real numbers R, the Laurent series representing f in $\{|z| > R\} \cup \{\infty\}$ has the form

$$f(z) = \sum_{n=\mu}^{\infty} a_n \left(\frac{1}{z}\right)^n.$$

Corollary 3.67. *If $f \in \mathbf{M}(D)$, then $f' \in \mathbf{M}(D)$. If in addition $\nu_c(f) \neq 0$ for $c \in D$, then*

$$\nu_c(f') = \nu_c(f) - 1.$$

Exercises

3.1. Determine the radius of convergence ρ of each of the following series:

$$\sum_{n=0}^{\infty} \frac{z^n}{n!}, \quad \sum_{n=1}^{\infty} \frac{z^n}{n^2}, \quad \sum_{n=1}^{\infty} \frac{z^n}{n}, \quad \sum_{n=0}^{\infty} n! z^n.$$

(1) For those cases with $\rho < +\infty$, determine the values $|z| = \rho$ for which the series converges.
(2) For the case with $\rho = +\infty$, can you conclude anything about convergence at infinity?

3.2. Find the radius of convergence of the power series

$$\sum_{n=0}^{\infty} a_n z^n,$$

where $a_0 = 0$, $a_1 = 1$, and $a_n = a_{n-1} + a_{n-2}$ for all $n > 1$.
(Hint: Multiply the series by $z^2 + z - 1$.)

3.3. Prove that if $|a_n| \leq M$ for $n \geq 0$, then the power series $\sum_{n=0}^{\infty} a_n z^n$ has radius of convergence $\rho \geq 1$.

3.4. Under the hypothesis that $\{a_n\}$ and $\{b_n\}$ are sequences of positive real numbers, prove that:

(a)

$$\overline{\lim_n}\, a_n b_n \leq \overline{\lim_n}\, a_n \, \overline{\lim_n}\, b_n,$$

provided the right side is not the indeterminate form $0 \times \infty$. Show by example that strict inequality may hold.

(b) If $\lim_n a_n$ exists, then the equality holds in (a) provided the right side is not indeterminate; that is, show that in this case,

$$\overline{\lim_n}\, a_n b_n = \lim_n a_n \,\overline{\lim_n}\, b_n .$$

3.5. Give a proof of the ratio test.

3.6. Give a proof of Theorem 3.18.

3.7. Formulate the appropriate version of Corollary 3.60 for the case $c = \infty \in D$.

3.8. Prove using power series expansions that $e^{-z} = \frac{1}{e^z}$.

3.9. (a) Give examples of algebraic functions that are neither polynomials nor rational functions.

(b) Show that the exponential function is a transcendental function.

3.10. Is it always true that $\operatorname{Log}(e^z) = z$? Support your answer with either a proof or a counterexample.

3.11. Find all zeros of $f(z) = 1 - \exp(\exp z)$.

3.12. (a) Show that both the sine and cosine (complex) functions are periodic with period 2π.

(b) Show that $\sin z = 0$ if and only if $z = \pi n$ for some $n \in \mathbb{Z}$.
(c) Show that $\cos z = 0$ if and only if $z = \frac{\pi}{2}(2n+1)$ for some $n \in \mathbb{Z}$.

3.13. (a) Find $\Re(\sin z)$, $\Im(\sin z)$, $\Re(\cos z)$, and $\Im(\cos z)$.

(b) Write $z = x + \imath y$ and prove that

$$|\sin z|^2 = \sin^2 x + \sinh^2 y$$

and

$$|\cos z|^2 = \cos^2 x + \sinh^2 y,$$

where

$$\cosh z = \frac{e^z + e^{-z}}{2} \quad \text{and} \quad \sinh z = \frac{e^z - e^{-z}}{2}$$

are the *hyperbolic trigonometric functions*.
(c) Derive the addition formulas for $\cosh(z+c)$ and $\sinh(z+c)$.
(d) Evaluate $D \sinh z$, $D \cosh z$, and $\cosh^2 z - \sinh^2 z$.

3.14. Find all the roots of $\cos z = 2$.

3.15. (a) What are all the possible values of $\imath^\imath$?

(b) Let a and $b \in \mathbb{C}$ with $a \neq 0$. Find necessary and sufficient conditions for a^b to consist of infinitely many distinct values.

(c) Let n be a positive integer. Find necessary and sufficient conditions for a^b to consist of n distinct values.

3.16. Let $\{k_n\}$ be a sequence in which every positive integer appears once and only once.

Let $\sum a_n$ be a series. Putting $a'_n = a_{k_n}$, we say that $\sum a'_n$ is a *rearrangement* of $\sum a_n$:

(a) Let $a_n \in \mathbb{R}$ and assume that $\sum a_n$ converges but $\sum |a_n|$ does not. Let $a \in \mathbb{R}$. Show that there is a rearrangement $\sum a'_n$ of $\sum a_n$ such that $a = \sum a'_n$.
(b) Show that $\sum a_n$ converges absolutely if and only if every rearrangement converges to the same sum.

3.17. Let $\{a_n\}$ be a real sequence. Show that

$$\overline{\lim_n}\,\{a_n\} = \sup\left\{\alpha; \alpha = \lim_n b_n\right\}$$

with $\{b_n\}$ a convergent subsequence of $\{a_n\}$ and that

$$\underline{\lim_n}\, a_n = \inf\left\{\alpha; \alpha = \lim_n b_n\right\}$$

with $\{b_n\}$ as above.

In this exercise a sequence $\{b_n\}$ with $\lim b_n = +\infty$ (similarly $-\infty$) is to be considered a convergent sequence.

3.18. Let $p(z) = a_n z^n + a_{n-1}z^{n-1} + \cdots + a_1 z + a_0$, $a_n \neq 0$, be a polynomial of degree $n \geq 1$. Consider p as a self-map of $\widehat{\mathbb{C}}$:

(a) Let $\alpha \in \widehat{\mathbb{C}}$. Show that there exists a $z \in \widehat{\mathbb{C}}$ such that $p(z) = \alpha$.[11]
(b) Let $z \in \widehat{\mathbb{C}}$ and $p(z) = \alpha \in \widehat{\mathbb{C}}$. Define appropriately $m_p(z)$, the *multiplicity* of α for p at z so that you can prove: for all $\alpha \in \widehat{\mathbb{C}}$,

$$\sum_{z \in \widehat{\mathbb{C}};\ p(z)=\alpha} m_p(z) = n. \tag{3.12}$$

(c) Relate $m_p(z)$ to $\nu_z(p)$ and $\nu_z(p')$.

Note: The integer $\sum_{z \in \widehat{\mathbb{C}};\ p(z)=\alpha} m_p(z)$ is the *topological degree* of the map $p : \widehat{\mathbb{C}} \to \widehat{\mathbb{C}}$.

[11] You may use, although other arguments are available, the fundamental theorem of algebra which will be established in Chap. 5.

3.19. Let $p = \frac{P}{Q}$ be a nonconstant rational map. It involves no loss of generality to assume, as we do, that P and Q do not have any common zeros. View, as in the case of polynomials, p as a self-map of $\widehat{\mathbb{C}}$:

(a) Show that p is surjective.
(b) Define the concepts of multiplicity at a point and topological degree for the rational map p so that (3.12) holds for some positive integer n.
(c) Determine n in terms of the zeros and poles of p and in terms of the degrees of the polynomials P and Q.

3.20. The *unit sphere* (with center at 0) $S^2 \subseteq \mathbb{R}^3$ is defined by

$$S^2 = \{(\xi, \eta, \zeta) \in \mathbb{R}^3;\ \xi^2 + \eta^2 + \zeta^2 = 1\}.$$

Show that *stereographic projection*

$$(\xi, \eta, \zeta) \mapsto \frac{\xi + \imath\,\eta}{1 - \zeta}$$

is a diffeomorphism from $S^2 - \{(0, 0, 1)\}$ onto $\mathbb{C}$ and that it extends to a diffeomorphism from S^2 onto $\widehat{\mathbb{C}}$ (that sends $(0, 0, 1)$ to ∞).

3.21. Justify the statement that stereographic projection takes circles to circles.

A *circle* on S^2 is the intersection of a plane in $\mathbb{R}^3$ with S^2. Such a circle is *maximal* if it is the intersection of S^2 with a plane through the point $(0, 0, 1)$ of $\mathbb{R}^3$.

Show that stereographic projection sets up a bijective correspondence between the set of maximal circles on S^2 and the set of *circles* through ∞ on $\widehat{\mathbb{C}}$; that is, straight lines in $\mathbb{C}$. Also show that stereographic projection sets up a bijective correspondence between the set of all circles on S^2 and the *circles in* $\widehat{\mathbb{C}}$: the union of the set of all circles in $\mathbb{C}$ and the set of all straight lines in $\mathbb{C}$.

The circles in $\widehat{\mathbb{C}}$ will play an important role in Chap. 8.

3.22. Show that stereographic projection preserves angles.

3.23. The formula

$$\tan z = \frac{\sin z}{\cos z}$$

defines a meromorphic function on $\mathbb{C}$. Show that it has simple poles at

$$z = (2k + 1)\frac{\pi}{2} \text{ for every integer } k$$

and is holomorphic elsewhere. Show that tan maps $\mathbb{C}$ onto $\mathbb{C} \cup \{\infty\}$:

(a) Show that $\tan z = \tan \zeta$ if and only if there exists an integer k such that $\zeta - z = \pi k$.
(b) Show that $z \mapsto \tan z$ is a holomorphic one to one map of

$$\left\{z \in \mathbb{C}; -\frac{\pi}{2} < \Re z < \frac{\pi}{2}\right\} \quad \text{onto} \quad \mathbb{C} - ((-\infty, -1) \cup (1, +\infty))$$

and of

$$\left\{z \in \mathbb{C}; -\frac{\pi}{2} < \Re z \leq \frac{\pi}{2}\right\} \text{ onto } \mathbb{C} \cup \{\infty\}.$$[12]

(c) Show that

$$\frac{\mathrm{d}}{\mathrm{d}z} \tan z = \frac{1}{\cos^2 z}.$$

3.24. One of the purposes of this exercise is to establish the beautiful formula (3.13). Verify each of the following assertions and/or answer the questions:

(a) The series

$$\frac{1}{1+z^2} = \sum_{k=0}^{\infty} (-1)^k z^{2k}$$

defines a holomorphic function on $|z| < 1$.

(b) Hence there is a holomorphic function

$$f(z) = \sum_{k=0}^{\infty} \frac{(-1)^k z^{2k+1}}{2k+1}$$

on $|z| < 1$ such that

$$f(0) = 0 \text{ and } f'(z) = \frac{1}{1+z^2}.$$

(c) Since tan is locally injective, there exists a multivalued inverse function arctan defined on $\mathbb{C} \cup \{\infty\}$ such that

$$\tan(\arctan z) = z \text{ for all } z \in \mathbb{C},$$

hence also $\tan(\arctan(z) + k\pi) = z$ for all $k \in \mathbb{Z}$ and all $z \in \mathbb{C}$. We can hence define the *principal branch of* arctan, Arctan, by requiring that

$$-\frac{\pi}{2} < \Re(\text{Arctan } z) \leq \frac{\pi}{2}.$$

(d) Show that

$$\arctan z = \frac{1}{2\imath} \log \frac{1+\imath z}{1-\imath z} \quad \text{and} \quad \text{Arctan } z = \frac{1}{2\imath} \text{Log} \frac{1+\imath z}{1-\imath z}.$$

[12]For this and the previous onto proof you will need either some of the results of the next exercise or something like Rouché's theorem, which is proven in Chap. 6.

(e) Let $g(z) = f(\tan z)$. Show that $g'(z) = 1$ for all z in a domain D. Describe D.
(f) Conclude that $f(z) = \text{Arctan}\, z$ for $|z| < 1$.
(g) Why does the Taylor series for Arctan at the origin not converge in a disc larger than $|z| < 1$?
(h) Show that Arctan 1 is given by $\sum_{k=0}^{\infty} \frac{(-1)^k}{2k+1}$, thus justifying

$$\pi = 4 \sum_{k=0}^{\infty} \frac{(-1)^k}{2k+1}. \tag{3.13}$$

3.25. (L'Hopital's Rule) Let f and g be two functions defined by convergent power series in a neighborhood of 0. Assume that $f(0) = 0 = g(0)$ and $g'(0) \neq 0$. Show that

$$\lim_{z \to 0} \frac{f(z)}{g(z)} = \frac{f'(0)}{g'(0)}.$$

3.26. The reader might be better prepared for this problem after studying Chap. 6.

Suppose the series $\sum_{-\infty}^{\infty} \alpha_j z^j$ and $\sum_{-\infty}^{\infty} \beta_j z^j$ converge for $1 < |z| < 3$ and $2 < |z| < 4$, respectively, and that they have the same sum for $2 < |z| < 3$. Does this imply that $\alpha_j = \beta_j$ for all j?

3.27. Let p be a polynomial of degree $n \geq 0$. We usually write

$$p(z) = \sum_{i=0}^{n} b_i z^i,$$

as a linear combination of the standard monomials z^i. In this exercise we introduce a second set of monomials that will help us to evaluate the infinite series

$$\sum_{k=0}^{\infty} p(k) z^k.$$

- Show that the series converges for $|z| < 1$.
- Define inductively (for $n \in \mathbb{Z}_{\geq 0})$) the monomials $z_{(n)}$, by setting

$$z_{(0)} = 1,$$

and for $n \in \mathbb{Z}_{\geq 0})$,

$$z_{(n)} = z_{(n-1)}(z - n + 1).$$

Show that there exists constants a_i such that

$$p(z) = \sum_{i=0}^{n} a_i z_{(i)}.$$

Evaluate these constants.

- Show that for all $n \in \mathbb{Z}_{\geq 0}$ and all $z \in \mathbb{C}$ such that $|z| < 1$,

$$S^{(n)}(z) = \sum_{k=n}^{\infty} k_{(n)} z^{k-n} = \frac{n!}{(1-z)^{n+1}}.$$

- Prove that for all $z \in \mathbb{C}$ such that $|z| < 1$,

$$\sum_{k=0}^{\infty} p(k) z^k = \sum_{k=0}^{\infty} \sum_{j=0}^{n} a_j k_{(j)} z^k = \sum_{j=0}^{n} a_j z^j \sum_{k=j}^{\infty} k_{(j)} z^{k-j}$$

$$= \sum_{j=0}^{n} a_j z^j S^{(j)}(z) = \frac{\sum_{j=0}^{n} a_j z^j (1-z)^{n-j} j!}{(1-z)^{n+1}}.$$

Chapter 4
The Cauchy Theory: A Fundamental Theorem

As with the theory of differentiation for complex-valued functions of a complex variable, the integration theory of such functions begins by mimicking and extending results from the theory for real-valued functions of a real variable, but again the resulting theory is substantially different, more robust, and more elegant. Specifically, a curve or path γ in $\mathbb{C}$ is a continuous *function* from a closed interval in $\mathbb{R}$ to $\mathbb{C}$. Thus the restriction of a complex-valued function f on $\mathbb{C}$ to the range of a curve has real and imaginary parts which can be viewed as real-valued functions of a real variable and thus integrated on the interval.[1] Adding the integral of the real part to $\imath$ times the integral of the imaginary part defines a complex-valued integral of a complex-valued function (i.e., $\int f = \int \Re f + \imath \int \Im f$). In fact, there are several useful ways to employ the ability to integrate a function of a real variable to define complex-valued integrals of a complex variable over certain paths. Among these integrals are those known as line integrals, complex line integrals, and integrals with respect to arc length. One can then use the integration theory of real variables to obtain an integration theory for complex-valued functions along curves in $\mathbb{C}$. This extends to a more general theory, the Cauchy theory, which constitutes a main portion of the fundamental theorem (Theorem 1.1). The integration theory depends not just on the integrated function being holomorphic but also upon the topology of the curve over which the integration is being carried out and the topology of the domain in which the curve lies. In the simplest situation Cauchy's theorem says that the integral of a holomorphic function over a simple closed curve lying in a convex domain is equal to zero.[2]

In this chapter, we lay the foundations for proving several of the equivalences in the fundamental theorem. Beginning in Sect. 4.1, we present a more or less self-contained treatment; our approach is through integration of closed forms over closed

[1]When suitable conditions on $\Re f$ and $\Im f$ hold.

[2]A simple closed curve is one whose initial and end points coincide and has no other self-intersections.

R.E. Rodríguez et al., *Complex Analysis: In the Spirit of Lipman Bers*, Graduate Texts in Mathematics 245, DOI 10.1007/978-1-4419-7323-8_4,

curves. Line integrals and differential forms are introduced in the next section. In the section following it, we emphasize the difference between exact and closed (locally exact) one-forms. Integration of closed forms along continuous, not necessarily rectifiable, paths is discussed next.[3] This is followed by sections on the winding number of a curve about a point, curve homotopy and simple connectivity, and properties of the winding number.

After that, we treat Goursat's form of Cauchy's theorem, which is reduced to establishing the following simple statement: the integral of a holomorphic function on a domain D over the boundary of a rectangle whose closure is contained in D is equal to zero. This result can hence be referred to as Cauchy's theorem for a rectangle. If we were willing to assume $\mathbf{C}^1$ in the definition of holomorphicity, then the simple statement would follow at once from Green's theorem. The weaker definition used complicates the proof but is needed to see the full power of holomorphicity. The chapter ends with an appendix on differential forms, and another appendix on a more classic approach to the Cauchy theory than the one we used, based on differential forms.

While the proofs of Goursat's theorem and some of the other main results become quite technical in places, the final conclusions are simple to state. This simplicity gives them a certain elegance and compactness.

The two chapters that follow present the core of the proofs of the fundamental theorem. In Chap. 5 we present key consequences of the initial Cauchy theory (viz., of Cauchy's theorem for a rectangle). This is followed by Chap. 6, where consequences related to holomorphic functions with isolated singularities are presented.

4.1 Line Integrals and Differential Forms

We recall the definitions of one-sided derivatives for functions of a real variable.

Definition 4.1. Let $[a,b]$ be a closed (finite) interval on $\mathbb{R}$ and let $g : [a,b] \to \mathbb{C}$ be a function. As in calculus, for $a \leq c < b$, we define

$$(D^+g)(c) = \lim_{\substack{h \to 0 \\ h > 0}} \frac{g(c+h) - g(c)}{h},$$

the *right-sided* derivative of g at c (whenever this limit exists).

Similarly, for $a < c \leq b$, we set

[3] This approach has the advantage of avoiding a discussion of rectifiability, but we lose the ability to integrate some non-closed forms.

$$(D^- g)(c) = \lim_{\substack{h \to 0 \\ h<0}} \frac{g(c+h) - g(c)}{h},$$

the *left-sided* derivative of g at c (whenever this limit exists), and for $a < c < b$, we let

$$g'(c) = (Dg)(c) = (D^+ g)(c) = (D^- g)(c),$$

the *derivative* of g at c, whenever the last two limits exist and are equal.

We say that g is *differentiable* on $[a,b]$ if g' exists on (a,b) and $(D^+ g)(a)$ and $(D^- g)(b)$ exist (these define $g'(a)$ and $g'(b)$, respectively); g is called *continuously differentiable* on $[a,b]$ if g' is continuous on $[a,b]$, in which case we will write $g \in \mathbf{C}^1_{\mathbb{C}}([a,b]) = \mathbf{C}^1([a,b])$ or, equivalently, $g' \in \mathbf{C}^0_{\mathbb{C}}([a,b]) = \mathbf{C}^0([a,b]) = \mathbf{C}_{\mathbb{C}}([a,b]) = \mathbf{C}([a,b])$. The spaces $\mathbf{C}^1_{\mathbb{R}}[a,b])$ and $\mathbf{C}_{\mathbb{R}}([a,b])$ are similarly defined.

Remark 4.2. The concepts we have been discussing are from real analysis; if $f : [a,b] \to \mathbb{C}$ is a complex-valued function, they apply to the two real-valued functions of a real variable given by $u = \Re f$ and $v = \Im f$, and we set $f' = u' + \imath\, v'$.

Definition 4.3. Let D be a domain in $\mathbb{C}$. A function $\gamma \in \mathbf{C}^1([a,b])$ (respectively, $\mathbf{C}^0([a,b])$) with $\gamma([a,b]) \subset D \subseteq \mathbb{C}$ will be called a *differentiable (respectively, continuous) path* or *curve in* D; we will say that γ is *parameterized* by $[a,b]$, and we will write

$$\gamma(t) = (x(t), y(t)) = z(t) = x(t) + \imath\, y(t) \text{ for } a \le t \le b.$$

The *image or range* of γ will be denoted by range γ.

A curve γ is called *closed* if $\gamma(a) = \gamma(b)$; a closed curve γ is *simple* or a *Jordan curve* if γ is one-to-one except at the end points of the interval $[a,b]$; to be precise, if $\gamma(t_1) = \gamma(t_2)$ for $a \le t_1 < t_2 \le b$, then $a = t_1$ and $t_2 = b$.

Definition 4.4. Let D be a domain in $\mathbb{C}$. A *differential (one-)form* ω *on* D is an expression

$$\omega = P\,\mathrm{d}x + Q\,\mathrm{d}y,$$

where $P = P(x,y)$ and $Q = Q(x,y)$ are continuous (complex-valued) functions on D, and $\mathrm{d}x$ and $\mathrm{d}y$ are symbols associated to the coordinate $z = x + \imath\, y$, called the *differentials* of x and y, respectively.

If γ is a differentiable path in D and ω is a differential form on D, then we define *the line* (or *path* or *contour*) *integral of* ω *along* γ by the formula

$$\int_\gamma \omega = \int_a^b [P(x(t), y(t))\, x'(t) + Q(x(t), y(t))\, y'(t)]\mathrm{d}t$$
$$= \int_a^b [p_1(x(t), y(t))\, x'(t) + q_1(x(t), y(t))\, y'(t)]\, \mathrm{d}t$$
$$+\imath \int_a^b [p_2(x(t), y(t))\, x'(t) + q_2(x(t), y(t))\, y'(t)]\, \mathrm{d}t,$$

where $P = p_1 + \imath\, p_2$ and $Q = q_1 + \imath\, q_2$.

Remark 4.5. The above definition involves again only concepts from real analysis, even though the paths and functions involve the complex numbers.

4.1.1 Reparameterization

If $t : [\alpha, \beta] \to [a, b]$ is a one-to-one, onto and differentiable function, and $\gamma : [a, b] \to \mathbb{C}$ is a differentiable path, then $\widetilde{\gamma} = \gamma \circ t$ is again a differentiable path, called a *reparameterization* of γ.

For all differential forms ω defined in a neighborhood of the range of the path γ and all reparameterizations $\widetilde{\gamma} = \gamma \circ t$ of γ, the following equalities hold readily from the definitions:

$$\int_{\widetilde{\gamma}} \omega = \int_\gamma \omega\ , \ \text{if}\ \ t'(u) \geq 0\ \ \text{for all}\ u \in [\alpha, \beta],$$

and

$$\int_{\widetilde{\gamma}} \omega = -\int_\gamma \omega\ , \ \text{if}\ \ t'(u) \leq 0\ \ \text{for all}\ \ u \in [\alpha, \beta].$$

Note that there are no other possibilities for the sign of the derivative of t, since t' cannot change signs, because t is one-to-one, and therefore either increasing or decreasing on $[\alpha, \beta]$.

Since for any closed interval $[a, b]$ there exists a one-to-one, onto and differentiable map $t : [0, 1] \to [a, b]$ with $t'(u) > 0$ for all $u \in [0, 1]$ (see Exercise 4.6), we can always assume, as we will when appropriate, that a given path γ is parameterized by $[0, 1]$.

4.1.2 Subdivision of Interval

Let $\gamma : [a, b] \to \mathbb{C}$ be a differentiable path and consider the *partition* of $[a, b]$ defined by

$$a = t_0 < t_1 < \cdots < t_{n+1} = b. \tag{4.1}$$

For a set T contained in the domain of γ, $\gamma|_T$, of course, denotes the restriction of γ to T. If we let

$$\gamma_j = \gamma|_{[t_j, t_{j+1}]}, \text{ for } j = 0, \ldots, n, \tag{4.2}$$

then each γ_j is a differentiable path, and

$$\int_\gamma \omega = \sum_{j=0}^{n} \int_{\gamma_j} \omega. \tag{4.3}$$

This observation allows us to generalize the notion of integral of a form over a differentiable path to a more general class of paths, as follows.

4.1.3 The Line Integral

Definition 4.6. Let $\gamma : [a,b] \to \mathbb{C}$ be a continuous path. We say that γ is a *piecewise differentiable path* (henceforth abbreviated *pdp*) if there exists a partition of $[a,b]$ of the form given in (4.1) such that each of the paths defined by (4.2) is differentiable. Then we use (4.3) to define the path integral $\int_\gamma \omega$.

Remark 4.7. The path integral over a pdp is well defined (independent of the partition) and agrees with our earlier definition for differentiable paths. The verification of these facts is left as an exercise.

Remark 4.8. There are three pictures in $\mathbb{R}^2$ that are naturally associated to each path $\gamma = x + \imath\, y$: the picture of the range of the curve and the graphs of the functions x and y. Figure 4.1 illustrates this for a curve whose image is the boundary of the rectangle with vertices (c,e), (d,e), (d,f), and (c,f).

Definition 4.9. Let γ denote a pdp parameterized by $[a,b]$ in a domain D. Then the path γ *traversed backward* is defined by

$$\gamma_{-}(t) = \gamma(-t + b + a) \text{ for all } t \in [a,b].$$

It follows that

$$\int_{\gamma_-} \omega = -\int_\gamma \omega$$

for all differential forms ω defined in a neighborhood of the range of γ.

Definition 4.10. If γ_1 and γ_2 are pdp's in D parameterized by $[0,1]$, with $\gamma_1(1) = \gamma_2(0)$, then a new pdp $\gamma_1 * \gamma_2$ in D may be defined by first traversing γ_1 and then continuing with γ_2, as follows:

$$\gamma_1 * \gamma_2(t) = \begin{cases} \gamma_1(2t), & \text{for } 0 \le t \le \frac{1}{2}, \\ \gamma_2(2t-1), & \text{for } \frac{1}{2} \le t \le 1. \end{cases}$$

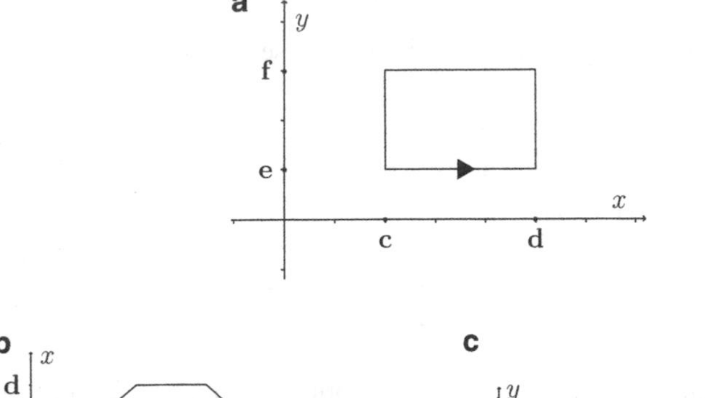

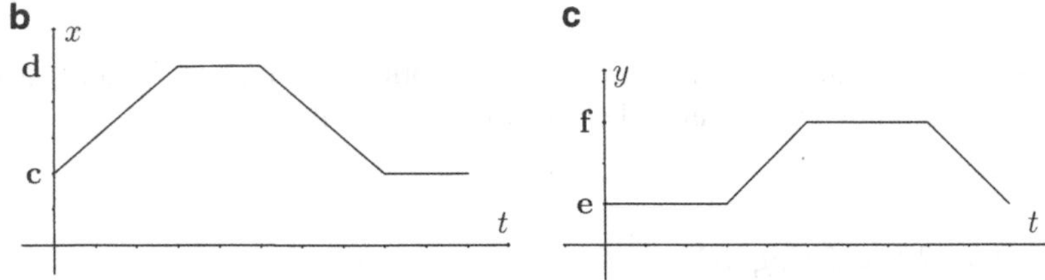

Fig. 4.1 Three figures for a curve (the boundary of a rectangle). (**a**) Picture of the curve. (**b**) The graph of x. (**c**) The graph of y

Thus

$$\int_{\gamma_1 * \gamma_2} \omega = \int_{\gamma_1} \omega + \int_{\gamma_2} \omega$$

for all differential forms ω defined in a neighborhood of the range of $\gamma_1 * \gamma_2$.

Lemma 4.11. *If D is a domain in $\mathbb{C}$, then any two points in D can be joined by a pdp in D.*

Proof. Fix $c \in D$ and let

$$E = \{z \in D; z \text{ can be joined to } c \text{ by a pdp in } D\}.$$

The set E is open in D, because if z denotes any point in E, then, since D is an open set, there is a small disc U with center at z contained in D, and any point w in U may be joined to z by a radial segment in U; the pdp consisting of this segment followed by the pdp joining z to c gives that w in E (see Definition 4.10). Similarly, $D - E$ is also open in D. Since D is connected and $c \in E$, we conclude that $E = D$. □

Definition 4.12. Although the concepts introduced here were used in previous chapters without explanation, this is a good place for formalizing them. Let D be a domain in $\mathbb{C}$

(1) We recall that a function f defined on D is *of class* $\mathbf{C}^p$ on D, with $p \in \mathbb{Z}_{\geq 0}$, if f has partial derivatives (with respect to x and y) up to and including order p, and these are continuous on D. We also say that f is a $\mathbf{C}^p$*-function* on D.

Of course $p = 0$ just means that f is continuous on D. The vector space of complex-valued functions of class $\mathbf{C}^p$ on D is denoted by $\mathbf{C}^p(D)$. We also say that f is of class $\mathbf{C}^\infty$ or *smooth* in D if it is of class $\mathbf{C}^p$ for all $p \in \mathbb{Z}_{\geq 0}$. In this case we also write $f \in \mathbf{C}^\infty(D)$.

(2) A differential form $\omega = P\mathrm{d}x + Q\mathrm{d}y$ is of class $\mathbf{C}^p$ on D if and only if P and Q are.

(3) For a given function f, we have the (real) partial derivatives f_x and f_y as well as the formal (complex) partial derivatives f_z and $f_{\bar{z}}$ introduced in Definition 2.37.

Similarly, we consider the differentials

$$\mathrm{d}z = \mathrm{d}x + \imath\,\mathrm{d}y \text{ and } \mathrm{d}\bar{z} = \mathrm{d}x - \imath\,\mathrm{d}y.$$

It follows that every differential form can be written in the following two ways:

$$P\mathrm{d}x + Q\mathrm{d}y = \frac{1}{2}[(P - \imath Q)\mathrm{d}z + (P + \imath Q)\mathrm{d}\bar{z}].$$

Remark 4.13. We recommend that all concepts and definitions that are formulated in terms of x and y be reformulated by the reader in terms of z and $\bar{z}$ (and vice versa).

(4) If f is a $\mathbf{C}^1$-function on D, then we define $\mathrm{d}f$, the *total differential of* f, by either of the two equivalent formulae:

$$\mathrm{d}f = f_x\,\mathrm{d}x + f_y\,\mathrm{d}y = f_z\,\mathrm{d}z + f_{\bar{z}}\,\mathrm{d}\bar{z}.$$

In addition to the differential operator d, we have two other important differential operators ∂ and $\bar{\partial}$ defined by

$$\partial f = f_z\,\mathrm{d}z \text{ and } \bar{\partial} f = f_{\bar{z}}\,\mathrm{d}\bar{z},$$

as well as the formula

$$d = \partial + \bar{\partial}.$$

We have defined the three differential operators on spaces of $\mathbf{C}^1$-functions. They can be also defined on spaces of $\mathbf{C}^1$-differential forms, and it follows from these definitions that, for example, on $\mathbf{C}^2$-functions the equality $d^2 = 0$ holds. We shall not need these extended definitions but will outline some facts concerning the exterior differential calculus in the first appendix to this chapter.

(5) A differential form ω is called *exact* if there exists a $\mathbf{C}^1$-function F on D (called a *primitive for* ω) such that $\omega = \mathrm{d}F$.

A primitive (if it exists) is unique up to addition of a constant, because $\mathrm{d}F = 0$ means $F_x = F_y = 0$, and this implies that F is constant on the connected set D.

By abuse of language we also say that a function F is a *primitive* for a function f if F is a primitive for the differential form $\omega = f(z)\mathrm{d}z$.

(6) A differential form ω on D is *closed* if it is locally exact; that is, if for each $c \in D$ there exists a neighborhood U of c in D such that $\omega|_U$ is exact.

4.2 The Precise Difference Between Closed and Exact Forms

While the definitions of exact and closed forms are straightforward, as is the fact that every exact differential is closed, an intuitive sense of the difference between the two properties may not immediately present itself. This is because these differences arise from the topology of the domain where the differential form is defined and from the behavior of the differential form along certain paths in that domain. A closed but not exact differential is given in Example 4.30. We will see that on a disc the two properties are equivalent, but situations where they are not equivalent are especially significant.

To understand this difference, we study the pairing that associates the complex number

$$\langle \gamma, \omega \rangle = \int_\gamma \omega$$

to a pdp γ in a domain D and a differential form ω on D (when the integral exists).

Lemma 4.14. *Let ω be a differential form on a domain D. Then ω is exact on D if and only if $\int_\gamma \omega = 0$ for all closed pdps γ in D.*

Proof. Assume that ω is exact. Then there exists a $\mathbf{C}^1$-function F on D with

$$\omega = F_x\,\mathrm{d}x + F_y\,\mathrm{d}y.$$

If γ is a pdp parameterized by $[a, b]$ joining two points P_1 to P_2 in D, then

$$\int_\gamma \omega = \int_a^b \left(F_x \frac{\mathrm{d}x}{\mathrm{d}t} + F_y \frac{\mathrm{d}y}{\mathrm{d}t} \right) \mathrm{d}t = \int_a^b \frac{\mathrm{d}F}{\mathrm{d}t}\,\mathrm{d}t = F(P_2) - F(P_1),$$

which equals zero if $P_1 = P_2$, as happens for every closed curve.

To prove the converse, let $Z_0 = (x_0, y_0)$ be a fixed point in D and let $Z = (x, y)$ be an arbitrary point in D. Let γ be a pdp in D joining Z_0 to Z and define

$$F(x, y) = \int_\gamma \omega.$$

To see that the function F is well defined on D, note that if γ_2 is another pdp in D joining Z_0 to Z, then $\gamma_2 * \gamma_-$ is a closed pdp in D, and it follows from the hypothesis and from Definitions 4.9 and 4.10 that

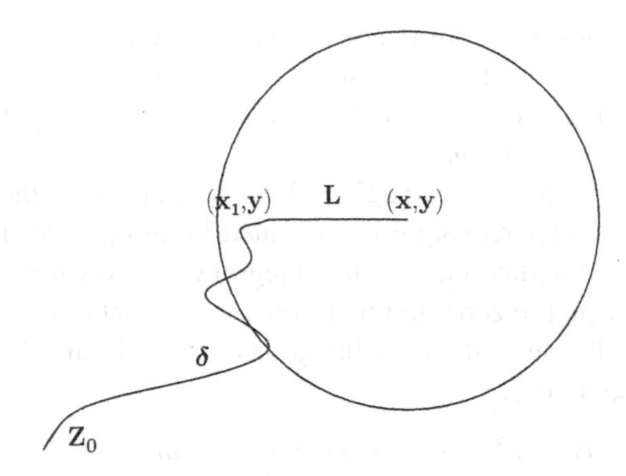

Fig. 4.2 The integration path for F

$$0 = \int_{\gamma_2 * \gamma_-} \omega = \int_{\gamma_2} \omega - \int_{\gamma} \omega.$$

We must show that $F \in \mathbf{C}^1(D)$ and $\mathrm{d}F = \omega$. Choose $\epsilon > 0$ so that $U = U_{(x,y)}(\epsilon) \subseteq D$; also choose $x_1 \neq x$ such that $(x_1, y) \in U$ and the (straight) segment L from (x_1, y) to (x, y) is contained in D. Let δ be any pdp in D from (x_0, y_0) to (x_1, y) (see Fig. 4.2), and assume that $\omega = P\,\mathrm{d}x + Q\,\mathrm{d}y$.

Then

$$F(x, y) = \int_{\delta} \omega + \int_{x_1}^{x} P(t, y)\,\mathrm{d}t.$$

It is now clear that $F_x = P$. Similarly, $F_y = Q$. □

For future use it is important to record the following result.

Corollary 4.15 (of proof). *If* $\mathrm{d}F$ *is an exact differential form on the domain* D *and* γ *is a pdp in* D *starting at* P_1 *and ending at* P_2, *then*

$$\int_{\gamma} \mathrm{d}F = F(P_2) - F(P_1).$$

Theorem 4.16. *Let* ω *be a differential form on an open disc* U. *Then* ω *is exact on* U *if and only if* $\int_{\gamma} \omega = 0$ *for all* γ *that are boundaries of rectangles contained in* U *with sides parallel to the coordinate axes.*

Proof. Repeat the appropriate argument in the proof of the last lemma with (x_0, y_0) the center of U, observing that any other point in the disc may be joined to the center by either a vertical segment, a horizontal segment, or two consecutive segments, one horizontal and one vertical. □

Corollary 4.17. *A differential form* ω *is closed on an open disc if and only if it is exact on the disc.*

Proof. Every exact form on a domain is closed, so we just need to show that the converse holds on an open disc. So assume ω is a closed form on the open disc U. By the theorem, it is enough to show that if R is any rectangle contained in U, with sides parallel to the coordinate axes, then $\int_{\partial R} \omega = 0$. Note that the rectangle R may be subdivided into smaller rectangles with sides parallel to the coordinate axes and such that each smaller rectangle is contained in an open set in U where ω is exact. It follows from the theorem that the integral of ω over the boundary of each smaller rectangle is equal to zero, and therefore the integral of ω over the boundary of R, being equal to the sum of all the integrals over the boundaries of the smaller rectangles, is also equal to zero. □

Corollary 4.18. *If ω is a differential form on a domain D, then ω is closed on D if and only if $\int_\gamma \omega = 0$ for all curves γ that are boundaries of rectangles contained in D with sides parallel to the coordinate axes.*

Definition 4.19. A region $R \subseteq \mathbb{R}^2$ is called (xy)-*simple* if it is bounded by a pdp and has the property that any horizontal or vertical line which has nonempty intersection with R intersects it an interval. Further, the set of values $a \in \mathbb{R}$ for which the vertical line $x = a$ has nonempty intersection with R is an interval, and the set of values $c \in \mathbb{R}$ for which the horizontal line $y = c$ has nonempty intersection with R is also an interval. Here an interval may consist of a single point.

In particular, there exist real numbers $c < d$ and functions h_1 and h_2 defined on the interval $[c, d]$ such that the region R may be described as follows:

$$R = \{(x, y); c \le y \le d, h_1(y) \le x \le h_2(y)\}.$$

A similar description may be given interchanging the roles of the two variables x and y.

Open discs and the interiors of rectangles and triangles are examples of (xy)-simple regions.

We recall[4] and establish a form of a theorem that will help us to further distinguish closed from exact differentials.

Theorem 4.20 (Green's Theorem). *Let R be an (xy)-simple region and let γ denote its boundary oriented counterclockwise (this means that R lies to the left of the oriented curves on its boundary). Consider a $\mathbf{C}^1$-form $\omega = P\,dx + Q\,dy$ on a region $D \supset R \cup \gamma$. Then*

$$\iint_R \left(\frac{\partial Q}{\partial x} - \frac{\partial P}{\partial y}\right) dx\,dy = \int_\gamma P\,dx + Q\,dy = \int_\gamma \omega.$$

[4]From calculus courses.

Proof. Using the notation introduced in the definition of (xy)-simple regions, we have

$$\begin{aligned}\iint_R \frac{\partial Q}{\partial x}\,\mathrm{d}x\,\mathrm{d}y &= \int_c^d \int_{h_1(y)}^{h_2(y)} \frac{\partial Q}{\partial x}\,\mathrm{d}x\,\mathrm{d}y \\ &= \int_c^d [Q(h_2(y), y) - Q(h_1(y), y)]\,\mathrm{d}y \\ &= \int_c^d Q(h_2(y), y)\,\mathrm{d}y + \int_d^c Q(h_1(y), y)\,\mathrm{d}y \\ &= \int_\gamma Q\,\mathrm{d}y.\end{aligned}$$

Similarly,

$$\iint_R -\frac{\partial P}{\partial y}\,\mathrm{d}x\,\mathrm{d}y = -\iint_R \frac{\partial P}{\partial y}\,\mathrm{d}y\,\mathrm{d}x = \int_\gamma P\,\mathrm{d}x. \qquad \square$$

Remark 4.21. (1) The theorem can be easily extended to any region that may be divided into a finite union of (xy)-simple regions and their boundaries (by cancelation of integrals over common boundaries oppositely oriented); for instance, the interior of a compact convex set.

(2) In terms of complex derivatives, the theorem can be restated as

$$\iint_R \left(\frac{\partial Q}{\partial z} - \frac{\partial P}{\partial \bar{z}}\right)\mathrm{d}z\,\mathrm{d}\bar{z} = \int_\gamma P\,\mathrm{d}z + Q\,\mathrm{d}\bar{z},$$

where $\mathrm{d}z\,\mathrm{d}\bar{z} = -2\imath\,\mathrm{d}x\,\mathrm{d}y$.

(3) In most real analysis courses and books (see, e.g., Theorem 5.12 of G.B. Folland *Advanced Calculus*, Prentice Hall, 2002), Green's theorem is given in the following form (we are now using complex notation).

Theorem 4.22 (Green's Theorem, Version 2). *Let K be a compact set in $\mathbb{C}$ which is the closure of its interior, with piecewise smooth positively oriented boundary ∂K. If f and g are $\mathbf{C}^1$-functions on a neighborhood of K, then*

$$\iint_K (g_z - f_{\bar{z}})\,\mathrm{d}z\,\mathrm{d}\bar{z} = \int_{\partial K} f(z)\,\mathrm{d}z + g(z)\,\mathrm{d}\bar{z}.$$

We can now characterize the closed $\mathbf{C}^1$-forms.

Theorem 4.23. *Suppose that $\omega = P\,\mathrm{d}x + Q\,\mathrm{d}y$ is a $\mathbf{C}^1$-differential form on a domain D. If ω is closed, then $P_x = Q_y$.*

***Conversely**, if D is an open disc, P and Q are $\mathbf{C}^1$-functions on D, and $P_x = Q_Y$, then $\omega = P\,\mathrm{d}x + Q\,\mathrm{d}y$ is closed (hence exact) on the disc.*

Proof. If ω is closed in the domain D, then near every point in D there exists a function F such that $\omega = \mathrm{d}F = F_x\,\mathrm{d}x + F_y\,\mathrm{d}y$. But ω is $\mathbf{C}^1$ and thus F is $\mathbf{C}^2$; therefore $P_y = F_{xy} = F_{yx} = Q_x$.

For the converse on a disc D, by Theorem 4.16 and Corollary 4.17, we need only show that $\int_\gamma \omega = 0$ for all paths γ in D that are boundaries of rectangles R with sides parallel to the coordinate axes and such that $R \cup \gamma \subset D$. But

$$\int_\gamma \omega = \iint_R \left(Q_x - P_y\right)\,\mathrm{d}x\,\mathrm{d}y = 0. \qquad \square$$

Corollary 4.24. *If $\omega = P\,\mathrm{d}x + Q\,\mathrm{d}y$ is a $\mathbf{C}^1$-form on a domain D, then ω is closed on D if and only if $P_y = Q_x$ in D.*

Proof. For any point in D, consider an open disc U centered at that point and contained in D, and apply the previous theorem to ω restricted to U. $\square$

Remark 4.25. Recall that

$$f(z)\,\mathrm{d}z = (u + \imath\, v)(\mathrm{d}x + \imath\,\mathrm{d}y) = (u\,\mathrm{d}x - v\,\mathrm{d}y) + \imath\,(u\,\mathrm{d}y + v\,\mathrm{d}x) = \omega_1 + \imath\,\omega_2,$$

with ω_1 and ω_2 real differentials. Thus

$$\int_\gamma f(z)\,\mathrm{d}z = \int_\gamma \omega_1 + \imath \int_\gamma \omega_2.$$

Further, $f(z)\,\mathrm{d}z$ is closed (respectively exact) if and only if both ω_1 and ω_2 are, and F_j is a primitive for ω_j $(j = 1, 2)$ if and only if $F_1 + \imath\, F_2$ is a primitive for $f(z)\,\mathrm{d}z$.

Lemma 4.26. *Let $f(z)\,\mathrm{d}z$ be of class $\mathbf{C}^1$ on a domain D. Then $f(z)\,\mathrm{d}z$ is closed on D if and only if f is holomorphic in D.*

Proof. By the above remarks and previous Corollary, $f(z)\,\mathrm{d}z$ is a closed form on D if and only if $u_y = -v_x$ and $v_y = u_x$ if and only if u and v satisfy CR if and only if f is holomorphic. $\square$

Lemma 4.27. *A $\mathbf{C}^1$-function F is a primitive for $f(z)\,\mathrm{d}z$ if and only if $F' = f$.*

Proof. The function F is a primitive for $f(z)\,\mathrm{d}z$ if and only if $\mathrm{d}F = F_z\,\mathrm{d}z + F_{\bar{z}}\,\mathrm{d}\bar{z} = f(z)\,\mathrm{d}z$ if and only if $F_{\bar{z}} = 0$ and $F_z = F' = f$. $\square$

We have now proven the following result that gives a preliminary characterization of certain closed forms.

Theorem 4.28. *The differential form $f(z)\,\mathrm{d}z$ is closed on a domain D if and only if $\int_\gamma f(z)\,\mathrm{d}z = 0$, for all boundaries γ of rectangles R contained in D[5] with sides parallel to the coordinate axes.*

If $f \in \mathbf{C}^1(D)$, then $f(z)\,\mathrm{d}z$ is closed if and only if f is holomorphic on D.

Remark 4.29. We shall see that the $\mathbf{C}^1$ assumption in the last part of the theorem is not needed.

Example 4.30. Not every closed form is exact. Let $D = \mathbb{C}_{\neq 0}$ and $\omega = \dfrac{\mathrm{d}z}{z}$.

(a) If $\gamma(t) = \mathrm{e}^{2\pi \imath t}$ for $t \in [0,1]$, then $\int_\gamma \omega = 2\pi \imath$. Thus ω is not exact on D.

(b) Since $f(z) = \dfrac{1}{z}$ is holomorphic and $\mathbf{C}^1$ on D, ω is closed on D.

Note that locally (in D) we have $\omega = \mathrm{d}F$, where F is a branch of the logarithm, and that we have just proved that there is no branch of the logarithm globally defined on D.

We have produced two real forms on $D = \mathbb{C}_{\neq 0}$, the real and imaginary parts of ω:

$$\frac{\mathrm{d}z}{z} = \frac{x\,\mathrm{d}x + y\,\mathrm{d}y}{x^2+y^2} + \imath\,\frac{-y\,\mathrm{d}x + x\,\mathrm{d}y}{x^2+y^2} = \mathrm{d}\log|z| + \imath\,\mathrm{d}\arg z = \mathrm{d}\log z.$$

The first of the two real forms is exact, the second closed but not exact on D. Note that $\mathrm{d}\arg z = \mathrm{d}\arctan\frac{y}{x}$ (for $x \neq 0$). Note also that $\arg z$ and $\arctan\frac{y}{x}$ are multivalued functions, whose differentials agree and are single-valued.

It is important to observe that if we change the domain D then ω may become exact; for instance, on the domain $\mathbb{C} - (-\infty, 0]$ the same form $\omega = \dfrac{\mathrm{d}z}{z}$ is exact because $\omega = \mathrm{dLog}$, where Log denotes the principal branch of the logarithm in this domain.

We have been working with the integral of any differential form over any pdp. We want to extend the definition to the integral over any *continuous* path γ but only for *closed* differential forms. The next result will lead us in that direction.

Definition 4.31. Let D be a domain in $\mathbb{C}$, $\gamma : [a,b] \to D$ be a continuous path in D, and $\omega = P\,\mathrm{d}x + Q\,\mathrm{d}y$ be a closed form in D. A *primitive for ω along γ* is a continuous function $f : [a,b] \to \mathbb{C}$ such that for all $t_0 \in [a,b]$ there exists a neighborhood N of $\gamma(t_0)$ in D and a primitive F for ω in N such that $F(\gamma(t)) = f(t)$ for all t in a neighborhood of t_0 in $[a,b]$.

[5]It should be emphasized that the rectangle R, not just its perimeter ∂R, is contained in D.

4.2.1 Caution

It is possible to have $t_1 \neq t_2$ with $\gamma(t_1) = \gamma(t_2)$ but $f(t_1) \neq f(t_2)$; that is, f need not be well defined on γ; of course, it is required to be well defined on $[a, b]$. See Exercise 4.8.

4.2.2 Existence and Uniqueness

Theorem 4.32. *If γ is a continuous path in a domain D and ω is a closed form on D, then there exists a primitive f of ω along γ. Furthermore, f is unique up to the addition of a constant.*

Proof. Suppose γ is parameterized by $[a, b]$ and $\omega = P\,\mathrm{d}x + Q\,\mathrm{d}y$.

Uniqueness: Suppose f and g are two primitives of ω along γ and let $t_0 \in [a, b]$. Then there exist primitives F and G of ω in a connected neighborhood U of $\gamma(t_0)$ in D such that

$$F(\gamma(t)) = f(t) \quad \text{and} \quad G(\gamma(t)) = g(t)$$

for t near t_0. Hence $F_x = G_x = P$ and $F_y = G_y = Q$ in U; thus $F - G$ is constant in U, and therefore $f - g$ is constant near t_0. We conclude that $f - g$ is a continuous and locally constant function on the connected set $[a, b]$, and thus $f - g$ is a constant function.

Existence: Given $t \in [a, b]$, there exists an interval $I(t) \subseteq [a, b]$ (open in $[a, b]$ and containing t) and an open set $U(\gamma(t)) \subseteq D$ such that ω has a primitive in $U(\gamma(t))$ and $\gamma(I(t)) \subset U(\gamma(t))$. Then

$$\bigcup_{t \in [a,b]} I(t)$$

is an open cover of $[a, b]$ and thus there exists a finite subcover

$$I_0 \cup I_1 \cup \cdots \cup I_n = [a, b],$$

with corresponding U_j. Without loss of generality we may assume $I_0 = [a_0, b_0)$, $I_j = (a_j, b_j)$ for $j = 1, \ldots, n-1$ and $I_n = (a_n, b_n]$, where

$$a_0 = a < a_1 < b_0 < a_2 < b_1 < \cdots < a_n < b_{n-1} < b_n = b.$$

Further, $\gamma(I_j) \subset U_j$ and ω has a primitive F_j on U_j for $j = 0, 1, \ldots, n$. Set $f(t) = F_0(\gamma(t))$ for $t \in I_0$. Having defined f on $I_0 \cup I_1 \cup \cdots \cup I_k$ for $0 \leq k < n$, we define f on I_{k+1} as follows. Let F_{k+1} be any primitive for ω in U_{k+1} and let F_k be the primitive for ω in U_k such that $f(t) = F_k(\gamma(t))$ for $t \in I_k$. Then F_{k+1} and F_k are two primitives for ω in $U_{k+1} \cap U_k$, and thus $F_{k+1} - F_k$ is constant on each connected component of $U_{k+1} \cap U_k$; in particular, let $F_{k+1} - F_k = c$ on the component containing $\gamma(I_{k+1} \cap I_k)$. Set $f(t) = F_{k+1}(\gamma(t)) - c$ for $t \in I_{k+1}$, then f is well defined on $I_{k+1} \cap I_k$. □

4.3 Integration of Closed Forms and the Winding Number

Consideration of the next example leads to the extension of the integral to more general paths. This leads in turn to the surprising result, given in Corollary 4.37, that certain integrals take on only integer values. This fact allows a precise definition corresponding to the intuitive idea of counting the number of times a curve winds around a point, the *winding number* of the curve with respect to a point.

Example 4.33. We use Theorem 4.32 to compute $\int_\gamma \omega$, where ω is a closed differential form in D and $\gamma : [a,b] \to D$ is a pdp in D.

Subdivide $[a,b] = I_0 \cup I_1 \cup \cdots \cup I_n$, where $I_j = [a_j, a_{j+1}]$, $a_0 = a$, and $a_{n+1} = b$, such that $\gamma_j = \gamma|_{I_j}$ is a differentiable path and ω has a primitive F_j in a neighborhood of $\gamma(I_j)$ for $j = 0, \ldots, n$. By the theorem, there exists a primitive f of ω along γ. Then

$$\int_\gamma \omega = \sum_{j=0}^{n} \int_{\gamma_j} \mathrm{d}F_j = f(b) - f(a).$$

Using the last equation, we extend the concept of a line integral to continuous paths.

Definition 4.34. Let ω be a *closed* differential form in D and $\gamma : [a,b] \to D$ be a *continuous* path in D. We define

$$\int_\gamma \omega = f(b) - f(a),$$

where f is a primitive of ω along γ.

Remark 4.35. The integral is well defined and agrees with the earlier definition for pdp's. Note that we have avoided any discussion of rectifiability[6] of the curve γ. We have however paid a price: we have not introduced the class of curves γ whose length is well defined.

Theorem 4.36. *For every $c \in \mathbb{C}$ and every continuous closed path γ in $\mathbb{C} - \{c\}$, the number*

$$\frac{1}{2\pi\imath} \int_\gamma \frac{\mathrm{d}z}{z - c} \in \mathbb{Z}.$$

Proof. We may assume $c = 0$. Let f be a primitive of $\frac{\mathrm{d}z}{z}$ along the curve γ. Then

[6]We leave it to the curious reader to consult other sources for the meaning of rectifiability.

$$\int_\gamma \frac{\mathrm{d}z}{z} = f(b) - f(a),$$

where $[a, b]$ parameterizes γ. Since $\gamma(a) = \gamma(b)$, we know from Example 4.30 that $f(b) - f(a)$ is just the difference between two branches of $\log z$ at the same point, hence of the form $2\pi\imath\, n$ with n in $\mathbb{Z}$. □

From Example 4.30 we obtain

Corollary 4.37. *For γ as in Theorem 4.36 with c=0, the value of* $\dfrac{1}{2\pi}\displaystyle\int_\gamma \frac{x\mathrm{d}y - y\mathrm{d}x}{x^2 + y^2}$ *is an integer.*

Definition 4.38. Let $c \in \mathbb{C}$ and let γ be a continuous closed path in $\mathbb{C} - \{c\}$. We define the *index* or *winding number of γ with respect to c* as the following integer:

$$I(\gamma, c) = \frac{1}{2\pi\imath}\int_\gamma \frac{\mathrm{d}z}{z - c}.$$

Example 4.39 (In polar coordinates). Let $r = g(\theta) > 0$, with $g \in \mathbf{C}^1(\mathbb{R})$. Let $n \in \mathbb{Z}_{>0}$ and define $\gamma(\theta) = g(\theta)\mathrm{e}^{\imath\theta}$, where $\theta \in [0, 2\pi n]$. Assume that $g(0) = g(2\pi n)$. Observe that the conditions on g imply that the curve γ winds around the origin n times in the counterclockwise direction and, as expected,

$$\begin{aligned} I(\gamma, 0) &= \frac{1}{2\pi\imath}\int_\gamma \frac{\mathrm{d}z}{z} = \frac{1}{2\pi\imath}\int_0^{2\pi n} \frac{d(g(\theta)\mathrm{e}^{\imath\theta})}{g(\theta)\mathrm{e}^{\imath\theta}} \\ &= \frac{1}{2\pi\imath}\int_0^{2\pi n} \frac{g'(\theta)\mathrm{e}^{\imath\theta} + \imath g(\theta)\mathrm{e}^{\imath\theta}}{g(\theta)\mathrm{e}^{\imath\theta}}\,\mathrm{d}\theta \\ &= \frac{1}{2\pi\imath}\int_0^{2\pi n} \left[\frac{g'(\theta)}{g(\theta)} + \imath\right]\mathrm{d}\theta = n. \end{aligned}$$

In general, let $\gamma : [a, b] \to \mathbb{C} - \{c\}$ be a continuous closed path and let f be a primitive of $\dfrac{\mathrm{d}z}{z - c}$ on γ. Then $f(t)$ agrees with a branch of $\log(\gamma(t) - c)$; that is,

$$\mathrm{e}^{f(t)} = \gamma(t) - c \quad \text{for all } t \in [a, b].$$

Hence

$$I(\gamma, c) = \frac{f(b) - f(a)}{2\pi\imath}.$$

We see that the $\mathbf{C}^1$ assumption on g is unnecessary as we will also be able to conclude using homotopy of curves discussed in the next section.

4.4 Homotopy and Simple Connectivity

In order to give the integration results the clearest formulation (see for instance Corollary 4.52), we introduce the topological concepts of homotopic curves and simply connected domains.

Definition 4.40. Let γ_0 and γ_1 be two continuous paths in a domain D, parameterized by $I = [0, 1]$ with the same end points; that is, $\gamma_0(0) = \gamma_1(0)$ and $\gamma_0(1) = \gamma_1(1)$. We say that γ_0 and γ_1 are *homotopic on D (with fixed end points)* if there exists a continuous function $\delta : I \times I \to D$ such that

(1) $\delta(t, 0) = \gamma_0(t)$ for all $t \in I$
(2) $\delta(t, 1) = \gamma_1(t)$ for all $t \in I$
(3) $\delta(0, u) = \gamma_0(0) = \gamma_1(0)$ for all $u \in I$
(4) $\delta(1, u) = \gamma_0(1) = \gamma_1(1)$ for all $u \in I$

We call δ a *homotopy with fixed end points* between γ_0 and γ_1; see Fig. 4.3, with $\gamma_u(t) = \delta(t, u)$, for fixed u in I.

Let γ_0 and γ_1 be two continuous closed paths in a domain D parameterized by $I = [0, 1]$; that is, $\gamma_0(0) = \gamma_0(1)$ and $\gamma_1(0) = \gamma_1(1)$. We say that γ_0 and γ_1 are *homotopic as closed paths on D* if there exists a continuous function $\delta : I \times I \to D$ such that

(1) $\delta(t, 0) = \gamma_0(t)$ for all $t \in I$
(2) $\delta(t, 1) = \gamma_1(t)$ for all $t \in I$
(3) $\delta(0, u) = \delta(1, u)$ for all $u \in I$

The map δ is called a *homotopy of closed curves or paths*; see Fig. 4.4, with $\gamma_u(t) = \delta(t, u)$ for fixed u in I.

A continuous closed path is *homotopic to a point* if it is homotopic to a constant path (as a closed path).

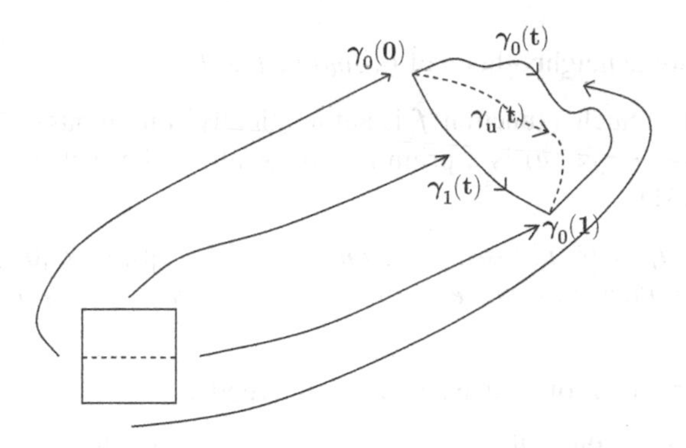

Fig. 4.3 Homotopy with fixed end points

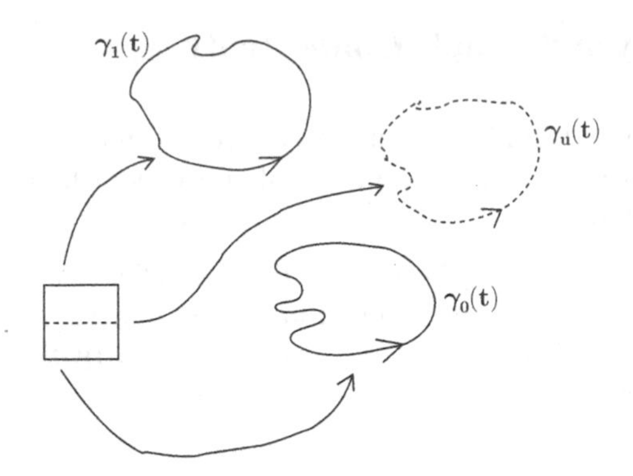

Fig. 4.4 Homotopy of closed paths

Example 4.41. Continuing with Example 4.39, it is easy to see that for continuous g, the path γ is homotopic as a closed path in $\mathbb{C} - \{0\}$ to the circle S^1 traversed n times in the positive direction. Thus both have the same winding number about the origin.

Remark 4.42. Note that the notion of being homotopic (in all its versions) depends on the domain D. For instance, the closed path $\gamma(t) = \exp(2\pi\imath t)$ for $0 \leq t \leq 1$ is homotopic to a point in $\mathbb{C}$ (set $\delta(t, u) = u\gamma(t)$), but it is not homotopic to a point in $\mathbb{C} - \{0\}$, as we will soon see (Remark 4.55).

Definition 4.43. Let I = [0, 1], let $\delta : I \times I \to D \subseteq \mathbb{C}$ be a continuous map, and let ω be a closed form on D.

A function $f : I \times I \to \mathbb{C}$ is said to be a *primitive* for ω along δ provided for every $(t_0, u_0) \in I \times I$, there exists a neighborhood V of $\delta(t_0, u_0)$ in D and a primitive F for ω on V such that

$$f(t, u) = F(\delta(t, u))$$

for all (t, u) in some neighborhood of (t_0, u_0) in $I \times I$.

Remark 4.44. (1) Such a function f is automatically continuous on $I \times I$.

(2) For fixed $u \in I$, $f(\cdot, u)$ is a primitive for ω along the path $t \mapsto \delta(t, u)$ (see Definition 4.31).

Theorem 4.45. *If ω is a closed form on D and $\delta : [0, 1] \times [0, 1] \to D$ is a continuous map, then a primitive f for ω along δ exists and is unique up to an additive constant.*

Proof. We leave the proof as an exercise for the reader. □

We now observe that all integrals of a closed form along homotopic paths coincide.

Theorem 4.46. *Let γ_0 and γ_1 be continuous paths in a domain D and let ω be a closed form on D. If γ_0 is homotopic to γ_1 with fixed end points, then*

$$\int_{\gamma_0} \omega = \int_{\gamma_1} \omega.$$

Proof. We assume that both paths are parameterized by the interval $I = [0, 1]$. Let $\delta : I \times I \to D$ be a homotopy between our two paths and let f be a primitive of ω along δ. Thus $u \mapsto f(0, u)$ is a primitive of ω along the constant curve $u \mapsto \delta(0, u) = \gamma_0(0)$, and hence $f(0, u)$ is a constant α independent of u. Similarly $f(1, u) = \beta \in \mathbb{C}$.

But then

$$\int_{\gamma_0} \omega = f(1, 0) - f(0, 0) = \beta - \alpha$$

and

$$\int_{\gamma_1} \omega = f(1, 1) - f(0, 1) = \beta - \alpha.$$

□

Remark 4.47. A similar result holds for two curves that are homotopic as closed paths (see Exercise 4.10).

Corollary 4.48. *If γ is homotopic to a point in D and ω is a closed form in D, then*

$$\int_{\gamma} \omega = 0.$$

This corollary motivates the following.

Definition 4.49. A region $D \subseteq \mathbb{C}$ is called *simply connected* if every continuous closed path in D is homotopic to a point in D.

Example 4.50. (1) The complex plane $\mathbb{C}$ is simply connected. More generally,
(2) Discs are simply connected: let $c \in \mathbb{C}$, and for $R \in (0, +\infty)$ set $D = U_c(R) = \{z \in \mathbb{C}; |z - c| < R\}$. Without loss of generality we assume $c = 0$ and $R = 1$. Let γ be a continuous closed path in D parameterized by $[0, 1]$, and define $\delta(t, u) = u\gamma(t)$.

Corollary 4.51. *If D is a simply connected domain and γ is a continuous closed path in D, then $\int_\gamma \omega = 0$ for all closed forms ω on D.*

We obtain the simplest formulation of the main result:

Corollary 4.52. *In a simply connected domain a differential form is closed if and only if it is exact.*

Remark 4.53. The property appearing in the last corollary actually characterizes simply connected domains. We will not prove nor use this fact.

An immediate corollary gives the existence of branches of the logarithm:

Corollary 4.54. *In every simply connected domain not containing the point* 0*, there exists a branch of* $\log z$.

Proof. The differential form $\omega = \dfrac{dz}{z}$ is closed and thus exact in the given domain. Hence there exists a holomorphic function F (on the same domain) such that $dF = \omega$. This function F is a branch of the logarithm. □

Remark 4.55. Annuli and punctured discs are not simply connected. To see this, let R_1, R, and R_2 be positive real numbers such that $R_1 < R < R_2$. For any complex number z_0, let A be the annulus $A = \{z; R_1 < |z - z_0| < R_2\}$, and let D denote the punctured disc $D = \{z; 0 < |z - z_0| < R_2\}$. The range of the continuous closed path $\gamma(t) = R\exp(2\pi\imath t)$, for $0 \leq t \leq 1$, is contained in A and in D, and $\int_\gamma \dfrac{dz}{z - z_0} = 2\pi\imath \neq 0$.

To give more examples of simple connectivity, we introduce

Definition 4.56. A region D is *convex* if every pair of points in D can be joined by a segment in D.

Note that convex implies simply connected, but the converse is not true.

4.5 More on the Winding Number

In Sect. 4.3 we defined the winding number $I(\gamma, c)$ of a curve γ with respect to a point c not on its range. In this section we will see that arguments involving the winding number allow us to draw strong conclusions about the behavior of a function defined in a disc (Theorem 4.57).

We begin with some **properties** of $I(\gamma, c)$ for γ a closed path and $c \notin$ range γ.

1. If γ_0 and γ_1 are homotopic as closed paths in $D = \mathbb{C} - \{c\}$, then $I(\gamma_0, c) = I(\gamma_1, c)$. □

   Proof. The differential $\omega = \dfrac{dz}{z - c}$ is closed on D. Thus $\displaystyle\int_{\gamma_0} \omega = \int_{\gamma_1} \omega$.

2. If $\gamma : [a, b] \to \mathbb{C}$ is a closed path, then $z \mapsto I(\gamma, z)$ is a locally constant function on the open set $E = \mathbb{C}$ − range γ; hence it is constant on each connected component of E.

   Proof. Let $c \in E$; we need to show that for all $h \in \mathbb{C}$ with $|h|$ sufficiently small, $I(\gamma, c + h) = I(\gamma, c)$. Without loss of generality, we assume $[a, b] = [0, 1]$.

   Choose δ_0 to be any positive number less than the distance of c to range γ, then

$$U(\gamma(t), \delta_0) \subset \mathbb{C} - \{c\}$$

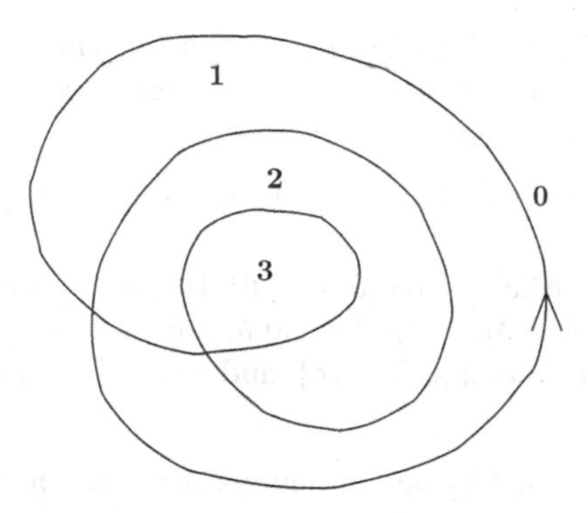

Fig. 4.5 Winding numbers

for all $t \in [0, 1]$.

If we fix any $h \in \mathbb{C}$ with $|h| < \delta_0$, then

$$\int_\gamma \frac{dz}{z-(c+h)} = \int_\gamma \frac{dz}{(z-h)-c} = \int_{\gamma'} \frac{dz'}{z'-c},$$

where $z' = z - h$ and $\gamma' = \gamma - h$; that is, $\gamma'(t) = \gamma(t) - h$ for all $t \in [0, 1]$. Now γ' is a closed path in $\mathbb{C} - \{c\}$, and γ' is homotopic to γ there (as closed paths), via the homotopy defined on $[0, 1] \times [0, 1]$ by

$$\delta(t, u) = \gamma(t) - uh, \quad \text{for } t, u \in [0, 1].$$

Thus $I(\gamma, c + h) = I(\gamma', c) = I(\gamma, c)$. □

3. If range $\gamma \subset D \subset \mathbb{C} - \{c\}$ with D simply connected, then $I(\gamma, c) = 0$.

   Proof. The differential $\dfrac{dz}{z-c}$ is closed in D, therefore exact. □

4. $I(\gamma, c) = 0$ for all c in the unbounded component of the complement of the range of γ.
5. Let $\gamma : t \mapsto Re^{\imath t}$, with $R > 0$ and $t \in [0, 2\pi]$. Then
   - $I(\gamma, 0) = 1$, by Example 4.39
   - $I(\gamma, z) = 1$ for $|z| < R$, by (2)
   - $I(\gamma, z) = 0$ for $|z| > R$, by (3)

In Fig. 4.5 we depict the range of a closed path γ. The numbers indicate the value of the winding number $I(\gamma, c)$ for a point c in the respective component of the complement of the range of γ.

We now show that if the image of the boundary of a disc under a continuous function f winds nontrivially around a point c, then f assumes the value c somewhere inside the disc. More precisely,

Theorem 4.57. *Let $f : \{z \in \mathbb{C} : |z| \leq R\} \to \mathbb{C}$ be a continuous map (with $R > 0$) and let $\gamma(\theta) = f(Re^{2\pi\imath\theta})$ for $\theta \in [0,1]$. If $c \notin$ range γ and $I(\gamma, c) \neq 0$, then there exists a z such that $|z| < R$ and $f(z) = c$.*

Proof. Assume $f(z) \neq c$ for all $|z| < R$. Then $f(z) \neq c$ for all $|z| \leq R$, because $c \notin$ range γ.

Define $\delta(\rho, \theta) = f(\rho Re^{2\pi\imath\theta})$ on $[0,1] \times [0,1]$. Then δ is continuous, $\delta(1,\theta) = \gamma(\theta)$, $\delta(0,\theta) = f(0)$, $\delta(\rho, 0) = \delta(\rho, 1)$, and $\delta(\rho,\theta) \in \mathbb{C} - \{c\}$ for all $\rho, \theta \in [0,1]$. Thus γ is homotopic to a point in $\mathbb{C} - \{c\}$, and hence $I(\gamma, c) = 0$; we have arrived at the needed contradiction. □

Definition 4.58. Let γ_1 and γ_2 be continuous paths parameterized by $[0,1]$. We define two new continuous paths, also parameterized by $[0,1]$:

$$\gamma_1\gamma_2 : t \mapsto \gamma_1(t)\gamma_2(t),$$
$$\gamma_1 + \gamma_2 : t \mapsto \gamma_1(t) + \gamma_2(t).$$

Note that the above definition of a product of two paths differs from the one given earlier in Definition 4.10, where the paths are traversed in succession but at twice the speed.

Theorem 4.59. *If γ_1 and γ_2 are continuous closed paths not passing through 0, then*

$$I(\gamma_1\gamma_2, 0) = I(\gamma_1, 0) + I(\gamma_2, 0).$$

Proof. Let $\omega = \dfrac{dz}{z}$ and $\gamma_j : [0,1] \to \mathbb{C} - \{0\}$. Choose continuous functions $f_j : [0,1] \to \mathbb{C}$ so that $e^{f_j(t)} = \gamma_j(t)$ for all $t \in I$. Then $e^{f_1(t)+f_2(t)} = \gamma_1\gamma_2(t)$. Thus $f = f_1 + f_2$ is a primitive of ω along $\gamma_1\gamma_2$, and

$$I(\gamma_1\gamma_2, 0) = \frac{f(1) - f(0)}{2\pi\imath} = \frac{f_1(1) - f_1(0)}{2\pi\imath} + \frac{f_2(1) - f_2(0)}{2\pi\imath}.$$

□

Theorem 4.60. *Let γ_1 and γ be continuous closed paths in $\mathbb{C}$ parameterized by $[0,1]$. Assume that*

$$0 < |\gamma_1(t)| < |\gamma(t)| \text{ for all } t \in [0,1].$$

Then

$$I(\gamma_1 + \gamma, 0) = I(\gamma, 0).$$

Proof. Note that

$$\gamma(t) + \gamma_1(t) = \gamma(t)\left(1 + \frac{\gamma_1(t)}{\gamma(t)}\right) = \gamma(t)\beta(t)$$

with $\beta(t) = 1 + \dfrac{\gamma_1(t)}{\gamma(t)}$ for all $t \in [0, 1]$, and thus

$$I(\gamma_1 + \gamma, 0) = I(\gamma, 0) + I(\beta, 0).$$

Now observe that

$$|\beta(t) - 1| < 1,$$

and thus β is a closed path in the simply connected domain $U(1, 1)$. Therefore $I(\beta, 0) = 0$. □

4.6 Cauchy Theory: Initial Version

The most important technical result of this chapter is the following.

Theorem 4.61 (Goursat's Theorem). *If f is a holomorphic function on a domain D, then $f(z)\,\mathrm{d}z$ is a closed differential form on D. Equivalently, if γ denotes the boundary of R oriented counterclockwise, where R is any rectangle in D with sides parallel to the coordinate axes, then $\int_\gamma \omega = 0$.*

This theorem has many significant consequences. To prove it we need some preliminaries. Recall that the only issue that needs to be addressed involves the smoothness of the function f; we are not assuming that the function has continuous partial derivatives. For $\mathbf{C}^1$-functions, we already have the result (see Theorem 4.28).

We follow a beautiful classic line of reasoning for the proof.

Definition 4.62. Let γ be a pdp parameterized by the unit interval $[0, 1]$ in a domain D and let f be a continuous function on D.

For such restricted paths, we define

(1) The *integral of f on γ with respect to arc length*

$$\int_\gamma f(z)\,|\mathrm{d}z| = \int_0^1 f(\gamma(t))\,\left|\gamma'(t)\right|\,\mathrm{d}t$$

(2) The *length of the curve γ*

$$L(\gamma) = \text{length of } \gamma = \int_\gamma |\mathrm{d}z|$$

The following results follow straightforwardly from these definitions.

Proposition 4.63. *Let γ be a pdp parameterized by the unit interval $[0, 1]$ in a domain D, and let f be a continuous function on D. Then*

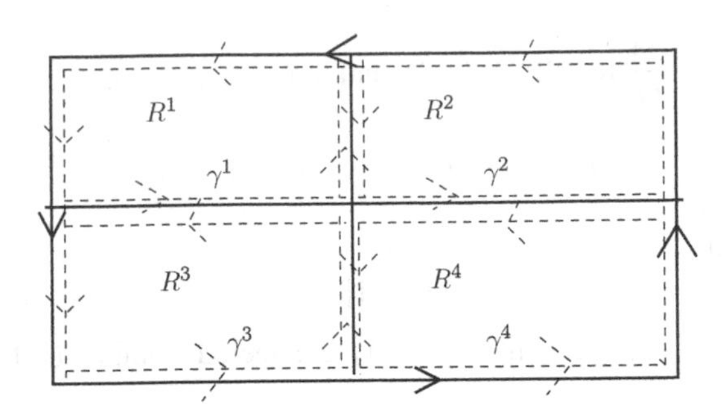

Fig. 4.6 The integrals along the common side of R^1 and R^2 are in opposite directions

$$\int_\gamma f(z)\ |\mathrm{d}z| = \int_{\gamma_-} f(z)\ |\mathrm{d}z|,$$

where γ_- denotes the pdp γ traversed backward (recall Definition 4.9), and

$$\left|\int_\gamma f(z)\,\mathrm{d}z\right| \le \int_\gamma |f(z)|\ |\mathrm{d}z| \le (\sup\{|f(z)| : z \in \text{range}\ \gamma\}) \cdot L(\gamma).$$

Proof of Theorem 4.61 (Due to E. Goursat). Let R be any rectangle in D with sides parallel to the coordinate axes, and let γ denotes the boundary of R oriented counterclockwise. We need to show that $\int_\gamma \omega = \int_\gamma f(z)\mathrm{d}z = 0$. (Even though the proof of this last statement is due to Goursat, the result is known as *Cauchy's theorem for a rectangle*.) Assume to the contrary that $\int_\gamma \omega = \alpha \neq 0$, and divide R into four congruent rectangles $R^1, \ldots, R^4$, with boundaries $\gamma^1, \ldots, \gamma^4$. Then

$$\alpha = \int_\gamma \omega = \sum_{i=1}^4 \int_{\gamma^i} \omega = \sum_{i=1}^4 \alpha^{(i)}.$$

The second equality follows from the fact that certain paths on the boundaries of the subrectangles have opposite directions, giving cancelations in the integrals (see Fig. 4.6).

It is clear that we must have

$$\left|\alpha^{(i)}\right| \ge \frac{|\alpha|}{4}$$

for at least one index i. Call the corresponding rectangle R_1, its boundary γ_1, and the corresponding integral α_1. By repeating this procedure, we obtain a sequence of closed rectangles

$$R \supset R_1 \supset \cdots \supset R_k \supset \cdots$$

with boundaries $\gamma_k = \partial R_k$ so that for $\alpha_k = \int_{\gamma_k} \omega$, we have

$$|\alpha_k| \geq \frac{|\alpha|}{4^k}.$$

Each rectangle R_k is a closed subset of D and also of $\mathbb{C}$; furthermore,

$$\lim_{k\to\infty} \text{Area}\,(R_k) = \lim_{k\to\infty} \frac{\text{Area}\,(R)}{4^k} = 0.$$

The Bolzano–Weierstrass theorem states that there exists a unique $c \in \bigcap_k R_k$. Since f is holomorphic at c,

$$f(z) = f(c) + f'(c)(z-c) + \varepsilon(z)\,|z-c|,$$

where

$$\lim_{z\to c} \varepsilon(z) = 0.$$

Let λ_k denote the length of a diagonal of R_k, then $\lambda_k = \frac{\lambda}{2^k}$, where λ is the length of a diagonal of R. Now

$$\alpha_k = \int_{\gamma_k} f(z)\,\mathrm{d}z = f(c)\int_{\gamma_k} \mathrm{d}z + f'(c)\int_{\gamma_k} (z-c)\,\mathrm{d}z + \int_{\gamma_k} |z-c|\,\varepsilon(z)\,\mathrm{d}z.$$

The first two integrals following the second equal sign are equal to zero since the integrands are exact forms ($\mathrm{d}z$ and $\frac{1}{2}\mathrm{d}(z-c)^2$, respectively). Thus we have

$$\alpha_k = \int_{\gamma_k} |z-c|\,\varepsilon(z)\,\mathrm{d}z.$$

Now on R_k we have

$$|z-c| \leq \lambda_k = 2^{-k}\,\lambda.$$

Given $\epsilon > 0$, there exists a $K \in \mathbb{Z}_{>0}$ such that $|\varepsilon(z)| < \epsilon$ for all z in R_K and thus also for all z in R_k and all $k \geq K$. Hence we have

$$|\alpha_k| \leq \int_{\gamma_k} |z-c|\,|\varepsilon(z)|\,|\mathrm{d}z| \leq 2^{-k}\,\lambda\,\epsilon\,L(\gamma_k) = \epsilon\,4^{-k}\,\lambda\,L(\gamma)$$

for $k \geq K$. We conclude that

$$|\alpha| \leq 4^k\,|\alpha_k| \leq \epsilon\,\lambda\,L(\gamma).$$

Since ϵ is arbitrary, we must have that $\alpha = 0$. □

Corollary 4.64 (Cauchy's Theorem). *If f is holomorphic on a domain D and γ is a continuous closed path in D that is homotopic to a point in D, then $\int_\gamma f(z)\,\mathrm{d}z=0$.*

Corollary 4.65. *If f is holomorphic in a domain D, then locally $f(z)\,\mathrm{d}z$ has a primitive in D.*

Remark 4.66. We have previously shown that (1) implies (4) in the fundamental theorem. We have just established another path for obtaining the same conclusion.

4.7 Appendix I: The Exterior Differential Calculus

To place the discussion of differential forms in its broader context, we outline some basic concepts of the exterior differential calculus for domains $D \subseteq \mathbb{C}$ using only complex coordinates $z = x + \imath y$ $(w = u + \imath v)$.[7] The reader is invited to translate all concepts under discussion to the real (rectangular) coordinates (x, y).[8] For every nonnegative integer n, let $\Omega^n(D)$ denote the space of n-forms on D. Since D is two dimensional over the reals, we need to consider only $n = 0, 1, 2$, because under any reasonable set of definitions $\Omega^n(D) = \{0\}$ for $n > 2$.

The set $\Omega^0(D)$ consists of smooth ($\mathbf{C}^\infty$-) functions on D. The space $\Omega^1(D)$ can be viewed as a module over $\Omega^0(D)$ generated by the 1-forms dz and $\mathrm{d}\bar{z}$; that is, expressions of the form $f(z)\mathrm{d}z + g(z)\mathrm{d}\bar{z}$ with f and g smooth, subject however to the appropriate transformation rule under complex coordinate changes (w is an injective holomorphic function of z):

$$F(w)\mathrm{d}w + G(w)\mathrm{d}\overline{w} = f(z)\mathrm{d}z + g(z)\mathrm{d}\bar{z},$$

where

$$f(z) = F(w(z))w'(z) \text{ and } g(z) = G(w(z))\overline{w'(z)},$$

and the general rule about exterior multiplication $\wedge$ of forms is given as follows:

$$\mathrm{d}z \wedge \mathrm{d}z = 0 = \mathrm{d}\bar{z} \wedge \mathrm{d}\bar{z},\ \mathrm{d}z \wedge \mathrm{d}\bar{z} = -\mathrm{d}\bar{z} \wedge \mathrm{d}z = -2\imath \mathrm{d}x \wedge \mathrm{d}y.$$

The set $\Omega^2(D)$ consists of elements of the form

$$f(z)\mathrm{d}z \wedge \mathrm{d}\bar{z} = F(w)\mathrm{d}w \wedge \mathrm{d}\overline{w},$$

where f and F are $\mathbf{C}^\infty$-functions with

$$f(z) = F(w(z))\left|w'(z)\right|^2,$$

where w is an injective holomorphic function of z.

[7] Usually w is an analytic function of z under appropriate circumstances.

[8] Allowing arbitrary $\mathbf{C}^\infty$ coordinate changes.

The space of one-forms $\Omega^1(D)$ decomposes as a direct sum

$$\Omega^1(D) = \Omega^{(1,0)}(D) \oplus \Omega^{(0,1)}(D),$$

where

$$\Omega^{(1,0)}(D) = \{\omega \in \Omega^1(D);\ \omega = f(z)\mathrm{d}z\},$$

and

$$\Omega^{(0,1)}(D) = \{\omega \in \Omega^1(D);\ \omega = g(z)\mathrm{d}\bar{z}\}.$$

Three differential operators, $\partial, \overline{\partial}$, and $d = \partial + \overline{\partial}$, act on forms; each of these follows the product rule; for example,

$$d(\omega_1 \wedge \omega_2) = d(\omega_1) \wedge \omega_2 + \omega_1 \wedge d(\omega_2).$$

Thus these operators are completely determined by their actions on functions (of z) and the distinguished one forms $\mathrm{d}z$ and $\mathrm{d}\bar{z}$, for example,

$$\partial(f(z)) = f_z \mathrm{d}z,\ \partial(\mathrm{d}z) = 0 = \partial(\bar{z}).$$

One easily check that

$$0 = \partial^2 = \partial\overline{\partial} + \overline{\partial}\partial = \overline{\partial}^2 \text{ and } d^2 = \partial\overline{\partial} + \overline{\partial}\partial.$$

The differential operators lead to three set of maps:

$$\{0\} \to \Omega^0(D) \overset{\partial}{\to} \Omega^{(1,0)}(D) \overset{\partial}{\to} \Omega^2(D) \overset{\partial}{\to} \{0\},$$

$$\{0\} \to \Omega^0(D) \overset{\overline{\partial}}{\to} \Omega^{(0,1)}(D) \overset{\overline{\partial}}{\to} \Omega^2(D) \overset{\overline{\partial}}{\to} \{0\},$$

and

$$\{0\} \to \Omega^0(D) \overset{d}{\to} \Omega^1(D) \overset{d}{\to} \Omega^2(D) \overset{d}{\to} \{0\}.$$

4.8 Appendix II: An Alternative Approach to the Cauchy Theory

There is an alternative approach to the Cauchy theory that does not begin with differential forms but rather ends with it. In the main body of the text we based the theory on integrating differential forms. The alternative development starts with integration of functions, mimicking as much as possible ideas from a classical treatment of undergraduate calculus. It is a matter of taste which development one uses or prefers. Both treatments yield the same or equivalent theorems. In this appendix we outline the alternative development leaving many details to the reader, essentially by listing the order of definitions and results we have already encountered and providing full statements for results, usually theorems and their corollaries, that appear different than in the main body of this book. Proofs are included as needed, particularly when they involve new ideas. This outline follows the approach that Bers used in many of his courses and is similar to the approach of Ahlfors in his text. We end this outline after we prove a strengthened version of Goursat's theorem (another version is in the next chapter). Since this approach does not rely on power series, we include a self-contained proof that the derivative of an analytic function is also analytic.

For the rest of this appendix $f = \Re f + \imath \Im f = u + \imath v$ is a fixed continuous function on a domain $D \subseteq \mathbb{C}$, and γ is a pdp in D, parameterized by the closed interval $[a, b] \subset \mathbb{R}$. At times we will put restrictions on each of these.

4.8.1 Integration of Functions

We start by considering four types of integrals, essentially mimicking and expanding slightly ideas from elementary calculus.

- For a complex-valued continuous function w defined on a real interval $[a, b]$ with $w(t) = u(t) + \imath\, v(t)$ (as usual, $u = \Re w$ and $v = \Im w$), we set

$$\int_a^b w(t)\,\mathrm{d}t = \int_a^b u(t)\,\mathrm{d}t + \imath \int_a^b v(t)\,\mathrm{d}t.$$

- Let $\gamma : [a, b] \to \mathbb{C}$ be a pdp and f a continuous function on the range of γ. The *complex line integral* of f over γ is

$$\int_\gamma f(z)\,\mathrm{d}z = \int_a^b f(\gamma(t))\,\gamma'(t)\,\mathrm{d}t.$$

  Hence the integral is reduced to the previous case.

- Under the same conditions for f and γ, we have *the integral with respect to arc length*

$$\int_\gamma f(z)\;|\mathrm{d}z| = \int_a^b f(\gamma(t))\;\left|\gamma'(t)\right|\;\mathrm{d}t.$$

  This integral and the next concept were encountered in our proof of Goursat's theorem.

- The *length of the pdp* γ is $L(\gamma) = \int_\gamma |\mathrm{d}z| = \int_a^b |\gamma'(t)|\,\mathrm{d}t$.

There are a number of properties that these integrals satisfy, including, for example, invariance under orientation preserving change of parameter. We do not discuss these here since they should be familiar to the reader who studied this chapter.

We have previously used, and find useful here, the calculation

$$\begin{aligned}\int_\gamma f(z)\,\mathrm{d}z &= \int_\gamma (u+\imath v)(\mathrm{d}x+\imath\,\mathrm{d}y)\\ &= \int_\gamma (u\mathrm{d}x - v\mathrm{d}y) + \imath(v\mathrm{d}x + u\mathrm{d}y)\\ &= \int_\gamma p\,\mathrm{d}x + q\,\mathrm{d}y.\end{aligned}$$

It is convenient to introduce, for appropriate c and $d \in \mathbb{C}$, the abbreviation

$$\int_c^d f(z)\,\mathrm{d}z$$

to denote the integral along the line segment L from c to d. Using the mean value integral theorem (from calculus), we easily compute that

$$\int_c^d f(z)\,\mathrm{d}z = (d-c)(\Re f(z_1) + \imath\Im f(z_2)),$$

where z_1 and $z_2 \in L$.

4.8.2 The Key Theorem

We proceed in an informal way to the key results. We fix a domain D, $z_0 \in D$, and $R > 0$ such that $U(z_0, R) \subseteq D$.

(1) Goursat's Theorem for Triangles: *If $f'(z)$ exists for all z in D, and T is any triangle contained in D (including its interior), then $\int_{\partial T} f(z)\mathrm{d}z = 0$.*

Proof. The proof is identical to the one of Goursat's theorem in this chapter. Since we are dealing with triangles, we join the midpoints of its sides to divide a triangle into four smaller triangles, each with perimeter equal to half the perimeter of the original triangle. □

(2) Strengthened Goursat's Theorem for Triangles: *If f is continuous on D, $f'(z)$ exists for all z in D except for a discrete subset, and T is a triangle contained in D (including its interior), then $\int_{\partial T} f(z)\mathrm{d}z = 0$.*

Proof. It suffices to show that $\left|\int_{\partial T} f(z)\mathrm{d}z\right| \le \epsilon$ for all $\epsilon > 0$. Let $M > 0$ be an upper bound for f on T. Since T is compact, there are only finitely many, say N, *bad* points in this set where f' does not exists; if $N = 0$ we are done, by Goursat's theorem for triangles, so assume $N > 0$. Divide T into smaller triangles $T_1, \ldots, T_k$ (several times as in the above proof) so that the perimeter of each triangle is of length $\leq \frac{\epsilon}{6MN}$. Now, as usual,

$$\int_{\partial T} f(z)\mathrm{d}z = \sum_{j=1}^{k} \int_{\partial T_j} f(z)\mathrm{d}z.$$

If T_j does not contain any bad points, then $\int_{\partial T_j} f(z)\mathrm{d}z = 0$ by Goursat's theorem for triangles. For a triangle T_j that contains a bad point, $\left|\int_{\partial T_j} f(z)\mathrm{d}z\right| \leq M\frac{\epsilon}{6MN}$. Since each bad point can be in at most six triangles and there are N bad points, the required estimate follows. □

(3) *If f is continuous on D, $f'(z)$ exists for all z in D except for a discrete subset, and D' is any convex subdomain of D, then there is $F : D' \to \mathbb{C}$ with $F'(z) = f(z)$.*

Proof. Fix $z_0 \in D'$, define $F(z) = \int_{z_0}^{z} f(t)\mathrm{d}t$ for z in D', and use the strengthened form of Goursat's theorem for triangles to conclude that $F'(z) = f(z)$. Specifically, by the strengthened form of Goursat's theorem for triangles, for all small $|h| > 0$ (such that the triangle with vertices z_0, z, and $z + h$, including its interior, is contained in D'), we obtain

$$F(z+h) = \int_{z_0}^{z+h} f(t)\mathrm{d}t = \int_{z_0}^{z} f(t)\mathrm{d}t + \int_{z}^{z+h} f(t)\mathrm{d}t.$$

Thus

$$\begin{aligned} F(z+h) &= F(z) + \int_{z}^{z+h} f(t)\mathrm{d}t \\ &= F(z) + h(\Re f(z_1) + \imath\, \Im f(z_2)), \end{aligned}$$

where z_1 and z_2 are on the line segment between z and $z + h$, and we conclude by the continuity of f at z that

$$F'(z) = \lim_{h\to 0} \frac{F(z+h) - F(z)}{h} = f(z).$$

□

(4) Fundamental Theorem of Calculus (FTC) (complex version): *If the function $F : D \to \mathbb{C}$ is complex differentiable and F' is continuous in D, then*

$$\int_{\gamma} F'(z)\mathrm{d}z = F(\gamma(b)) - F(\gamma(a))$$

for all pdp's $\gamma : [a, b] \to D$.

Proof. Use the definition of the complex line integral, CR and FTC (real version)). □

(5) *If for every convex subdomain D' contained in D there is an $F : D' \to \mathbb{C}$ with $F' = f$, then $\int_\gamma f(z)\mathrm{d}z = 0$ for any closed pdp γ in D'.*

Proof. $\int_\gamma f(z)\mathrm{d}z = \int_\gamma F'(z)\mathrm{d}z = 0$ by FTC. □

(6) *If γ is a closed pdp in $\mathbb{C}$ and z_0 is not in its image, then*

$$I(\gamma, z_0) = \frac{1}{2\pi\imath}\int_\gamma \frac{\mathrm{d}z}{z - z_0} \in \mathbb{Z}.$$

We have established this fact for any closed curve, not just a pdp, because of work with closed differential forms and logarithms. We now show that these ingredients are unnecessary if we restrict to pdp.

Proof. For $\gamma : [a, b] \to \mathbb{C}$, set

$$h(t) = \int_a^t \frac{\gamma'(s)}{\gamma(s) - z_0}\,\mathrm{d}s$$

for $a \le t \le b$. Then h is continuous in $[a, b]$,

$$h(a) = 0,\ h(b) = \int_\gamma \frac{\mathrm{d}z}{z - z_0},\ \text{and } h'(t) = \frac{\gamma'(t)}{\gamma(t) - z_0}$$

at the points t where $\gamma'(t)$ is continuous. Let

$$g(t) = \mathrm{e}^{-h(t)}(\gamma(t) - z_0).$$

Then g is continuous in $[a, b]$, $g'(t) = 0$ except at a finite number of points t, and hence g is constant:

$$g(t) = g(a) = \gamma(a) - z_0$$

for all t in $[a, b]$. Therefore

$$\mathrm{e}^{h(t)} = \frac{\gamma(t) - z_0}{\gamma(a) - z_0}.$$

But then

$$\mathrm{e}^{h(b)} = \frac{\gamma(b) - z_0}{\gamma(a) - z_0} = 1$$

since γ is closed, and therefore there exists k in $\mathbb{Z}$ such that

$$\int_\gamma \frac{\mathrm{d}z}{z - z_0} = h(b) = 2\pi\imath k.$$ □

The reader may notice that part of our argument here is very similar to one used in the main body of our book under different circumstances (see the proof of Theorem 3.38).

Many properties of the index can be developed. Among them:
For the curve $\gamma(t) = z_0 + R\mathrm{e}^{2\pi\imath t}$, $0 \le t \le 1$, *and* $|z - z_0| \neq R$,

$$\int_\gamma \frac{\mathrm{d}\tau}{\tau - z} = \begin{cases} 2\pi\imath, & \text{if } |z - z_0| < R, \\ 0, & \text{if } |z - z_0| > R. \end{cases}$$

(7) *If* f' *exists in* D *and* $\{|z - z_0| \le R\} \subset D$, *then*

$$f(z) = \frac{1}{2\pi\imath} \int_{|\tau - z_0| = R} \frac{f(\tau)}{\tau - z} \mathrm{d}\tau$$

for all z *such that* $|z - z_0| < R$.

In the last formula, $\int_{|\tau - z_0| = R}$ represents integration over the path $\gamma(t) = z_0 + R\mathrm{e}^{2\pi\imath t}$, $0 \le t \le 1$.

Proof. Fix z such that $|z - z_0| < R$ and define the function $g : D \to \mathbb{C}$ as follows:

$$g(\tau) = \begin{cases} \dfrac{f(\tau) - f(z)}{\tau - z}, & \text{for } \tau \neq z, \\ f'(z), & \text{for } \tau = z. \end{cases}$$

Then g is continuous in D, differentiable except perhaps at z, and it follows from (4.8.2) above that there exists a complex differentiable $G : \{|z - z_0| \le R\} \to \mathbb{C}$ with $G' = g$. Thus by (4.8.2) that

$$0 = \int_{|\tau - z_0| = R} g(\tau)\mathrm{d}\tau = \int_{|\tau - z_0| = R} \frac{f(\tau) - f(z)}{\tau - z} \mathrm{d}\tau,$$

and the desired integral formula follows. □

(8) It remains (under this approach) to be shown that the derivative of a holomorphic function is again holomorphic. This follows from the next lemma, whose similarity to Exercise 4.11 should be noted.

Lemma 4.67. *If* γ *is a pdp,* g *a continuous function on the range of* γ, *and* $n \in \mathbb{Z}_{>0}$, *then the function defined by*

$$F_n(z) = \int_\gamma \frac{g(\tau)}{(\tau - z)^n} \mathrm{d}\tau, \quad z \in \mathbb{C} - \text{range } \gamma, \tag{4.4}$$

is analytic in each of the regions in the complement of the range of γ *and satisfies*

$$F_n' = nF_{n+1}$$

there.

Proof. We first prove that F_1 is continuous at z_0 in Δ, a component of the complement of the range of γ. Choose $\delta > 0$ such that $U(z_0, \delta) \subseteq \Delta$. For z such that $|z - z_0| < \frac{\delta}{2}$, we have $|\tau - z| > \frac{\delta}{2}$ for all τ in range γ; hence

$$\begin{aligned} |F_1(z) - F_1(z_0)| &= \left| (z - z_0) \int_\gamma \frac{g(\tau)\,d\tau}{(\tau - z)(\tau - z_0)} \right| \\ &\le |z - z_0| \int_\gamma \frac{|g(\tau)|\,|d\tau|}{\frac{\delta}{2} \cdot \delta} \\ &= |z - z_0| \frac{2}{\delta^2} \int_\gamma |g(\tau)|\,|d\tau|, \end{aligned}$$

and the last term goes to zero when $z \to z_0$.

We also obtain, for all small positive values of $|z - z_0|$,

$$\frac{F_1(z) - F_1(z_0)}{z - z_0} = \int_\gamma \frac{g(\tau)\,d\tau}{(\tau - z)(\tau - z_0)}.$$

Thus

$$\begin{aligned} F_1'(z_0) &= \lim_{z \to z_0} \int_\gamma \frac{g(\tau)\,d\tau}{(\tau - z)(\tau - z_0)} \\ &= \int_\gamma \frac{g(\tau)\,d\tau}{(\tau - z_0)^2} = F_2(z_0). \end{aligned}$$

The general case is proved by induction: assume that n is at least equal to two, that F_{n-1} is analytic for any function g continuous on the range of γ, and that $F_{n-1}'(z) = (n-1)F_n(z)$ for all z in the complement of the range of γ.

Note that the function $\widetilde{g}(\tau) = \dfrac{g(\tau)}{\tau - z_0}$ is continuous on the range of γ, and therefore the function

$$\widetilde{F_{n-1}}(z) = \int_\gamma \frac{\widetilde{g}(\tau)\,d\tau}{(\tau - z)^{n-1}}$$

is analytic in the complement of range γ and satisfies

$$\widetilde{F_{n-1}}'(z) = (n-1)\widetilde{F_n}(z)$$

there.

We need to show that F_n, defined in (4.4), is analytic and satisfies $F_n'(z) = nF_{n+1}(z)$ in the complement of the range of γ. To do this, first observe that

$$\begin{aligned}\frac{1}{(\tau - z)^n} - \frac{1}{(\tau - z_0)^n} &= \frac{(\tau - z_0)^n - (\tau - z)^n}{(\tau - z)^n(\tau - z_0)^n} \\ &= \frac{(\tau - z_0)^{n-1}(\tau - z + z - z_0) - (\tau - z)^n}{(\tau - z)^n(\tau - z_0)^n} \\ &= \frac{1}{(\tau - z)^{n-1}(\tau - z_0)} - \frac{1}{(\tau - z_0)^n} + \frac{z - z_0}{(\tau - z)^n(\tau - z_0)},\end{aligned}$$

and then conclude that

$$\begin{aligned}F_n(z) - F_n(z_0) &= \int_\gamma g(\tau)\left[\frac{1}{(\tau - z)^n} - \frac{1}{(\tau - z_0)^n}\right] \\ &= \int_\gamma g(\tau)\left[\frac{1}{(\tau - z)^{n-1}(\tau - z_0)} - \frac{1}{(\tau - z_0)^n} + \frac{z - z_0}{(\tau - z)^n(\tau - z_0)}\right] \\ &= \widetilde{F_{n-1}}(z) - \widetilde{F_{n-1}}(z_0) + (z - z_0)\int_\gamma \frac{g(\tau)\,d\tau}{(\tau - z)^n(\tau - z_0)}.\end{aligned}$$

Thus $\lim_{z \to z_0} F_n(z) = F_n(z_0)$, and F_n is continuous at z_0. In addition we obtain, as before,

$$\frac{F_n(z) - F_n(z_0)}{z - z_0} = \frac{\widetilde{F_{n-1}}(z) - \widetilde{F_{n-1}}(z_0)}{z - z_0} + \int_\gamma \frac{g(\tau)\,d\tau}{(\tau - z)^n(\tau - z_0)},$$

and hence

$$\begin{aligned}F_n'(z_0) &= \widetilde{F_{n-1}}'(z_0) + \int_\gamma \frac{g(\tau)\,d\tau}{(\tau - z_0)^{n+1}} \\ &= (n-1)\widetilde{F_n}(z_0) + \int_\gamma \frac{g(\tau)\,d\tau}{(\tau - z_0)^{n+1}} \\ &= (n-1)\int_\gamma \frac{g(\tau)\,d\tau}{(\tau - z_0)^{n+1}} + \int_\gamma \frac{g(\tau)\,d\tau}{(\tau - z_0)^{n+1}} = nF_{n+1}(z_0). \qquad \square\end{aligned}$$

(9) *If f is continuous on D, and for all $R > 0$ such that $\{|z - z_0| \le R\} \subset D$, the equality*

$$f(z) = \frac{1}{2\pi\imath}\int_{|\tau - z_0| = R} \frac{f(\tau)}{\tau - z}\,d\tau$$

holds for all z such that $|z - z_0| < R$, then $f^{(n)}(z)$ exists for all z in D and for all $n \in \mathbb{Z}_{\geq 0}$. In particular, if $f'(z)$ exists for all z in D, then the conclusion holds.

Proof. Fix z_0 in D and choose $R > 0$ such that

$$f(z) = \frac{1}{2\pi\imath} \int_{|\tau - z_0| = R} \frac{f(\tau)}{\tau - z}\, \mathrm{d}\tau$$

for all z such that $|z - z_0| < R$. It follows from Lemma 4.67 that then

$$f'(z) = \frac{1}{2\pi\imath} \int_{|\tau - z_0| = R} \frac{f(\tau)}{(\tau - z)^2} \mathrm{d}\tau,$$

for all z such that $|z - z_0| < R$ and, more generally, that for each $n \geq 0$ and z such that $|z - z_0| < R$:

$$f^{(n)}(z) = \frac{n!}{2\pi\imath} \int_{|\tau - z_0| = R} \frac{f(\tau)}{(\tau - z)^{n+1}} \mathrm{d}\tau.$$

□

This material suffices to produce the Cauchy estimates and Morera's and Liouville's theorems, as done in the next chapter.

Exercises

4.1. Evaluate the line or contour integral $\int_C |z|\, \mathrm{d}z$ directly from the definition if

(1) C is a straight line segment from $-\imath$ to $\imath$.
(2) C is the left half of the unit circle traversed from $-\imath$ to $\imath$.
(3) C is the right half of the unit circle traversed from $-\imath$ to $\imath$.

4.2. Evaluate the line or contour integral $\int_C x \mathrm{d}z$ directly from the definition when C is the line segment from 0 to $i + 1$.

4.3. Let $z_0 \in \mathbb{C}$. Evaluate the line or contour integral $\int_C (z - z_0)^m \mathrm{d}z$ directly from the definition, where C is the circle centered at z_0 with radius $r > 0$ and

(1) m is an integer, $m \geq 0$.
(2) m is an integer, $m < 0$.

4.4. Evaluate the line or contour integral $\int_\gamma z^3 \mathrm{d}z$ directly from the definition over the path $\gamma(t), 0 \leq t \leq 1$, where

(1) $\gamma(t) = 1 + it$.
(2) $\gamma(t) = \mathrm{e}^{-\pi\imath t}$.

(3) $\gamma(t) = e^{\pi \imath t}$.
(4) $\gamma(t) = 1 + \imath t + t^2$.

Evaluate the integrals $\int_\gamma \bar{z}\,dz$ and $\int_\gamma \frac{1}{z}\,dz$ over the same paths.

4.5. Let D_1 and D_2 be simply connected plane domains whose intersection is nonempty and connected. Prove that their intersection and their union are both simply connected.

4.6. Show that for any closed interval $[a, b]$ there exists a one-to-one, onto, and differentiable map $t : [0, 1] \to [a, b]$ with $t'(u) > 0$ for all u.

4.7. Establish Remark 4.7.

4.8. Verify that $f(t) = 2\pi \imath\, t$ for $0 \le t \le 1$ is a primitive of the closed form

$$\omega = \frac{d}{z} \text{ along the path } \gamma(t) = \exp(2\pi \imath t),\ 0 \le t \le 1,$$

on the domain $D = \mathbb{C}_{\neq 0}$. Observe that $f(0) \neq f(1)$ and $\gamma(0) = \gamma(1)$; see the warning after Definition 4.31.

4.9. Let P and Q be smooth functions on a domain $D \subseteq \mathbb{C}$. Find necessary and sufficient conditions for the form $P dz + Q d\bar{z}$ to be closed.

4.10. Let D be a domain in $\mathbb{C}$. We have studied the pairing $\int_\gamma \omega$, where γ is a closed path in D and ω is a closed differential form in D. Show that

(1) $\int_\gamma \omega$ depends only on the homotopy class of the closed path γ; that is, if we replace γ by a closed path γ' homotopic to γ (as closed paths), then the integral is unchanged.
(2) (Only for those who know some algebraic topology) $\int_\gamma \omega$ depends only on the homology class of the closed path γ; that is, if we replace γ by a closed path γ' homologous to γ, then the integral is unchanged.
(3) $\int_\gamma \omega$ depends only on the cohomology class of the closed form ω; that is, if we replace ω by $\omega + df$ with $f \in \mathbf{C}^2(D)$, then the integral is unchanged.

4.11. In this exercise we consider "differentiation under the integral sign."

(1) Let $\gamma : [a, b] \to \mathbb{C}$ denote a pdp, and let φ : range $\gamma \to \mathbb{C}$ be a continuous function.
Define $g : D = \mathbb{C} - \text{range}\ \gamma \to \mathbb{C}$ by

$$g(z) = \int_\gamma \frac{\varphi(u)}{u - z}\, du.$$

Show that g has derivatives of all orders $n \ge 0$ and that

$$g^{(n)}(z) = n! \int_\gamma \frac{\varphi(u)}{(u-z)^{n+1}}\,\mathrm{d}u$$

for all n in $\mathbb{Z}_{\geq 0}$. Thus, in particular, g is holomorphic on D.

(2) Let $z_0 \in \mathbb{C}$, $R > 0$ and $\gamma(t) = z_0 + R\exp(2\pi\imath t)$ for $0 \leq t \leq 1$. Use the first part of this exercise to show that

$$\frac{1}{2\pi\imath}\int_\gamma \frac{\mathrm{d}z}{z-w} = \begin{cases} 1, \text{ if } |z-w| < R, \\ 0, \text{ if } |z-w| > R. \end{cases}$$

Chapter 5
The Cauchy Theory: Key Consequences

This chapter is devoted to some immediate consequences of the fundamental result for the Cauchy theory, Theorem 4.61, of the last chapter. Although the chapter is very short, it includes proofs of many of the implications of the fundamental theorem in complex function theory (Theorem 1.1). We point out that these relatively compact proofs of a host of major theorems result from the work put into Chap. 4 and earlier chapters.

The appendix to this chapter contains a version of Cauchy's integral formula for smooth (not necessarily holomorphic) functions.

5.1 Consequences of the Cauchy Theory

We begin with a technical strengthening of Theorem 4.61 allowing functions that are holomorphic on a domain except on a line segment. It will lead to Cauchy's integral formula, once described as the most beautiful theorem in complex variables.

Theorem 5.1 (Goursat's Theorem, Strengthened Version). *If f is continuous in a domain D and holomorphic except possibly on a line segment in D, then $f(z)\,\mathrm{d}z$ is closed in D.*

Proof. Without loss of generality, D is the unit disc, and the line segment is all or part of the real axis in D.

We must show that the integral $\int_\gamma f(z)\,\mathrm{d}z$ vanishes whenever γ is the (positively oriented) boundary of an open rectangle R whose closure is contained in D and whose sides are parallel to the coordinate axes.

There are three possibilities for such rectangles:

(1) The closure of R does not intersect the real axis.
(2) The closure of R has one side on $\mathbb{R}$.
(3) (The interior of) R intersects $\mathbb{R}$.

R.E. Rodríguez et al., *Complex Analysis: In the Spirit of Lipman Bers*, Graduate Texts in Mathematics 245, DOI 10.1007/978-1-4419-7323-8_5,

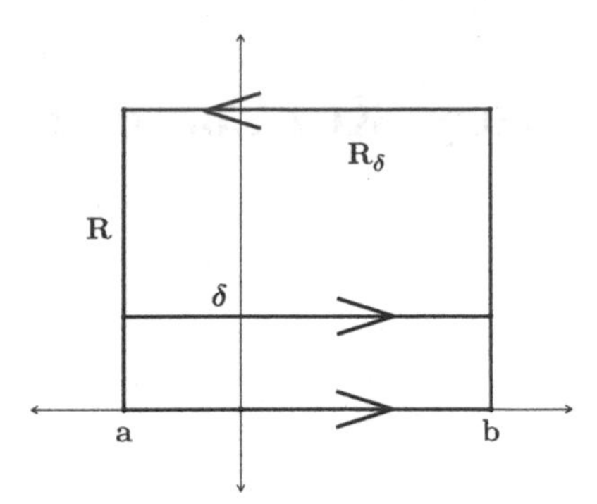

Fig. 5.1 The rectangles R and R_δ

In case (1), there is nothing to do by the results of the last chapter. In case (3), we reduce to the case of two rectangles as in case (2). Thus it suffices to consider a rectangle of type (2). Assume that the rectangle R lies in the upper semi-disc with one side on $\mathbb{R}$ from a to $b > a$. (The possibility of R in the lower half disc is handled similarly.) Let R_δ be the rectangle R with the portion below height δ chopped off, for $\delta > 0$ sufficiently small (see Fig. 5.1), and to be chosen later. Then the difference of the integrals over the boundary of R and the boundary of R_δ is an integral over an appropriate rectangle:

$$\begin{aligned}\int_{\partial R} f(z)\,\mathrm{d}z - \int_{\partial R_\delta} f(z)\,\mathrm{d}z &= \int_a^b f(x,0)\,\mathrm{d}x + \imath \int_0^\delta f(b,y)\,\mathrm{d}y \\ &\quad + \int_b^a f(x,\delta)\,dx + \imath \int_\delta^0 f(a,y)\,\mathrm{d}y \\ &= \int_a^b (f(x,0) - f(x,\delta))\,\mathrm{d}x \\ &\quad + \imath \int_0^\delta (f(b,y) - f(a,y))\,\mathrm{d}y.\end{aligned}$$

Now given $\epsilon > 0$, there exists a δ with $0 < \delta < \epsilon$ such that

$$|z - c| < \delta \text{ implies that } |f(z) - f(c)| < \epsilon$$

for all z and c in R (by the uniform continuity of f on R). Also there exists an $M > 0$ such that $|f(z)| \leq M$ for all $z \in R$. Thus

$$\left|\int_{\partial R} f(z)\,\mathrm{d}z - \int_{\partial R_\delta} f(z)\,\mathrm{d}z\right| \leq \epsilon(b - a) + 2M\epsilon.$$

Since ϵ is arbitrary, this tells us that

$$\int_{\partial R} f(z)\,\mathrm{d}z - \int_{\partial R_\delta} f(z)\,\mathrm{d}z = 0.$$

But we know that $\int_{\partial R_\delta} f(z)\,\mathrm{d}z = 0$; hence also $\int_{\partial R} f(z)\,\mathrm{d}z = 0$. □

We apply this strengthened theorem to obtain

Theorem 5.2 (Cauchy's Integral Formula). *If f is holomorphic on a domain D and γ is a continuous closed path homotopic to a point in D, then for all $c \in \mathbb{C} - \text{range}\,\gamma$, we have*

$$\frac{1}{2\pi\imath}\int_\gamma \frac{f(z)}{z-c}\,\mathrm{d}z = I(\gamma, c)\cdot f(c). \tag{5.1}$$

Proof. Define, for $z \in D$,

$$g(z) = \begin{cases} \dfrac{f(z)-f(c)}{z-c}, & \text{if } z \neq c, \\ f'(c), & \text{if } z = c. \end{cases}$$

Then g is continuous on D and holomorphic except (possibly) at c and thus, by Theorem 5.1, $g(z)\,\mathrm{d}z$ is closed in D. It follows that

$$0 = \int_\gamma g(z)\,\mathrm{d}z = \int_\gamma \frac{f(z)-f(c)}{z-c}\,\mathrm{d}z.$$

Thus

$$\int_\gamma \frac{f(z)}{z-c}\,\mathrm{d}z = f(c)\int_\gamma \frac{\mathrm{d}z}{z-c}.$$

□

Example 5.3. Let D be a domain in $\mathbb{C}$, f a holomorphic function defined on D, and $c \in D$. Choose $R > 0$ such that $\mathrm{cl}U(c, R) \subset D$ and let $\gamma(\theta) = c + R\mathrm{e}^{2\pi\imath\theta}$, for $0 \le \theta \le 1$. Then $I(\gamma, w) = 1$ for $|w - c| < R$ and $I(\gamma, w) = 0$ for $|w - c| > R$, by property (4) of the winding numbers given in Sect. 4.5. Thus

(1)

$$\frac{1}{2\pi\imath}\int_\gamma \frac{f(z)}{z-w}\,\mathrm{d}z = f(w) \text{ for } |w-c| < R \text{ and}$$

(2)

$$\frac{1}{2\pi\imath}\int_\gamma \frac{f(z)}{z-w}\,\mathrm{d}z = 0 \text{ for } |w-c| > R.$$

Remark 5.4. Equation (5.1) gives the amazing result that the value of a holomorphic function at a point interior to a circle (or eventually to any simple closed curve) is completely determined by the values of the function on the boundary circle. The function must, of course, be holomorphic in a neighborhood of the closed disc bounded by the circle. The requirement that the function be analytic in a neighborhood of the disc can be relaxed considerably, but this topic is not discussed in the book. It follows from (5.1) and Exercise 4.11 or from the next theorem that a holomorphic function has derivatives of all orders. We prove a more general result next.

Theorem 5.5 (Taylor (Power) Series Expansions for Holomorphic Functions). *If f is holomorphic in the open disc $\{|z| < R\}$, with $R \in (0, +\infty]$, then f has a power series expansion at each point in this disc. In particular, there exists a power series $\sum_{k=0}^{\infty} a_k z^k$ with radius of convergence $\rho \geq R$ such that*

$$f(z) = \sum_{k=0}^{\infty} a_k z^k \text{ for } |z| < R.$$

Proof. It suffices to establish the particular claim. Choose $0 < r_0 < R$ and define $\gamma(\theta) = r_0 e^{2\pi\imath\theta}$ for $0 \leq \theta \leq 1$. Then $|z| = r < r_0$ implies $I(\gamma, z) = 1$. We start for such z with

$$f(z) = \frac{1}{2\pi\imath} \int_\gamma \frac{f(t)}{t - z} \, dt,$$

by Theorem 5.2. Our plan is to write the integrand as a power series in z and then interchange the order of the operations (integration and summation). Now

$$\frac{1}{t-z} - \frac{1}{t} \frac{\left(\frac{z}{t}\right)^{n+1}}{1 - \frac{z}{t}} = \frac{1}{t} \frac{1 - \left(\frac{z}{t}\right)^{n+1}}{1 - \frac{z}{t}} = \frac{1}{t} \sum_{k=0}^{n} \left(\frac{z}{t}\right)^k.$$

Hence

$$\frac{1}{t-z} = \frac{1}{t} \sum_{k=0}^{n} \left(\frac{z}{t}\right)^k + \frac{1}{t} \frac{\left(\frac{z}{t}\right)^{n+1}}{1 - \frac{z}{t}}$$

and

$$\begin{aligned} f(z) &= \frac{1}{2\pi\imath} \left[\sum_{k=0}^{n} z^k \int_\gamma \frac{f(t)}{t^{k+1}} \, dt + \int_\gamma \frac{f(t)}{t} \frac{\left(\frac{z}{t}\right)^{n+1}}{1 - \frac{z}{t}} \, dt \right] \\ &= \sum_{k=0}^{n} a_k z^k + R_n, \end{aligned}$$

where

$$a_k = \frac{1}{2\pi\imath}\int_\gamma \frac{f(t)}{t^{k+1}}\,\mathrm{d}t\,, \text{ and } R_n = \frac{1}{2\pi\imath}\int_\gamma \frac{f(t)}{t}\,\frac{\left(\frac{z}{t}\right)^{n+1}}{1-\frac{z}{t}}\,\mathrm{d}t.$$

On the range of γ we have $t = r_0 e^{2\pi\imath\theta}$, $|t| = r_0$, and $|\mathrm{d}t| = 2\pi r_0 \mathrm{d}\theta$. Let

$$M(r_0) = \sup\{\left|f(r_0 e^{2\pi\imath\theta})\right| : 0 \le \theta \le 1\}$$

and observe that $\left|1 - \frac{z}{t}\right| \ge 1 - \frac{r}{r_0}$. Hence

$$\begin{aligned} |R_n| &\le \frac{1}{2\pi}\int_0^1 \frac{M(r_0)}{r_0}\,\frac{\left(\frac{r}{r_0}\right)^{n+1}}{1-\frac{r}{r_0}}\, r_0\, 2\pi\, \mathrm{d}\theta \\ &= M(r_0)\frac{\left(\frac{r}{r_0}\right)^{n+1}}{1-\frac{r}{r_0}} \to 0 \text{ as } n \to \infty. \end{aligned}$$

We conclude that $f(z) = \sum_{k=0}^{\infty} a_k z^k$ for $|z| < R$; further,

$$a_k = \frac{1}{2\pi\imath}\int_\gamma \frac{f(t)}{t^{k+1}}\,\mathrm{d}t$$

is independent of r_0 for $0 < r_0 < R$. We have also obtained the estimates

$$|a_k| \le \frac{M(r_0)}{r_0^k} \tag{5.2}$$

and therefore

$$|a_k|^{\frac{1}{k}} \le \frac{M(r_0)^{\frac{1}{k}}}{r_0}.$$

Thus (we have obtained a second proof that) the radius of convergence ρ of $\sum a_k z^k$ satisfies $\rho \ge r_0$ for all $r_0 < R$; in particular, $\rho \ge R$. □

We will make further use of (5.2) shortly. Note that the theorem (once again) allows us to interchange the orders of operations:

$$2\pi\imath f(z) = \int_\gamma f(t)\sum_{k=0}^{\infty}\frac{z^k}{t^{k+1}}\,\mathrm{d}t = \sum_{k=0}^{\infty} z^k \int_\gamma \frac{f(t)}{t^{k+1}}\,\mathrm{d}t.$$

Corollary 5.6. *A function f is holomorphic in an open set D if and only if f has a power series expansion at each point of D. For a holomorphic function f on D, the Taylor series expansion of f at $c \in D$*

$$f(z) = \sum_{k=0}^{\infty} a_k (z-c)^k$$

has radius of convergence

$$\rho \geq \sup\{r > 0;\ U(c,r) \subseteq D\}.$$

Remark 5.7. For $c = 0$, the above Taylor series is also called a *Maclaurin* series.

Corollary 5.8. *If f is holomorphic on a domain D, then f is $\mathbf{C}^\infty$ in D, and for each $n \in \mathbb{Z}_{\geq 0}$, $f^{(n)}$ is holomorphic on D.*

Corollary 5.9 (Cauchy's Generalized Integral Formula). *Let f be holomorphic on a domain D containing $\operatorname{cl} U(c,R)$ for some $c \in D$ and $R > 0$. If $\gamma(\theta) = c + R\,e^{\imath\theta}$ for $0 \leq \theta \leq 2\pi$, then for $n = 0, 1, 2, \ldots,$*

$$f^{(n)}(c) = \frac{n!}{2\pi\imath} \int_\gamma \frac{f(t)}{(t-c)^{n+1}} \mathrm{d}t.$$

Proof. Recall that for $n = 0, 1, 2, \ldots,$

$$a_n = \frac{1}{2\pi\imath} \int_\gamma \frac{f(t)}{(t-c)^{n+1}} \,\mathrm{d}t = \frac{f^{(n)}(c)}{n!}.$$

□

Theorem 5.10 (Morera's Theorem). *If $f \in \mathbf{C}^0(D)$ and $f(z)\,\mathrm{d}z$ is closed on D, then f is holomorphic on D.*

Proof. Since the differential form $\omega = f(z)\,\mathrm{d}z$ is locally exact, for each point $c \in D$ there is a neighborhood U of c in D and a primitive F of ω in U. That is, there is a $\mathbf{C}^1$-function F on U with $F_z = f$ and $F_{\bar{z}} = 0$; thus F is holomorphic on U and so is its derivative f, by Corollary 5.8. Since being holomorphic is a local property, f is holomorphic on D. □

An immediate consequence of Morera's theorem together with Theorem 5.1 is

Corollary 5.11. *If f is continuous in D and holomorphic except possibly on a line segment in D, then f is holomorphic in D.*

We have by now established the following important

Theorem 5.12. *Let f be a complex-valued function defined on an open set D in $\mathbb{C}$. Then the following conditions are equivalent:*

(a) f *is holomorphic on* D.
(b) f *is* $\mathbf{C}^1$ *and satisfies* CR *on* D.
(c) f *is* $\mathbf{C}^0$ *and* $f(z)\,dz$ *is closed on* D.
(d) f *is* $\mathbf{C}^0$ *on* D *and holomorphic except possibly on a line segment in* D.
(e) f *has a power series expansion at each point in* D.

Remark 5.13. As a consequence of the theorem, the space $\mathbf{H}(D)$ defined in Chap. 3 (see Definition 3.57) consists precisely of the holomorphic functions on D, and a meromorphic function (an element of $\mathbf{M}(D)$, see Definition 3.59) is locally the ratio of two holomorphic functions.

Recall that we have established the estimates (5.2) from the Cauchy integral formula. An immediate consequence is the following result.

Corollary 5.14 (Cauchy's Inequalities). *Let* $c \in \mathbb{C}$ *and assume that*

$$f(z) = \sum_{n=0}^{\infty} a_n (z-c)^n$$

has radius of convergence $\rho > 0$. *Then*

$$a_n = \frac{f^{(n)}(c)}{n!}$$

and

$$|a_n| = \left|\frac{f^{(n)}(c)}{n!}\right| \leq \frac{M(r)}{r^n}, \tag{5.3}$$

for all $0 < r < \rho$, *where*

$$M(r) = \sup\{|f(z)|\,;\; |z-c| = r\}.$$

Theorem 5.15 (Liouville's Theorem). *A bounded entire function is constant.*

Proof. Observe that the Taylor series expansion $\sum_{n=0}^{\infty} a_n z^n$ of the entire function f at the origin has radius of convergence equal to ∞ and that the estimates (5.3) hold for all $r > 0$. Since there exists $M > 0$ such that $|f(z)| \leq M$ for all $z \in \mathbb{C}$, we obtain

$$|a_n| \leq \frac{M}{r^n}$$

for all positive r and all $n \in \mathbb{Z}_{n \geq 0}$; taking limit as r goes to ∞ shows that $a_n = 0$ for all $n \geq 1$. □

Theorem 5.16 (Fundamental Theorem of Algebra). *If* P *is a polynomial of degree* $n \geq 1$, *then there exist* $a_1, \ldots, a_n \in \mathbb{C}$ *and* $b \in \mathbb{C}_{\neq 0}$ *such that*

$$P(z) = b \prod_{j=1}^{n} (z - a_j) \text{ for all } z \in \mathbb{C}.$$

Proof. It suffices to show that P has a root. If not, $\frac{1}{P}$ is an entire function. It is also bounded since $\lim_{z \to \infty} \frac{1}{P(z)} = 0$ and thus must be constant. □

5.2 Cycles and Homology

In order to prove the general version of Cauchy's integral formula, we will need a more general form of Cauchy's theorem that deals with integrals over *cycles* that are *homologous* to zero. However, except for the next section and Sect. 9.3.3, we develop the subsequent chapters without reference to the material in this section.

Definition 5.17. A *cycle* γ is a finite sequence of continuous closed paths in the complex plane.

If γ is a cycle, we refer to the continuous closed paths $\gamma_1, \gamma_2, \ldots, \gamma_n$ that make up γ as its *component curves*, and we write $\gamma = (\gamma_1, \gamma_2, \ldots, \gamma_n)$. Note that the component curves of a cycle need not be distinct; we also mention that the order of the component curves will not be relevant in our considerations. We consider the *range of* γ to be the union of the ranges of its components.

We extend the notion of the integral of a function over a single closed path to the integral over a cycle as follows.

Definition 5.18. If $\gamma = (\gamma_1, \gamma_2, \ldots, \gamma_n)$ is a cycle, then for any holomorphic function f defined on a domain D such that range $\gamma \subset D$, we set

$$\int_\gamma f(z)\,dz = \int_{\gamma_1} f(z)\,dz + \cdots + \int_{\gamma_n} f(z)\,dz. \tag{5.4}$$

We can extend the notion of the index of a point with respect to a path to the index of a point with respect to a cycle as follows.

Definition 5.19. The *index of a cycle* $\gamma = (\gamma_1, \gamma_2, \ldots, \gamma_n)$ *with respect to a point* $c \in \mathbb{C} -$ range γ is denoted by $I(\gamma, c)$ and defined by

$$I(\gamma, c) = I(\gamma_1, c) + \cdots + I(\gamma_n, c). \tag{5.5}$$

Definition 5.20. A cycle γ with range contained in a domain $D \subseteq \mathbb{C}$ is said to be *homologous to zero in* D if $I(\gamma, c) = 0$ for every $c \in \mathbb{C} - D$.

Observe that if a continuous closed path γ is homotopic to a point in D, then the cycle (γ) with the single component γ is homologous to zero in D. However, the two notions are different, see Exercise 5.3.

With these definitions and some work,[1] we can obtain the most general forms of Cauchy's theorem and integral formula.

Theorem 5.21 (Cauchy's Theorem and Integral Formula: General Form). *If f is analytic in a domain $D \subseteq \mathbb{C}$ and γ is a cycle homologous to zero in D, then*

(a) $\int_\gamma f(z)\,\mathrm{d}z = 0$.
(b) For all $c \in D - \operatorname{range}\gamma$, we have (5.1).

Proof. If

$$E = \{z \in \mathbb{C} - \operatorname{range}\gamma;\, I(\gamma, z) = 0\},$$

then the set E is open in $\mathbb{C}$ and contains the unbounded component of the complement of the range of γ in $\mathbb{C}$, because it contains the unbounded component of the complement of the range of each component curve of γ, as we saw in Sect. 4.5. Moreover $E \supset (\mathbb{C} - D)$, since γ is homologous to zero in D.

Define $g : D \times D \to \mathbb{C}$ by

$$g(w,z) = \begin{cases} \dfrac{f(z) - f(w)}{z - w} & \text{for } z \neq w, \\ \\ f'(w) & \text{for } z = w. \end{cases}$$

The function g is continuous in $D \times D$, and for fixed $z \in D$, $g(\cdot, z)$ is holomorphic on D. Furthermore, for all $c \in D - \operatorname{range}\gamma$, we have

$$\begin{aligned} \int_\gamma g(c,z)\,\mathrm{d}z &= \int_\gamma \frac{f(z) - f(c)}{z - c}\,\mathrm{d}z \\ &= \int_\gamma \frac{f(z)}{z-c}\mathrm{d}z - f(c)\int_\gamma \frac{\mathrm{d}z}{z-c} \\ &= \int_\gamma \frac{f(z)}{z-c}\,\mathrm{d}z - f(c)\,2\pi\imath\, I(\gamma, c). \end{aligned} \tag{5.6}$$

We define next

$$h(w) = \begin{cases} \int_\gamma g(w,z)\,\mathrm{d}z \text{ for } w \in D, \\ \\ \int_\gamma \dfrac{f(z)\,\mathrm{d}z}{z - w} \quad \text{for } w \in E. \end{cases}$$

Noting that $D \cup E = \mathbb{C}$, we see from (5.6) that for $w \in D \cap E$,

[1] We are following a course outlined by J. D. Dixon, *A brief proof of Cauchy's integral formula*, Proc. Amer. Math. Soc. **29** (1971), 625–626.

$$\int_\gamma g(w,z)\,\mathrm{d}z = \int_\gamma \frac{f(z)}{z-w}\,\mathrm{d}z,$$

because $I(\gamma, w) = 0$, and thus h is a well-defined function on the plane.

The set E contains the complement of a large disc, and the function h is clearly bounded there. By Exercise 4.11, h is complex differentiable; thus h is a bounded analytic function in $\mathbb{C}$ and hence constant, by Liouville's theorem. Since $\lim_{z\to\infty} h(w) = 0$, h is the zero function. In particular,

$$\int_\gamma g(w,z)\,\mathrm{d}z = 0$$

for all $w \in D - \text{range } \gamma$, and (b) follows from (5.6).

We now fix a point $c \in D - \text{range } \gamma$ and apply part (b) to the analytic function defined on D by $z \mapsto (z-c)f(z)$ and the cycle γ, to obtain

$$I(\gamma, w)(w-c)f(w) = \frac{1}{2\pi\imath}\int_\gamma \frac{(z-c)f(z)}{z-w}\mathrm{d}z$$

for all $w \in D - \text{range } \gamma$. We obtain part (a) by evaluating the last equation at $w = c$. □

Remark 5.22. 1. A topologist would develop the concept of homology in much more detail using chains and cycles. However, for our purposes, the above definitions suffice.

2. To help the reader with some of the problems of the last chapter, we review a standard definition from algebraic topology: two cycles $\gamma = (\gamma_1, \ldots, \gamma_n)$ and $\delta = (\delta_1, \ldots, \delta_m)$ with ranges contained in a domain D are *homologous in D* if the cycle with components $(\gamma_1, \ldots, \gamma_n, \delta_{1\_}, \ldots, \delta_{m\_})$ is homologous to zero in D, where $\delta_{i\_}$ is the curve δ_i traversed backward (see Definition 4.9).
3. It is also standard to define the next relation between curves, that we have not had any reason to use. Two non-closed paths γ_1 and γ_2 (with ranges contained in D) are *homologous in D* if they have the same initial point and the same end point and the cycle $(\gamma_1 * \gamma_{2\_})$ is homologous to zero in D; note that the only component of this cycle is the closed path $\gamma_1 * \gamma_{2\_}$.
4. The notions of a cycle $\gamma = (\gamma_1, \gamma_2, \ldots, \gamma_n)$ and the sum $\gamma_1 + \gamma_2 + \cdots + \gamma_n$ of its components, as in Definition 4.58, are different and should not be confused.
5. A domain D in $\mathbb{C}$ is simply connected if and only if $I(\gamma, c) = 0$ for all cycles γ in D and all $c \in \mathbb{C} - D$.

5.3 Jordan Curves

We recall that the continuous closed path $\gamma : [0, 1] \to \mathbb{C}$ is a *simple closed path* or a *Jordan curve* whenever $\gamma(t_1) = \gamma(t_2)$ with $0 \le t_1 < t_2 \le 1$ implies $t_1 = 0$ and $t_2 = 1$.

In this case, the range of γ is a homeomorphic image of the unit circle S^1. To see this, we define

$$h(e^{2\pi \imath t}) = \gamma(t)$$

and note that h maps S^1 onto the range of γ. Observe that h is well defined, continuous, and injective. Since the circle is compact, h is a homeomorphism.

Theorem 5.23 (Jordan Curve Theorem[2]). *If γ is a simple closed path in $\mathbb{C}$, then*

(a) $\mathbb{C} -$ range γ *has exactly two connected components, one of which is bounded.*
(b) Range γ *is the boundary of each of these components, and*
(c) $I(\gamma, c) = 0$ *for all c in the unbounded component of the complement of the range of γ. $I(\gamma, c) = \pm 1$ for all c in the bounded component of the complement of the range of γ. The choice of sign depends only on the choice of direction for traversal on γ.*

Definition 5.24. For a simple closed path γ in $\mathbb{C}$ we define the *interior of γ, $i(\gamma)$*, to be the bounded component of $\mathbb{C} -$ range γ and the *exterior of γ, $e(\gamma)$*, to be the unbounded component of $\mathbb{C} -$ range γ.

If $I(\gamma, c) = +1$ (respectively -1) for c in $i(\gamma)$ then we say that γ is a *Jordan curve* with *positive* (respectively *negative*) *orientation*.

We shall not prove the above theorem. It is a deep result. In all of our applications, it will be obvious that our Jordan curves have the above properties.

Remark 5.25. Another important (and nontrivial to prove) property of Jordan curves is the fact that the interior of a Jordan curve is always a simply connected domain in $\mathbb{C}$. If we view the Jordan curve as lying on the Riemann sphere $\widehat{\mathbb{C}}$, then each component of the complement of its range is simply connected.

This property allows us to prove the following result.

Theorem 5.26 (Cauchy's Theorem (Extended Version)). *Let $\gamma_0, \ldots, \gamma_n$ be $n+1$ positively oriented Jordan curves. Assume that*

$$\text{range } \gamma_j \subset e(\gamma_k) \cap i(\gamma_0)$$

[2] For a proof see the appendix to Ch. IX of J. Dieudonné, *Foundations of Modern Analysis*, Pure and Applied Mathematics, vol. X, Academic Press, 1960 or Chap. 10 of J. R. Munkres, *Topology (Second Edition)*, Dover, 2000.

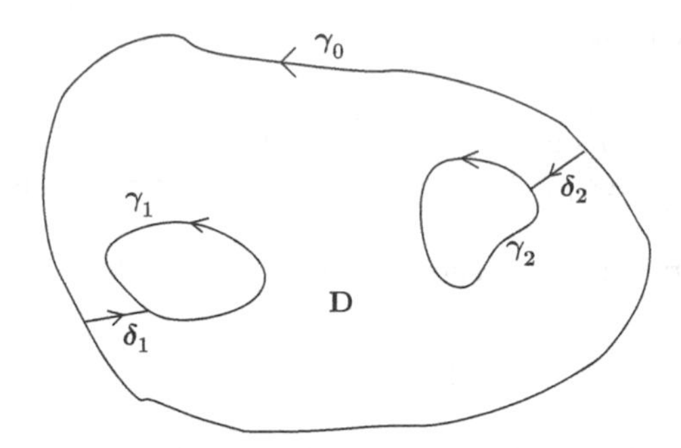

Fig. 5.2 Jordan curves and the domain they define

for all $1 \leq j \neq k \leq n$, *see Fig. 5.2. If* f *is a holomorphic function on a neighborhood* N *of the closure of the domain*

$$D = i(\gamma_0) \cap e(\gamma_1) \cap \cdots \cap e(\gamma_n),$$

then

$$\int_{\gamma_0} f(z)\, \mathrm{d}z = \sum_{k=1}^{n} \int_{\gamma_k} f(z)\, \mathrm{d}z.$$

Proof. Adjoin nonintersecting curves δ_j in D from γ_0 to γ_j for $j = 1, \ldots, n$, as in Fig. 5.2. Then the cycle

$$\delta = (\gamma_0, \delta_1 * \gamma_{1\_} * \delta_{1\_}, \ldots, \delta_n * \gamma_{n\_} * \delta_{n\_})$$

is homologous to zero in N. Thus Theorem 5.21 implies that

$$\left(\int_{\gamma_0} + \sum_{k=1}^{n} \left(\int_{\delta_k} + \int_{\gamma_{k\_}} + \int_{\delta_{k\_}} \right) \right) f(z)\, \mathrm{d}z = 0,$$

and the result follows by noting that the integral over each δ_k is canceled by the corresponding integral over $\delta_{k\_}$. □

An immediate consequence is

Theorem 5.27 (Cauchy's Integral Formula (Extended Version)). *With the hypotheses as in the extended version of Cauchy's Theorem 5.26, we have*

$$2\pi\imath f(c) = \int_{\gamma_0} \frac{f(z)}{z-c}\, \mathrm{d}z - \sum_{k=1}^{n} \int_{\gamma_k} \frac{f(z)}{z-c}\, \mathrm{d}z$$

for all $c \in D$.

Proof. We can apply Theorem 5.2 to the function f, using the neighborhood N of Theorem 5.26 and the cycle δ constructed in its proof, since δ is homologous to zero in N and $I(\delta, c) = +1$. As before, the integral over each δ_k is canceled by the corresponding integral over δ_{k_-}. □

5.4 The Mean Value Property

The next concept applies in a broader context than that of holomorphic functions, as we will see in Chap. 9.

Definition 5.28. Let f be a function defined on a domain D in $\mathbb{C}$. We say that f has the *mean value property* (MVP) if for each $c \in D$ there exists $r_0 > 0$ with $U(c, r_0) \subseteq D$ and

$$f(c) = \frac{1}{2\pi}\int_0^{2\pi} f(c + r\,\mathrm{e}^{\imath\theta})\,\mathrm{d}\theta \text{ for all } 0 \leq r < r_0. \tag{5.7}$$

Remark 5.29. A holomorphic function f on a domain D has the MVP (with $r_0 =$ the distance of $c \in D$ to ∂D). Hence so do its real and imaginary parts.

Theorem 5.30 (Maximum Modulus Principle). *Suppose f is a continuous complex-valued function defined on a domain D in $\mathbb{C}$ that has the MVP. If $|f|$ has a relative maximum at a point $c \in D$, then f is constant in a neighborhood of c.*

Proof. The result is clear if $f(c) = 0$. If $f(c) \neq 0$, replacing f by $\mathrm{e}^{-\imath\theta} f$ for some $\theta \in \mathbb{R}$, we may assume that $f(c) > 0$. Write $f = u + \imath\, v$ and choose $r_0 > 0$ such that

(1) $\mathrm{cl}U(c, r_0) \subset D$.
(2) Equation (5.7) holds.
(3) $|f(z)| \leq f(c)$ for $z \in \mathrm{cl}\, U(c, r_0)$.

If we define

$$M(r) = \sup\{|f(z)|\,;\,|z - c| = r\} \text{ for } 0 \leq r \leq r_0,$$

then it follows from (3) that

$$M(r) \leq f(c) \text{ for } 0 \leq r \leq r_0.$$

Since the MVP implies that

$$f(c) = \frac{1}{2\pi}\int_0^{2\pi} f(c + r\,\mathrm{e}^{i\theta})\,\mathrm{d}\theta \text{ for } 0 \leq r < r_0,$$

we also have $f(c) \leq M(r)$, and we conclude that $f(c) = M(r)$ for $0 \leq r \leq r_0$. Now observe that

$$\frac{1}{2\pi}\int_0^{2\pi} M(r)\,\mathrm{d}\theta = M(r) = f(c) = \frac{1}{2\pi}\int_0^{2\pi} u(c + r\,\mathrm{e}^{\imath\theta})\,\mathrm{d}\theta, \tag{5.8}$$

where the last equality holds because $f(c)$ is real. Also note that

$$M(r) - u(c + r\,\mathrm{e}^{\imath\theta}) \geq 0, \tag{5.9}$$

from the definition of $M(r)$. But from (5.8) we obtain that

$$\int_0^{2\pi} \left[M(r) - u(c + r\mathrm{e}^{\imath\theta})\right]\mathrm{d}\theta = 0,$$

and hence we must have equality in (5.9) for all θ. Finally,

$$M(r) \geq (u^2(c + r\,\mathrm{e}^{\imath\theta}) + v^2(c + r\,\mathrm{e}^{\imath\theta}))^{\frac{1}{2}} = (M(r)^2 + v^2(c + r\,\mathrm{e}^{\imath\theta}))^{\frac{1}{2}}$$

which implies that $v(c + r\,\mathrm{e}^{\imath\theta}) = 0$ for $0 \leq r \leq r_0$ and $0 \leq \theta \leq 2\pi$. Therefore $f(z) = u(z) = M(|z|) = f(c)$ for all $|z| \leq r_0$. □

It is now easy to deduce that *if f is a nonconstant holomorphic function on a bounded domain that extends to a continuous function on the closure of the domain, then $|f|$ assumes its maximum on the boundary of that domain*. More is true, as seen in

Corollary 5.31. *Suppose D is a bounded domain and $f \in \mathbf{C}^0(\mathrm{cl}D)$ satisfies the* MVP *in D. If*

$$M = \sup\{|f(z)|\,; z \in \partial D\},$$

then

(a) $|f(z)| \leq M$ *for all* $z \in D$.
(b) *If* $|f(c)| = M$ *for some* $c \in D$, *then* f *is constant in* D.

Proof. If

$$M' = \sup\{|f(z)|\,; z \in \mathrm{cl}D\},$$

then

$$M \leq M' < +\infty.$$

We know that there exists a c in $\mathrm{cl}D$ such that $|f(c)| = M'$. If $c \in D$, then f is constant in a neighborhood of c by the MMP. Let

$$D' = \{z \in D;\ |f(z)| = M'\}.$$

The set D' is closed and open in D; hence, if nonempty, it is all of D. In this latter case f is constant on D, since it is locally constant and D is connected, and thus also constant on $\text{cl}D$. Then $M = M'$, and (a) and (b) are trivially true. On the other hand, if $D' = \emptyset$, then $M = M'$; (a) follows and (b) is trivially true. □

In particular, since a function that is holomorphic in a domain satisfies the MVP, we have

Corollary 5.32 (The Maximum Modulus Principle for Analytic Functions). *If f is a nonconstant holomorphic function on a domain D, then $|f|$ has no relative maximum in D.*

Further, if D is bounded and f is continuous on the boundary of D, then $|f|$ assumes its maximum on the boundary of D.

Remark 5.33. By studying the proof of Theorem 5.30, one can prove the *maximum principle*, an interesting result that may be stated as follows.

Suppose f is a continuous real-valued function defined on a domain D in $\mathbb{C}$ that satisfies the MVP. If f has a relative maximum at a point $c \in D$, then f is constant in a neighborhood of c.

Similarly, the *minimum principle* asserts that *a continuous real-valued function defined on a domain D in $\mathbb{C}$ that satisfies the MVP on D and has a relative minimum at a point $c \in D$ must be constant in a neighborhood of c* (apply the maximum principle to the negative of the function).

An important consequence of Corollary 5.31 is

Theorem 5.34 (Schwarz's Lemma). *If f is a holomorphic function defined on $U(0,1)$ satisfying $|f(z)| < 1$ for $|z| < 1$ and $f(0) = 0$, then $|f(z)| \leq |z|$ for $|z| < 1$ and $|f'(0)| \leq 1$.*

Furthermore, if $|f(c)| = |c|$ for some c with $0 < |c| < 1$ or if $|f'(0)| = 1$, then there exists a $\lambda \in \mathbb{C}$ with $|\lambda| = 1$ such that $f(z) = \lambda z$ for all $|z| < 1$; that is, f is a rotation around zero, with angle $\text{Arg}(\lambda)$.

Proof. Using the Taylor series expansion for f at 0, we can write $f(z) = \sum_{n=1}^{\infty} a_n z^n$; this power series has radius of convergence $\rho \geq 1$, by Theorem 5.5. Then the function defined by

$$g(z) = \begin{cases} \dfrac{f(z)}{z}, & \text{for } 0 < |z| < 1, \\ a_1 = f'(0), & \text{for } z = 0, \end{cases}$$

satisfies

$$g(z) = \sum_{n=1}^{\infty} a_n z^{n-1} \quad \text{for } |z| < 1$$

and is holomorphic on $U(0,1)$. Now for any r with $0 < r < 1$ and any z with $|z| = r$, we have

$$|g(z)| = \left|\frac{f(z)}{z}\right| \leq \frac{1}{r};$$

by the MMP, the same inequality holds for all $|z| < r$. Hence $|g(z)| \leq 1$ for all $|z| < 1$, or, equivalently,

$$|f(z)| \leq |z| \text{ for all } |z| < 1 \text{ and } |f'(0)| \leq 1.$$

If $|g(c)| = 1$ for some c with $|c| < 1$, then g is constant, again by the MMP. □

Remark 5.35. Schwarz's lemma implies that a holomorphic function from the unit disc $U(0,1)$ to itself fixing zero either decreases the distance to the origin for all points in the disc or maintains it for all points. We will return to a natural geometric generalization of this fact in Sect. 8.4.4.

Remark 5.36. In Schwarz's lemma, the strong hypothesis $|f(z)| < 1$ for all $|z| < 1$ can be replaced by the weaker hypothesis that $|f(z)| \leq 1$ for all $|z| < 1$. Under this weaker hypothesis we conclude the stronger one, for otherwise there exists a c in $U(0,1)$ with $|f(c)| = 1$, and by the MMP, the function f must then be constant—obviously impossible.

5.5 Appendix: Cauchy's Integral Formula for Smooth Functions

In Sect. 4.2, we proved a simple version of Green's theorem (Theorem 4.20) and stated another version as Theorem 4.22. As a consequence of the second version, we now prove

Theorem 5.37 (Cauchy's Integral Formula for Smooth Functions). *Let K be a compact set in $\mathbb{C}$ that is the closure of its interior, with piecewise smooth positively oriented boundary ∂K. If f is a $\mathbf{C}^1$-function on a neighborhood of K and c is a point in the interior of K, then*

$$f(c) = \frac{1}{2\pi\imath}\left[\int_{\partial K} \frac{f(z)}{z-c}\,\mathrm{d}z + \iint_K \frac{f_{\bar{z}}(z)}{z-c}\,\mathrm{d}z\mathrm{d}\bar{z}\right]. \tag{5.10}$$

Proof. Choose $\epsilon > 0$ such that the closure of the ball $U(c,\epsilon)$ is contained in the interior of K, and let $K_\epsilon = K - U(c,\epsilon)$. We apply Green's theorem to the smooth differential form $\dfrac{f(z)}{z-c}\,\mathrm{d}z$ on K_ϵ and obtain

$$\int_{\partial K_\epsilon} \frac{f(z)}{z-c}\mathrm{d}z = -\iint_{K_\epsilon} \frac{f_{\bar{z}}(z)}{z-c}\,\mathrm{d}z\mathrm{d}\bar{z}. \tag{5.11}$$

But ∂K_ϵ consists of two parts: ∂K and the clockwise oriented circle with center at c and radius ϵ. Hence

$$\int_{\partial K_\epsilon} \frac{f(z)}{z-c}\,dz = \int_{\partial K} \frac{f(z)}{z-c}\,dz - \imath \int_0^{2\pi} f(c+\epsilon e^{\imath\theta})\,d\theta.$$

Letting $\epsilon \to 0$ in (5.11) yields (5.10). □

Remark 5.38.

- For holomorphic functions (5.10) reduces to (5.1).
- The last result shows, once again, that the Cauchy theory is a consequence of Green's theorem for $\mathbf{C}^1$-functions that satisfy CR.

Exercises

5.1. Let D be a domain in $\mathbb{C}$. Prove that the following conditions are equivalent.

(1) D is simply connected.
(2) $\widehat{\mathbb{C}} - D$ is connected.
(3) For each holomorphic function f on D such that $f(z) \neq 0$ for all $z \in D$, there exists a $g \in \mathbf{H}(D)$ such that $f = e^g$ (g is a logarithm of f).
(4) For each holomorphic function f on D such that $f(z) \neq 0$ for all $z \in D$ and for each positive integer n there exists an $h \in \mathbf{H}(D)$ such that $f = h^n$.

How unique are the functions g and h?

5.2. Let f be analytic in a simply connected domain D and let γ be a closed pdp in D. Set $\beta = f \circ \gamma$. Show that $I(\beta, c) = 0$ for all $c \in \mathbb{C}$, $c \notin f(D)$.

5.3. (a) Show that if a continuous closed curve γ is homotopic to a point in a domain D, then the cycle (γ) is homologous to zero in D.
(b) Is the converse true?
Hint: Consider the curve γ of Fig. 5.3, with $D = \mathbb{C} - \{p, q\}$, and verify that $I(\gamma, p) = 0 = I(\gamma, q)$.

5.4. Let f be a holomorphic function on $|z| < 1$ with $|f(z)| < 1$ for all $|z| < 1$.

(1) Prove the *invariant form* of Schwarz's lemma (also known as the Schwarz–Pick lemma):

$$\frac{|f'(z)|}{1-|f(z)|^2} \leq \frac{1}{1-|z|^2} \quad \text{for all } |z| < 1.$$

Hint: Use properties of Möbius transformation (developed subsequently) to reduce to the standard Schwarz lemma.
(2) Find necessary and sufficient conditions for equality in the last equation.
(3) If $f\left(\frac{1}{2}\right) = \frac{1}{3}$, find a sharp upper bound for $\left|f'\left(\frac{1}{2}\right)\right|$.

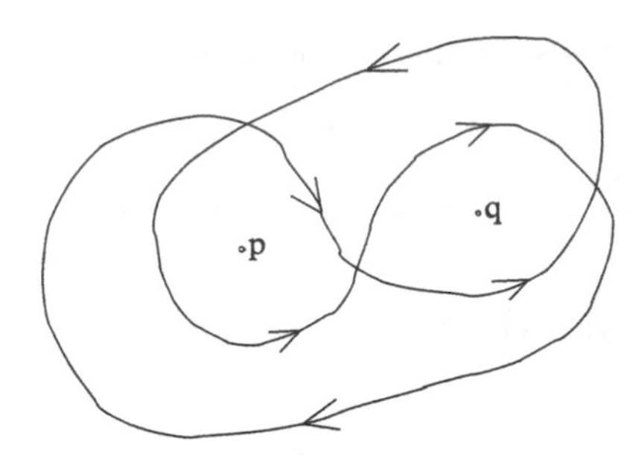

Fig. 5.3 Homologous to zero but not homotopic to a point?

5.5. Let f be a holomorphic function on $U(0, R)$, $R > 0$. Assume there exist an $M \in \mathbb{R}_{>0}$ such that $|f(z)| \leq M$ for all $z \in U(0, R)$ and an $n \in \mathbb{Z}_{\geq 0}$ such that

$$0 = f(0) = f'(0) = \cdots = f^{(n)}(0).$$

(1) Prove that

$$|f(z)| \leq M \left(\frac{|z|}{R}\right)^{n+1} \quad \text{for all } z \in U(0, R),$$

and

$$\left|\frac{f^{(n+1)}(0)}{(n+1)!}\right| \leq \frac{M}{R^{n+1}}.$$

(2) Assume that either $\left|f^{(n+1)}(0)\right| = (n+1)! \frac{M}{R^{n+1}}$ or $|f(c)| = M \left(\frac{|c|}{R}\right)^{n+1}$ for some c with $0 < |c| < R$. Prove that there exists an $\alpha \in \mathbb{C}$, $|\alpha| = 1$ such that $f(z) = \alpha M \left(\frac{z}{R}\right)^{n+1}$ for all $z \in U(0, R)$.

5.6. Let f be a holomorphic function on the punctured plane $\mathbb{C}_{\neq 0}$. Assume that there exist a positive constant C and a real constant M such that

$$|f(z)| \leq C\, |z|^M \text{ for } 0 < |z| < \frac{1}{2}.$$

Show that $z = 0$ is either a pole or a removable singularity for f, and find sharp bounds for $\nu_0(f)$ the order of f at 0.

5.7. Prove by use of Schwarz's lemma that every one-to-one holomorphic mapping of a disc onto another disc is given by a fractional linear transformation. Here the term "disc"is also meant to include half planes (with ∞ adjoined to the boundary), and a fractional linear transformation is a map of the form $T(z) = \frac{az+b}{cz+d}$, where a, b, c, d are complex numbers with $ad - bc \neq 0$.

Formulate and prove a corresponding theorem for one-to-one surjective holomorphic self mappings of $1 < |z| < +\infty$.

5.8. Let f be an entire function with $|f(z)| \leq a\,|z|^b + c$ for all z, where a, b, and c are positive constants. Prove that f is a polynomial of degree at most b.

5.9. Let f be an entire function such that $f(0) = 0$ and that $\Re(f(z)) \to 0$ as $|z| \to \infty$. Show that f is identically 0.

5.10. Let D be a bounded domain in $\mathbb{C}$. Let $f : \text{cl}D \to \mathbb{C}$ be a nonconstant continuous function, which is analytic in D and satisfies $|f(z)| = 1$ for all $z \in \partial D$. Show that $f(z_0) = 0$ for some $z_0 \in D$.

5.11. Prove the maximum and the minimum principles stated in Remark 5.33.

Furthermore, if D is a bounded domain in $\mathbb{C}$ and f is a continuous real-valued function on the closure of D and satisfies the MVP in D, with $m \leq f \leq M$ on ∂D for some real constants m and M, show that then $m \leq f \leq M$ on D.

5.12. Let f be a nonconstant holomorphic function on the disc $U(0, R)$, $0 < R \leq +\infty$. For $0 \leq r < R$, let

$$M(r) = \sup\{|f(z)|\,;\ |z| = r\}.$$

Show that M is a strictly increasing function on the open interval $(0, R)$.

Chapter 6
Cauchy Theory: Local Behavior and Singularities of Holomorphic Functions

In this chapter we use the Cauchy theory to study functions that are holomorphic on an annulus, and analytic functions with isolated singularities. We describe a classification for isolated singularities. Functions that are holomorphic on an annulus have *Laurent series* expansions, an analogue of power series expansions for holomorphic functions on discs. Holomorphic functions with a finite number of isolated singularities in a domain can be integrated using the *residue theorem*, an analogue of the Cauchy integral formula. We discuss local properties of these functions. The study of zeros and poles of meromorphic functions leads to a theorem of Rouché that connects the number of zeros and poles to an integral. The theorem is not only aesthetically pleasing in its own right but also allows us to give alternate proofs of many important results. In the penultimate section of this chapter we illustrate the use of complex function theory in the evaluation of real definite integrals. The appendix discusses Cauchy principal values, a way to integrate functions with certain singularities.

6.1 Functions Holomorphic on An Annulus

Theorem 6.1 (Laurent Series Expansion). *Let $c \in \mathbb{C}$ and let f be holomorphic in the annulus*

$$A = \{z \in \mathbb{C};\ 0 \leq R_1 < |z-c| < R_2 \leq +\infty\}.^{1}$$

Then

$$f(z) = \sum_{n=-\infty}^{\infty} a_n (z-c)^n \text{ for all } z \in A, \tag{6.1}$$

[1]We are including here the cases of *degenerate* annuli: those with $R_1 = 0$ and/or $R_2 = +\infty$.

R.E. Rodríguez et al., *Complex Analysis: In the Spirit of Lipman Bers*, Graduate Texts in Mathematics 245, DOI 10.1007/978-1-4419-7323-8_6,

where the series converges uniformly and absolutely on compact subsets of A, and

$$a_n = \frac{1}{2\pi\imath}\int_{\gamma_r}\frac{f(t)}{(t-c)^{n+1}}\,\mathrm{d}t \quad \textit{for } R_1 < r < R_2,$$

with

$$\gamma_r(\theta) = c + r\mathrm{e}^{\imath\theta}, 0 \le \theta \le 2\pi.$$

This series is called a *Laurent* series for f. It is uniquely determined by f and A.

Proof. Without loss of generality we assume $c = 0$. Consider two concentric circles $\gamma_{r_j} = \{z; |z| = r_j\}$ $(j = 1,2)$, bounding a smaller annulus

$$R_1 < r_1 < |z| < r_2 < R_2.$$

If for $j \in \{1,2\}$, we let

$$f_j(z) = \frac{1}{2\pi\imath}\int_{\gamma_{r_j}}\frac{f(t)}{t-z}\,\mathrm{d}t,$$

then it follows from the extended version of Cauchy's integral formula (Theorem 5.26) that

$$f(z) = \frac{1}{2\pi\imath}\int_{\gamma_{r_2}}\frac{f(t)}{t-z}\,\mathrm{d}t - \frac{1}{2\pi\imath}\int_{\gamma_{r_1}}\frac{f(t)}{t-z}\,\mathrm{d}t = f_2(z) - f_1(z).$$

Since f_2 can be extended to be a holomorphic function in the disc $\{z; |z| < r_2\}$, by Exercise 4.11, we obtain

$$f_2(z) = \sum_{n=0}^{\infty} a_n z^n, \text{ with } a_n = \frac{1}{2\pi\imath}\int_{\gamma_{r_2}}\frac{f(t)}{t^{n+1}}\,\mathrm{d}t;$$

furthermore, this power series converges for $|z| < R_2$, and a_n is independent of r_2, since any two circles centered at zero and contained in A are homotopic in A. As for f_1, note that

$$-\frac{1}{(t-z)} = \frac{1}{(z-t)} = \frac{1}{z}\frac{1}{(1-\frac{t}{z})} = \frac{1}{z}\sum_{k=0}^{\infty}\left(\frac{t}{z}\right)^k$$

for $|t| = r_1$ and $|z| > r_1$. Thus

$$-f_1(z) = \frac{1}{2\pi\imath}\frac{1}{z}\sum_{k=0}^{\infty}\int_{\gamma_{r_1}}\frac{f(t)}{z^k}t^k\,\mathrm{d}t = \sum_{k=0}^{\infty} z^{-k-1}\frac{1}{2\pi\imath}\int_{\gamma_{r_1}} f(t)t^k\,\mathrm{d}t.$$

Letting $-k-1=n$ we obtain

$$-f_1(z)=\sum_{n\le -1} a_n z^n, \quad \text{with } a_n=\frac{1}{2\pi\imath}\int_{\gamma_{r_1}}\frac{f(t)}{t^{n+1}}\,\mathrm{d}t,$$

and a_n is independent of r_1, since, as before, any two circles contained in A and centered at zero are homotopic in A.

Observe that f_1 can be extended to be a holomorphic function in $|z|>R_1$, including the point ∞, and $f_1(\infty)=0$. □

Corollary 6.2. *If f is holomorphic in A, then $f=f_2-f_1$ where f_2 is holomorphic in $|z-c|<R_2$ and f_1 is holomorphic in $|z-c|>R_1$ (including the point at ∞). The functions f_j are unique if we require that $f_1(\infty)=0$.*

Proof. Once again, without loss of generality, we assume $c=0$.

Existence: Already done.

Uniqueness: Suppose $f=f_2-f_1=g_2-g_1$ with appropriate f_i and g_j. Then $0=(f_2-g_2)-(f_1-g_1)$ in A, the function (f_2-g_2) is holomorphic in $|z|<R_2$, and the function (f_1-g_1) is holomorphic in $|z|>R_1$ and vanishes at ∞. Define

$$h(z)=\begin{cases}(f_1-g_1)(z), \text{ if } |z|>R_1,\\ (f_2-g_2)(z), \text{ if } |z|<R_2.\end{cases}$$

The function h is well defined and holomorphic on $\mathbb{C}\cup\{\infty\}$, and vanishes at ∞. Hence it is identically zero. □

Example 6.3. The rational function

$$f(z)=\frac{1}{z(1-z))(2-z)}$$

is holomorphic on three annuli centered at zero. On each of these it has a different Laurent series expansion of the form $\sum_{n=-\infty}^{+\infty} a_n z^n$; the uniqueness of expansion is valid only if we specify both the function and annulus.

(1) On $A_1=\{0<|z|<1\}$: we do not need to evaluate the coefficients a_n by the integration process described in the proof of the last theorem. On this annulus, we have

$$\frac{1}{1-z}=1+z+z^2+\cdots+z^n+\cdots \text{ (valid for } |z|<1),$$

and

$$\frac{1}{2-z}=\frac{1}{2\left(1-\frac{z}{2}\right)}=\frac{1}{2}\left(1+\frac{z}{2}+\frac{z^2}{2^2}+\cdots+\frac{z^n}{2^n}+\cdots\right) \text{ (valid for } |z|<2).$$

Thus

$$
\begin{aligned}
f(z) &= \frac{1}{2z}\left(1+z+z^2+\cdots+z^n+\cdots\right)\left(1+\frac{z}{2}+\frac{z^2}{2^2}+\cdots+\frac{z^n}{2^n}+\cdots\right) \\
&= \frac{1}{2}z^{-1}+\frac{3}{4}+\frac{7}{8}z+\cdots+\left(1-\frac{1}{2^{n+2}}\right)z^n+\cdots \quad \text{for } 0<|z|<1.
\end{aligned}
$$

There is another, perhaps simpler, method for computing the series. We use the partial fraction decomposition of the function

$$
f(z) = \frac{A}{z}+\frac{B}{1-z}+\frac{C}{2-z};
$$

the constants are computed to be

$$
A=\frac{1}{2},\ B=1,\ C=-\frac{1}{2},
$$

and we use the respective geometric series for the three fractions to obtain once again:

$$
\begin{aligned}
f(z) =&\frac{1}{2}z^{-1}+\left(1+z+z^2+\cdots+z^n+\cdots\right) \\
&-\frac{1}{4}\left(1+\frac{z}{2}+\frac{z^2}{2^2}+\cdots+\frac{z^n}{2^n}+\cdots\right) \\
=&\sum_{n=-1}^{+\infty}\left(1-\frac{1}{2^{n+2}}\right)z^n \quad \text{for } 0<|z|<1.
\end{aligned}
$$

(2) On $A_2=\{1<|z|<2\}$: we use again the partial fraction decomposition of f. We need to know in this case that

$$
\frac{1}{1-z}=\frac{-1}{z\left(1-\frac{1}{z}\right)}=-\frac{1}{z}\left(1+\frac{1}{z}+\frac{1}{z^2}+\cdots+\frac{1}{z^n}+\cdots\right) \quad \text{for } |z|>1
$$

and

$$
\frac{1}{2-z}=\frac{1}{2\left(1-\frac{z}{2}\right)}=\frac{1}{2}\left(1+\frac{z}{2}+\frac{z^2}{2^2}+\cdots+\frac{z^n}{2^n}+\cdots\right) \quad \text{for } |z|<2.
$$

Therefore

$$
f(z) =\frac{1}{2}z^{-1}-\left(z^{-1}+z^{-2}+\cdots+z^{-n}+\cdots\right)
$$

$$-\frac{1}{4}\left(1+\frac{z}{2}+\frac{z^2}{2^2}+\cdots+\frac{z^n}{2^n}+\cdots\right)$$

$$=-\sum_{n=-\infty}^{-2} z^n-\frac{1}{2z}-\sum_{n=0}^{\infty}\frac{1}{2^{n+2}}z^n \text{ for } 1<|z|<2.$$

(3) On $A_3=\{2<|z|\}$: in this case we use

$$\frac{1}{1-z}=\frac{-1}{z\left(1-\frac{1}{z}\right)}=-\frac{1}{z}\left(1+\frac{1}{z}+\frac{1}{z^2}+\cdots+\frac{1}{z^n}+\cdots\right) \text{ for } |z|>1$$

and

$$\frac{1}{2-z}=\frac{-1}{z\left(1-\frac{2}{z}\right)}=-\frac{1}{z}\left(1+\frac{2}{z}+\frac{2^2}{z^2}+\cdots+\frac{2^n}{z^n}+\cdots\right) \text{ for } |z|>2.$$

Then

$$f(z)=\frac{1}{2}z^{-1}-\left(z^{-1}+z^{-2}+\cdots+z^{-n}+\cdots\right)$$

$$+\frac{1}{2}\left(z^{-1}+2z^{-2}+2^2z^{-3}+\cdots+2^nz^{-n-1}+\cdots\right)$$

$$=\sum_{n=-\infty}^{-3}\left(-1+2^{-n-2}\right)z^n \text{ for } |z|>2.$$

6.2 Isolated Singularities

In this section we study the behavior of a function holomorphic in a punctured disc.

Definition 6.4. We consider the special case of functions holomorphic on a degenerate annulus with $R_1=0$ and $R_2\in(0,+\infty]$; that is, we fix a point $c\in\mathbb{C}$ and a holomorphic function f on the punctured disc $A=\{z\in\mathbb{C};\ 0<|z-c|<R_2\}$. In this case c is called an *isolated singularity* of f.

We know that then f has a Laurent series expansion (6.1). There are three possibilities for the coefficients $\{a_n\}_{n\in\mathbb{Z}_{<0}}$; we now analyze each possibility.

1. If $a_n=0$ for all n in $\mathbb{Z}_{<0}$, then f has a *removable singularity* at $z=c$, and f can be extended to be a holomorphic function in the disc $|z-c|<R_2$ by defining $f(c)=a_0$.

Shortly (in Theorem 6.7), we will establish a useful criterion for proving that such an isolated singularity is removable.

2. Only finitely many, at least one, nonzero coefficients with negative indices appear in the Laurent series; that is, there exists N in $\mathbb{Z}_{>0}$ such that $a_{-n} = 0$ for all $n > N$ and $a_{-N} \neq 0$. We can hence write

$$f(z) = \sum_{n=-N}^{-1} a_n(z-c)^n + \sum_{n=0}^{\infty} a_n(z-c)^n$$

for $0 < |z-c| < R_2$. In this case, $\sum_{n=-N}^{-1} a_n(z-c)^n$ is called the *principal or singular part of f at c*. Observe that

$$\lim_{z \to c} (z-c)^N f(z) = a_{-N} \neq 0,$$

and N is characterized by this property (that the limit exists and is different from zero). Note that in this case the function $z \mapsto (z-c)^N f(z)$ has a removable singularity at c, and hence can be extended to be holomorphic in the disc $|z-c| < R_2$. Therefore f is meromorphic in the disc $|z-c| < R_2$, and f has a pole of order N at $z = c$.

3. Infinitely many nonzero coefficients with negative indices appear in the Laurent series. Then c is called an *essential* singularity of f.

We extend the definition of isolated singularity to include the point at ∞.

Definition 6.5. A function f holomorphic in a deleted neighborhood of ∞ (see Definition 3.52) has an *isolated singularity at* ∞ if $g(z) = f\left(\frac{1}{z}\right)$ has an isolated singularity at $z = 0$.

If f has a pole of order N at ∞, *the principal part of f at ∞* is the polynomial $\sum_{n=1}^{N} a_n z^n$, where the Laurent series expansion for g near zero is given by

$$g(z) = f\left(\frac{1}{z}\right) = \sum_{n=1}^{N} a_n z^n + \sum_{n=0}^{\infty} a_n z^{-n}$$

for $|z| > 0$ small.

Example 6.6. We have

$$\exp\left(\frac{1}{z}\right) = \sum_{n=0}^{\infty} \frac{z^{-n}}{n!}$$

for $|z| > 0$. Here $R_1 = 0$ and $R_2 = +\infty$; 0 is an essential singularity of the function and ∞ is a removable singularity ($f(\infty) = 1$).

Theorem 6.7. *Let $c \in \mathbb{C}$ and f be a holomorphic function on $A = \{0 < |z - c| < R_2\}$. Assume that* (6.1) *is the Laurent series of f on the punctured disc A. If there exist an $M > 0$ and $0 < r_0 < R_2$ such that*

$$|f(z)| \le M \text{ for } 0 < |z - c| < r_0,$$

then f has a removable singularity at $z = c$.

Proof. We know that $a_n = \frac{1}{2\pi\imath}\int_{\gamma_r} \frac{f(t)}{(t-c)^{n+1}}\,dt$ for all $n \in \mathbb{Z}$, where $\gamma_r(\theta) = c + r\,e^{\imath\theta}$, for $\theta \in [0, 2\pi]$ and $0 < r < r_0$. We estimate $|a_n| \le \frac{M}{r^n}$. For $n < 0$, we let $r \to 0$ and conclude that $a_n = 0$. □

Theorem 6.8 (Casorati–Weierstrass). *If f is holomorphic on $\{0 < |z - c| < R_2\}$ and has an essential singularity at $z = c$, then for all $w \in \mathbb{C}$ the function*

$$g(z) = \frac{1}{f(z) - w}$$

is unbounded in any punctured neighborhood of $z = c$. Therefore the range of f restricted to any such neighborhood is dense in $\mathbb{C}$.

Proof. Assume that, for some $w \in \mathbb{C}$, the function g is unbounded in some punctured neighborhood N of $z = c$. Then there is $\varepsilon > 0$ such that $N = U(c, \varepsilon) - \{c\}$, and, for any $M > 0$, there exists a $z \in N$ such that $|g(z)| > M$; that is, such that $|f(z) - w| < \frac{1}{M}$. Thus w is a limit point of $f(N)$ and the last statement in the theorem is proved. Now it suffices to prove that for all $w \in \mathbb{C}$ and all $\varepsilon > 0$, the function g is unbounded in $U(c, \varepsilon) - \{c\}$. If g were bounded in such a neighborhood, it would have a removable singularity at $z = c$ and thus would extend to a holomorphic function on $U(c, \varepsilon)$; therefore f would be meromorphic there. □

A much stronger result can be established. We state it without proof.[2]

Theorem 6.9 (Picard). *If f is holomorphic in $0 < |z - c| < R_2$ and has an essential singularity at $z = c$, then there exists a $w_0 \in \mathbb{C}$ such that for all $w \in \mathbb{C} - \{w_0\}$, $f(z) = w$ has infinitely many solutions in $0 < |z - c| < R_2$.*

Example 6.10. The function $\exp\left(\frac{1}{z}\right)$ shows the above theorem is sharp, with $c = 0$ and $w_0 = 0$.

[2]For a proof, see Conway's book listed in the bibliography.

We now have a complete description of the behavior of a holomorphic function near an isolated singularity.

Theorem 6.11. *Assume that f is a holomorphic function in a punctured disc $U' = U(c, R) - \{c\}$ around the isolated singularity $c \in \mathbb{C}$. Then*

(1) *c is a removable singularity if and only if f is bounded in U' if and only if $\lim_{z \to c} f(z)$ exists and is finite.*

(2) *c is a pole of f if and only if $\lim_{z \to c} f(z) = \infty$, in the sense of Definition 3.52.*

(3) *c is an essential singularity of f if and only if $f(U')$ is dense in $\mathbb{C}$.*

Proof. (1) follows from Theorem 6.7 and the definition of removable singularity.

The Casorati–Weierstrass Theorem 6.8 shows that if c is an essential singularity then $f(U')$ is dense in $\mathbb{C}$. We complete the proof by showing (2). If c is a pole of f of order $N \geq 1$, then we know from the Laurent series expansion for f that there exists $0 < r < R$ and a holomorphic function $g : U(c, r) \to \mathbb{C}$ such that $f(z) = \dfrac{g(z)}{(z - c)^N}$ on $U(c, r) - \{c\}$ and $g(c) \neq 0$. By continuity of g we may assume that $|g(z)| \geq M > 0$ for all z in $U(c, r)$. Then $|f(z)| \geq \dfrac{M}{|z - c|^N}$, and it follows that $\lim_{z \to c} f(z) = \infty$. Conversely, if $\lim_{z \to c} f(z) = \infty$, then c is not a removable singularity of f since for every $M > 0$ there exists $\delta > 0$ such that

$$|f(z)| \geq M \text{ for } 0 < |z - c| < \delta,$$

and the Casorati–Weierstrass theorem implies that c is not an essential singularity of f. Thus it must be a pole. □

Example 6.12. For an entire function $f(z) = \sum_{n=0}^{\infty} a_n z^n$ (we know that its radius of convergence $\rho = +\infty$), there are two possibilities:

(a) Either there exists an N such that $a_n = 0$ for all $n > N$, in which case f is a polynomial of degree $\leq N$. If $\deg f = N \geq 1$, then f has a pole of order N at ∞, and $f(z) - f(0) = \sum_{i=1}^{N} a_i z^i$ is the principal part of f at ∞. If $\deg f = 0$, then f is constant, of course.

(b) Or f has an essential singularity at ∞.

We can now establish the following result.

Theorem 6.13. *Let $f : \mathbb{C} \cup \{\infty\} \to \mathbb{C} \cup \{\infty\}$. Then,*

(a) *If f is holomorphic, it is constant, and*

(b) *If f is meromorphic, it is a rational function.*

Proof. If f is holomorphic on $\mathbb{C} \cup \{\infty\}$, a compact set, it must be bounded. Since it is also an entire function, it must be constant, by Liouville's theorem.

If f is meromorphic on $\mathbb{C} \cup \{\infty\}$, its set of poles must be finite, being isolated points in a compact set. Denote them by $z_1, \ldots, z_k$, which may include ∞. If this set is empty, we are in the previous case. Otherwise, let $N_j > 0$ be the order of the pole at z_j and let $P_j(z)$ be the principal part of f at z_j. Then

$$P_j(z) = \begin{cases} \sum_{n=1}^{N_j} \frac{a_{n,j}}{(z-z_j)^n}, & \text{if } z_j \in \mathbb{C}, \\ \sum_{n=1}^{N_j} a_{n,j} z^n, & \text{if } z_j = \infty. \end{cases}$$

It follows that $f - \sum_{j=1}^{k} P_j$ is holomorphic in $\mathbb{C} \cup \{\infty\}$ and therefore constant. The result follows. □

6.3 Residues

If f is a holomorphic function in a deleted neighborhood U of c in $\mathbb{C} \cup \{\infty\}$, then $\omega = f(z)\,\mathrm{d}z$ defines a holomorphic differential form on U with an isolated singularity at c.

Definition 6.14. If the holomorphic function f has an isolated singularity at $c \in \mathbb{C}$, with Laurent series expansion (6.1) on

$$|z - c| < R_2 \text{ for some } R_2 > 0,$$

we define the *residue of the differential form* $\omega = f(z)\,\mathrm{d}z$ *at* c by the formula

$$\operatorname{Res}(f(z)\,\mathrm{d}z, c) = \operatorname{Res}(\omega, c) = a_{-1}.$$

Remark 6.15. Note that if f has a simple pole at c (or more generally, if $\nu_c(f) \geq -1$), then

$$\operatorname{Res}(f(z)\,\mathrm{d}z, c) = \lim_{z \to c} (z - c) f(z).$$

Our next result will give an alternate way to define the residue of the differential form $f(z)\,\mathrm{d}z$, and will show its invariance. It is an analogue of the Cauchy integral formula for the case of an isolated singularity.

Theorem 6.16. *Let A denote the annulus $R_1 < |z - c| < R_2$. If γ is a closed path in A and if f is holomorphic in A with Laurent series $f(z) = \sum_{n=-\infty}^{+\infty} a_n (z - c)^n$, then*

$$\frac{1}{2\pi\imath} \int_\gamma f(z)\,\mathrm{d}z = I(\gamma, c)\, a_{-1}.$$

In the special case that $R_1 = 0$ and $I(\gamma, c) = 1$, we have

$$\frac{1}{2\pi\imath}\int_\gamma f(z)\,\mathrm{d}z = \mathrm{Res}(f(z)\,\mathrm{d}z, c).$$

Proof. Write

$$g(z) = \sum_{n\neq -1} a_n(z-c)^n \text{ and } f(z) = \frac{a_{-1}}{z-c} + g(z).$$

The function g has a primitive in the annulus A; namely,

$$\sum_{n\neq -1} \frac{1}{n+1} a_n\,(z-c)^n.$$

Since $\int_\gamma g(z)\,\mathrm{d}z = 0$, the result follows. □

Theorem 6.17 (Residue Theorem). *Let f be holomorphic in a domain $D \subseteq \mathbb{C}$ except for isolated singularities at $z_1, \ldots, z_n$ in D. If γ is a positively oriented Jordan curve homotopic to a point in D and all z_j are in the interior of γ, then*

$$\int_\gamma f(z)\,\mathrm{d}z = 2\pi\imath \sum_{j=1}^{n} \mathrm{Res}(f(z)\,\mathrm{d}z, z_j).$$

Proof. Put a small positively oriented circle around each z_j and use the extended version of Cauchy's Theorems 5.26and 6.16. □

Let c and $b \in \mathbb{C}$. As above assume that c is an isolated singularity of a holomorphic function f defined in a deleted neighborhood U of c. Let h be a holomorphic function defined in a deleted neighborhood of b with values in U, and let γ be a positively oriented circle centered at b with sufficiently small radius. Then

$$\begin{aligned}\mathrm{Res}((f\circ h)(z)\,\mathrm{d}h(z), b) &= \frac{1}{2\pi\imath}\int_\gamma f(h(z))h'(z)\,\mathrm{d}z \\ &= \frac{1}{2\pi\imath}\int_{h(\gamma)} f(w)\,\mathrm{d}w \\ &= I(h(\gamma), c)\,\mathrm{Res}(f(w)\,\mathrm{d}w, c). \qquad (6.2)\end{aligned}$$

The above invariance property of residues allows us to extend its definition to holomorphic differentials with an isolated singularity at infinity. If ∞ is an isolated singularity of the holomorphic function f with Laurent series expansion $f(z) =$

$\sum_{n=-\infty}^{+\infty} a_n z^n$ for $|z| > R$, then we define the *residue of the differential form* $\omega = f(z)\,dz$ *at* ∞ by the formula

$$\text{Res}(\omega, \infty) = -a_{-1}.$$

To justify the last definition (its consistency with previously defined concepts), we show that it satisfies the invariance property (6.2). This is in agreement with our earlier convention: to obtain invariants at $z = \infty$ we change the variable from z to $w = \frac{1}{z}$, and then use the invariants defined for $w = 0$.

Thus, in this case, we have

$$\text{Res}(f(z)\,dz, \infty) = \text{Res}\left(f\left(\frac{1}{w}\right) d\left(\frac{1}{w}\right), 0\right).$$

But

$$f\left(\frac{1}{w}\right) d\left(\frac{1}{w}\right) = \left(\sum_{n=-\infty}^{+\infty} a_n w^{-n}\right)\left(-\frac{1}{w^2}\,dw\right)$$
$$= \sum_{k=-\infty}^{+\infty} (-a_{-k-2}) w^k\,dw.$$

6.4 Zeros and Poles of Meromorphic Functions

In this section we study several consequences of the residue theorem.

Let D be a domain in $\mathbb{C}$ and $f : D \to \widehat{\mathbb{C}}$ be a meromorphic function. This means that f is holomorphic except for isolated singularities in D, which are removable or poles (see Sect. 3.5). We have denoted the field of meromorphic functions on D by $\mathbf{M}(D)$.

Note that if γ is a positively oriented Jordan curve in D which is homotopic to a point in D, then the interior $i(\gamma)$ of γ is contained in D. Furthermore, the number of zeros and the number of poles in the interior of γ of any nonconstant meromorphic function in $\mathbf{M}(D)$ are both finite, since range $\gamma \cup i(\gamma)$ is a compact set. The next result counts the difference between these two numbers.

We recall from Definition 3.54 that the meromorphic function f has order or multiplicity $\nu_c(f)$ at $c \in D$, if there exists a holomorphic function g defined near c such that $g(c) \neq 0$ and $f(z) = (z - c)^{\nu_c(f)}\, g(z)$ near c.

Theorem 6.18 (The Argument Principle). *Let D be a domain in $\mathbb{C}$ and let $f \in \mathbf{M}(D)$. Suppose γ is a positively oriented Jordan curve in D which is homotopic to a point in D. Let $c \in \mathbb{C}$ and assume that $f(z) \neq c$ and $f(z) \neq \infty$ for all z in range γ. Then*

$$\frac{1}{2\pi\imath}\int_\gamma \frac{f'(z)}{f(z)-c}\,\mathrm{d}z = \sum_{z\in i(\gamma)} \nu_z(f-c) = Z - P,$$

where Z is the number of zeros of the function $f - c$ inside γ (counting multiplicities) and P is the number of poles of the function f inside γ (counting multiplicities).

Proof. If $F(z) = \dfrac{f'(z)}{f(z)-c}$ for $z \in D$, then $F \in \mathbf{M}(D)$, and we claim that F has only simple poles, at the zeros and poles of $f - c$, and that

$$\operatorname{Res}(F(z)\,\mathrm{d}z, d) = \nu_d(f-c) \text{ for all } d \in D. \tag{6.3}$$

The theorem then follows immediately from (6.3) and the residue theorem. To verify our claim, it suffices to assume that $c = 0 = d$. If we set $n = \nu_0(f)$, then $f(z) = z^n g(z)$ with g holomorphic near 0 and $g(0) \neq 0$. It follows that

$$f'(z) = nz^{n-1}g(z) + z^n g'(z)$$

and hence $\dfrac{f'(z)}{f(z)} = \dfrac{n}{z} + \dfrac{g'(z)}{g(z)}$ near 0. Thus $\dfrac{f'(z)}{f(z)}\,\mathrm{d}z$ has residue n at zero. □

Remark 6.19. The name "argument principle" attached to the previous result may be explained in the following way: let D be a domain in $\mathbb{C}$ and let $f \in \mathbf{M}(D)$ be a nonconstant function. Assume γ is a positively oriented Jordan curve in D which is homotopic to a point in D. Let $c \in \mathbb{C}$ and suppose that $f(z) \neq c$ and $f(z) \neq \infty$ for all z in range γ. If Z denotes the number of zeros of the function $f - c$ inside γ (counting multiplicities) and P denotes the number of poles of f inside γ (counting multiplicities), then the argument of $f - c$ increases by $2\pi(Z - P)$ upon traversing γ. Indeed, note that $\dfrac{f'}{f-c} = (\log(f-c))'$ and recall that

$$\log(f-c) = \log|f-c| + \imath\arg(f-c).$$

Therefore,

$$\int_\gamma \frac{f'(z)}{f(z)-c}\,\mathrm{d}z = \int_\gamma d\log|f(z)-c| + \imath\int_\gamma d\arg(f(z)-c).$$

The first integral on the rightmost side of the equation equals zero because $z \mapsto \log|f(z)-c|$ is a well-defined (single-valued) function on the range of γ. The second integral on the rightmost side equals the change in the argument of $f - c$ as one traverses γ.

Corollary 6.20. *Let f be a nonconstant holomorphic function on a neighborhood of $c \in \mathbb{C}$, $\alpha = f(c)$, and $m = \nu_c(f - \alpha)$. Then there exist $r > 0$ and $\varepsilon > 0$ such that for all $\beta \in \mathbb{C}$ with $0 < |\beta - \alpha| < \varepsilon$, $f - \beta$ has exactly m simple zeros in $0 < |z - c| < r$.*

Proof. Observe that $m \geq 1$. Choose a positively oriented circle γ around c such that $f - \alpha$ vanishes only at c in cl $i(\gamma)$ and $f'(z) \neq 0$ for all $z \in$ cl $i(\gamma) - \{c\}$. If we consider the curve $\gamma_1 = f \circ \gamma$, it follows from Theorem 6.18 that

$$I(\gamma_1, \alpha) = \frac{1}{2\pi\imath} \int_{\gamma} \frac{f'(z)}{f(z) - \alpha}\, dz = m.$$

Let $w = f(z)$ and conclude that for every β not in the range of γ_1, we have

$$\frac{1}{2\pi\imath} \int_{\gamma} \frac{f'(z)}{f(z) - \beta}\, dz = \frac{1}{2\pi\,\imath} \int_{\gamma_1} \frac{1}{w - \beta}\, dw = I(\gamma_1, \beta).$$

Now, there exists a $\delta > 0$ such that $|f(z) - \alpha| \geq \delta$ for all z in range γ. Hence $|\beta - \alpha| < \delta$ implies that, for all z in range γ,

$$|f(z) - \beta| = |(f(z) - \alpha) - (\alpha - \beta)| \geq |f(z) - \alpha| - |\alpha - \beta| > 0.$$

Thus $f - \beta$ does not vanish on range γ for such β. Since $I(\gamma_1, \beta)$ is constant on each connected component of the complement of the range of γ_1 in $\mathbb{C}$, there is an $\varepsilon > 0$ such that

$$|\beta - \alpha| < \varepsilon < \delta \Rightarrow I(\gamma_1, \beta) = m.$$

If $\beta \neq \alpha$ and $|\beta - \alpha| < \varepsilon$, then all the zeros of $f - \beta$ in $i(\gamma)$ are simple (since f' is not zero near, but not necessarily at, c), and therefore $f - \beta$ has m simple zeros in $i(\gamma) - \{c\} = \{z \in \mathbb{C}; 0 < |z - c| < r\}$ for some positive value of r. □

Corollary 6.21. *A nonconstant holomorphic function is an open mapping.*

Proof. If $f : D \to \mathbb{C}$ is holomorphic on a domain D and is not a constant, we obtain from Corollary 6.20 that for any α in $f(D)$ there exists $\epsilon > 0$ such that $U(\alpha, \epsilon) \subseteq f(D)$, and the result follows. □

Corollary 6.22. *An injective holomorphic function is a diffeomorphism from its domain onto its image.*

Remark 6.23. Corollary 6.20 characterizes when a holomorphic function f is locally injective: f is injective in a neighborhood of c if and only if $m = \nu_c(f - f(c)) = 1$. Together with Corollaries 6.21 and 6.22, we see that this condition is equivalent to being a local diffeomorphism. In particular, in the case that $m = 1$, f defines a bi-holomorphic map (a bijective holomorphic function; the local inverse of f is also holomorphic) between neighborhoods of c and $f(c)$. See also the discussion in the next section.

Theorem 6.24 (Rouché's Theorem). *Let f and g be holomorphic functions on a domain D. Let γ be a positively oriented Jordan curve with* cl $i(\gamma)$ *contained in D, and assume that $|f| > |g|$ on* range γ. *Then $Z_{f+g} = Z_f$, where Z_f denotes the number of zeros of f in $i(\gamma)$.*

Proof. It follows from Theorem 6.18 that $Z_f = I(f \circ \gamma, 0)$. Now apply Theorem 4.60 with $\gamma_1 = g \circ \gamma$ and $\gamma_2 = f \circ \gamma$. □

Example 6.25. Rouché's theorem is very useful in locating zeros of a holomorphic function, as this example shows. Let $h(z) = z^5 + z^4 + 6z + 1$. Then

$$\left|z^4 + 6z + 1\right| \leq 29 < 32 = \left|z^5\right|$$

for all $|z| = 2$, and

$$\left|z^5 + z^4 + 1\right| \leq 3 < 6 = |6z|$$

for all $|z| = 1$. Therefore h has its five zeros contained in $|z| < 2$, and four of them are contained in $\{z \in \mathbb{C}; 1 < |z| < 2\}$.

Theorem 6.26 (Integral Formula for the Inverse Function). *Let $R > 0$. Suppose f is holomorphic on $|z| < R$, $f(0) = 0$, $f'(z) \neq 0$ for $|z| < R$, and $f(z) \neq 0$ for $0 < |z| < R$. For $0 < r < R$, let γ_r be the positively oriented circle of radius r about 0, and let $m = \min |f|$ on γ_r. Then*

$$g(w) = \frac{1}{2\pi\imath} \int_{\gamma_r} \frac{t f'(t)}{f(t) - w}\, \mathrm{d}t$$

defines a holomorphic function in $|w| < m$ with

$$f(g(w)) = w \;\; on \;\; |w| < m$$

and

$$g(f(z)) = z \;\; for\; z \in i(\gamma_r) \cap f^{-1}(|w| < m).$$

Proof. Observe that $m > 0$, and fix w_0 with $|w_0| < m$. Then on the circle γ_r we have

$$|f(z)| \geq m > |w_0|\,.$$

Thus f and $(f - w_0)$ have the same number of zeros in $i(\gamma_r)$, by Rouché's theorem, and hence $f(z) - w_0 = 0$ has a unique solution z_0 in $i(\gamma_r)$. Therefore it suffices to show the following:

(1) $g(w_0) = z_0$ if $f(z_0) = w_0$.
(2) g is a holomorphic function on the disc $|w| < m$.

To verify (1), note that it follows from the residue theorem that

$$g(w_0) = \operatorname{Res}(F(s)\,\mathrm{d}s, z_0), \;\text{ where } F(s) = \frac{s\, f'(s)}{f(s) - w_0} \;\text{ for } |s| < R.$$

Thus

$$\begin{aligned} g(w_0) &= \lim_{s\to z_0}(s-z_0)\frac{s\,f'(s)}{f(s)-f(z_0)} \\ &= \lim_{s\to z_0}\frac{(2s-z_0)f'(s)+(s^2-sz_0)f''(s)}{f'(s)} \\ &= z_0, \end{aligned}$$

where the second equality follows from l'Hopital's rule (see Exercise 3.25).

Alternatively, to avoid the use of l'Hopital's rule, we change the previous series of equalities to

$$\begin{aligned} g(w_0) &= \lim_{s\to z_0}(s-z_0)\frac{s\,f'(s)}{f(s)-f(z_0)} \\ &= \lim_{s\to z_0}(s-z_0)\frac{sf'(s)}{(s-z_0)\left(f'(z_0)+\frac{f''(z_0)}{2}(s-z_0)+\cdots\right)} \\ &= z_0. \end{aligned}$$

To show (2) we note that $|f(t)| > |w|$ on γ_r and hence

$$\frac{1}{f(t)-w} = \frac{1}{f(t)\left[1-\frac{w}{f(t)}\right]} = \frac{1}{f(t)}\sum_{n=0}^{\infty}\left(\frac{w}{f(t)}\right)^n.$$

Thus

$$g(w) = \frac{1}{2\pi\imath}\sum_{n=0}^{\infty} w^n \int_{\gamma_r}\frac{t\,f'(t)}{f(t)^{n+1}}\,dt.$$

Since $\left|\int_{\gamma_r}\frac{t\,f'(t)}{f(t)^{n+1}}\,dt\right| \le \frac{M}{m^{n+1}}$ for some constant M that is independent of n, the last power series has radius of convergence $\ge m$. □

6.5 Local Properties of Holomorphic Maps

In this section we describe the behavior of an analytic function f near any point z_0 in its domain of definition D, using results from the previous section. We use the following standard notation:

$$z = x + \imath\, y,\ w = s + \imath\, t = f(z) = u(x, y) + \imath\, v(x, y) \quad \text{for } z \in D.$$

Proposition 6.27. *Let D be a domain in $\mathbb{C}$, z_0 a point in D, and f a function holomorphic on D. Then the following properties hold:*

(1) If $f'(z_0) \neq 0$, then f defines a homeomorphism of some neighborhood of z_0 onto some neighborhood of $f(z_0)$.

(2) If there exists $n \in \mathbb{Z}_{\geq 1}$ such that

$$0 = f'(z_0) = \cdots = f^{(n)}(z_0) \text{ and } f^{(n+1)}(z_0) \neq 0,$$

then f is $n+1$ to 1 near z_0.

(3) If $f'(z_0) \neq 0$, then angles between tangent vectors to curves at z_0 are preserved, and infinitesimal lengths at z_0 are multiplied by $|f'(z_0)|$.[3] *More generally, if $0 = f'(z_0) = \cdots = f^n(z_0)$ and $f^{(n+1)}(z_0) \neq 0$ for some n in $\mathbb{Z}_{\geq 1}$, then angles between tangent vectors to curves at z_0 are multiplied by $n+1$.*

(4) Conversely, if $g \in \mathbf{C}^1(D)$ preserves angles, then $g \in \mathbf{H}(D)$.

(5) The change in infinitesimal areas is given by multiplication by $|f'(z_0)|^2$.

Proof. Let $w_0 = f(z_0)$. We proceed to establish the various parts of the theorem.

(1) The condition implies that $\nu_{z_0}(f(z) - w_0) = 1$, and it follows from Corollary 6.20 that there exist $r > 0$ and $\varepsilon > 0$ such that for all $w \in \mathbb{C}$ with $0 < |w - w_0| < \varepsilon$, $f - w_0$ has exactly one simple zero in $0 < |z - z_0| < r$. In other words, f is injective near z_0. Now use Corollary 6.22 to conclude.

(2) Let $g(z) = f(z) - w_0$. It is enough to prove that g is $n+1$ to 1 near z_0. But

$$g(z_0) = 0 = g'(z_0) = \cdots = g^{(n)}(z_0), \text{ and } g^{(n+1)}(z_0) \neq 0,$$

and therefore, for $|z - z_0|$ small, we may write

$$\begin{aligned} g(z) &= \sum_{m \geq n+1} a_m (z - z_0)^m \quad (\text{where } a_{n+1} \neq 0) \\ &= (z - z_0)^{n+1} \sum_{k=0}^{\infty} a_{k+n+1} (z - z_0)^k \\ &= (z - z_0)^{n+1} (h(z))^{n+1} \\ &= (g_1(z))^{n+1}, \end{aligned}$$

where h and g_1 are holomorphic functions near z_0, $g_1(z_0) = 0$, and $(g_1)'(z_0) = h(z_0) \neq 0$, since $(h(z_0))^{n+1} = a_{n+1}$. The existence of h is a consequence of

[3] Both claims are stated in standard shorthand form. The first statement is reformulated in the proof that follows; the second can be reformulated as $|f(z) - f(z_0)| = |f'(z_0)|\ |z - z_0| + \epsilon(z)$, where $\lim_{z \to z_0} \epsilon(z) = 0$.

Exercise 5.1. By (1), g_1 is a homeomorphism from a neighborhood of z_0 to a neighborhood of 0. Since $p(z) = z^{n+1}$ is clearly $n+1$ to 1 from a neighborhood of 0 to a neighborhood of 0, and since $g = p \circ g_1$, it follows that g is $n+1$ to 1 from a neighborhood of z_0 to a neighborhood of $g(z_0) = 0$, as claimed.

(3) Let us write

$$f'(z_0) = \lim_{z \to z_0} \frac{f(z) - f(z_0)}{z - z_0} = \lim_{z \to z_0} \frac{\Delta w}{\Delta z}.$$

Assume first that $f'(z_0) \neq 0$. Then

$$f'(z_0) = \rho\, e^{\imath\theta}, \quad \text{for some } \rho > 0,\ \theta \in \mathbb{R}.$$

If $z : [0,1] \to D$ is a $\mathbf{C}^1$-curve with $z(0) = z_0$ and $z'(0) \neq 0$, then $w = f \circ z$ is a $\mathbf{C}^1$-curve with $w(0) = w_0$, $w'(0) \neq 0$. Furthermore, if we denote $\Delta z = z - z_0$ (for z close to but different from z_0) and $\Delta w = f(z) - w_0$, then $\Delta w = \Delta z \frac{\Delta w}{\Delta z}$ implies that

$$\arg \Delta w = \arg \Delta z + \arg \frac{\Delta w}{\Delta z},$$

which together with

$$\lim_{z \to z_0} \arg \frac{\Delta w}{\Delta z} = \theta$$

imply that

$$\arg w'(0) = \arg z'(0) + \arg f'(z_0).$$

All uses of the multi-valued arg function need to be interpreted appropriately; we leave it to the reader to do so. The assertion about lengths means that the ratio of the length of Δw to the length of Δz tends to $|f'(z_0)|$ as z tends to z_0. This follows immediately from

$$\lim_{z \to z_0} \left| \frac{\Delta w}{\Delta z} \right| = \rho.$$

The argument for the case with vanishing derivative is almost identical to the one used to establish (2) and is hence left to the reader.

(4) Let $z : [0,1] \to D$ be a $\mathbf{C}^1$-curve with $z'(t) \neq 0$ for all t. Then

$$w = g \circ z : [0,1] \to g(D)$$

is also a $\mathbf{C}^1$-curve and

$$w'(t) = g_z\, z'(t) + g_{\bar{z}}\, \overline{z'(t)}.$$

Since g preserves angles, $\arg \frac{w'(t)}{z'(t)}$ must be independent of $\arg z'(t)$. But

$$\frac{w'(t)}{z'(t)} = g_z + g_{\overline{z}}\, \frac{\overline{z'(t)}}{z'(t)},$$

and therefore $g_{\overline{z}} \equiv 0$.

(5) We compute the Jacobian of the holomorphic map f at z_0:

$$\begin{aligned} J(f)(z_0) &= \begin{vmatrix} u_x(z_0) & v_y(z_0) \\ v_x(z_0) & v_y(z_0) \end{vmatrix} = u_x(z_0)v_y(z_0) - u_y(z_0)v_x(z_0) \\ &= u_x^2(z_0) + v_x^2(z_0) = \left|f'(z_0)\right|^2. \end{aligned}$$

□

Remark 6.28. The above property (2) of holomorphic mappings is also a consequence of Corollary 6.20. Much of the above discussion for (1) and (2), as well as the next corollary, are slight amplifications of the material in the previous section. We have collected them in Proposition 6.27 together with (3), (4), and (5) so as to have the complete local behavior description in one place.

Corollary 6.29. *A holomorphic function f is injective near a point z_0 in its domain if and only if $f'(z_0) \neq 0$ if and only if f is a homeomorphism near z_0.*

6.6 Evaluation of Definite Integrals

The residue theorem is a powerful tool for the evaluation of many definite integrals. We illustrate this with a few examples. For the third example, the reader might want to look at the appendix to the chapter.

1. The first integral to be evaluated is

$$\int_{-\infty}^{\infty} \frac{1}{x^4+1}\, \mathrm{d}x.$$

This method will work for the evaluation of integrals of the form $\int_{-\infty}^{+\infty} F(x)$ $\mathrm{d}x$, where F is a rational function with no singularities on $\mathbb{R}$ satisfying $\nu_\infty(F) \geq 2$.

We will obviously want to integrate $F(z)\,\mathrm{d}z = \dfrac{1}{z^4+1}\,\mathrm{d}z$. To apply the residue theorem we must carefully choose the path of integration. Let $R > 1$; choose γ_R to be the portion on $\mathbb{R}$ from $-R$ to $+R$ followed by the upper half of the circle $|z| = R$, as in Fig. 6.1.

Since

$$z^4 + 1 = (z - \mathrm{e}^{\frac{\pi\imath}{4}})(z - \mathrm{e}^{\frac{3\pi\imath}{4}})(z + \mathrm{e}^{\frac{\pi\imath}{4}})(z + \mathrm{e}^{\frac{3\pi\imath}{4}}),$$

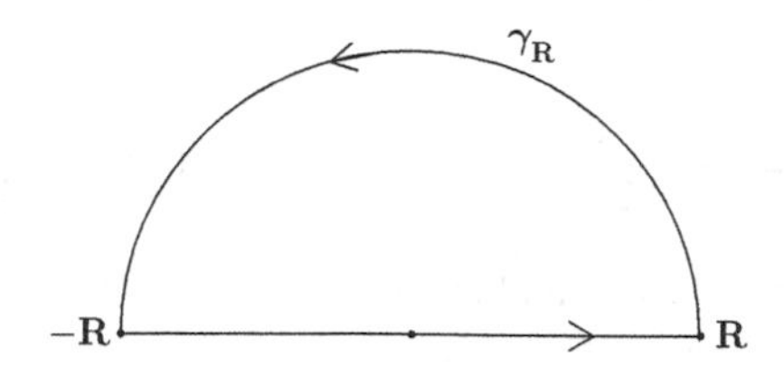

Fig. 6.1 The path of integration for first example

the function F has (possibly) nonzero residues only at these four roots of unity, and we conclude from the residue theorem that

$$\int_{\gamma_R} \frac{1}{z^4+1}\,\mathrm{d}z = 2\pi\imath\left(\operatorname{Res}\left(F, \mathrm{e}^{\frac{\pi\imath}{4}}\right) + \operatorname{Res}\left(F, \mathrm{e}^{\frac{3\pi\imath}{4}}\right)\right).$$

The residues are easy to compute:

$$\operatorname{Res}(F, \mathrm{e}^{\frac{\pi\imath}{4}}) = \left.\frac{1}{(z^2+\imath)(z+\mathrm{e}^{\frac{\pi\imath}{4}})}\right|_{z=\mathrm{e}^{\frac{\pi\imath}{4}}} = \frac{-1}{2(\sqrt{2}-\imath\sqrt{2})}$$

and

$$\operatorname{Res}(F, \mathrm{e}^{\frac{3\pi\imath}{4}}) = \left.\frac{1}{(z^2-\imath)(z+\mathrm{e}^{\frac{3\pi\imath}{4}})}\right|_{z=\mathrm{e}^{\frac{3\pi\imath}{4}}} = \frac{1}{2(\sqrt{2}+\imath\sqrt{2})}.$$

Next we estimate the absolute value of the integral over the semicircle $\{z;\ \Im z \geq 0,\ |z| = R\}$:

$$\left|\int_0^{\pi} \frac{R\,\imath\,\mathrm{e}^{\imath\,\theta}}{R^4\,\mathrm{e}^{4\,\imath\,\theta}+1}\,\mathrm{d}\theta\right| \leq \frac{\pi R}{R^4-1} \to 0 \text{ as } R \to +\infty.$$

We conclude that $\displaystyle\int_{-\infty}^{\infty} \frac{1}{x^4+1}\,\mathrm{d}x = \frac{\sqrt{2}\,\pi}{2}.$

2. A second class of integrals that can be evaluated by the residue theorem consists of those of the form

$$I = \int_0^{2\pi} Q(\cos\theta, \sin\theta)\,\mathrm{d}\theta,$$

where Q is a rational function of two variables with no singularities on the unit circle $S^1 = \{z; |z| = 1\}$. To apply the residue theorem, we express I as an integral of a holomorphic function over the unit circle. We use the change of variables

$$z = \mathrm{e}^{i\theta}, \text{ from where } \mathrm{d}z = \mathrm{e}^{\imath\,\theta}\,\imath\,\mathrm{d}\theta = \imath\,z\,\mathrm{d}\theta$$

and

$$\cos\theta = \frac{e^{\imath\theta} + e^{-\imath\theta}}{2} = \frac{z + z^{-1}}{2}, \; \sin\theta = \frac{e^{\imath\theta} - e^{-\imath\theta}}{2\imath} = \frac{z - z^{-1}}{2\imath}.$$

Example 6.30. Let $0 < b < a$ and evaluate

$$\begin{aligned} I &= \int_0^{2\pi} \frac{1}{a + b\cos\theta}\, d\theta = \int_{|z|=1} \frac{1}{(\imath z)\left(a + b\left(z + \frac{1}{z}\right)\frac{1}{2}\right)}\, dz \\ &= \int_{|z|=1} \frac{-2\imath}{bz^2 + 2az + b}\, dz \\ &= 2\pi\imath \sum_{|z|<1} \operatorname{Res}\left(\frac{-2\imath}{bz^2 + 2az + b}, z\right). \end{aligned}$$

The denominator of the integrand in the last integral is a quadratic polynomial in z with precisely one root inside the unit circle (the product of the roots is $+1$). We thus have one residue to evaluate and conclude that

$$I = 2\pi(a^2 - b^2)^{-\frac{1}{2}}.$$

3. The next type of integral to be discussed here is

$$I = \int_{-\infty}^{\infty} Q(x)e^{\imath x}dx,$$

where Q is a rational function with (at least) a simple zero at infinity and, in general, with no singularities on $\mathbb{R}$.

We illustrate with a more complicated example, where $Q(z) = \frac{1}{z}$ has a simple pole at the origin. Here the ordinary integral is replaced by its *principal value* (pr. v.) defined below:

$$\text{pr. v.} \int_{-\infty}^{\infty} \frac{e^{\imath x}}{x}\, dx = \lim_{\substack{\delta \to 0+ \\ R_1 \to +\infty \\ R_2 \to +\infty}} \left(\int_{\delta}^{R_2} + \int_{-R_1}^{-\delta}\right) \frac{e^{\imath x}}{x}\, dx.$$

We must choose a nice contour for integration; start with large X_1, X_2, and Y, and small δ, all four positive. Our closed path γ has several segments:

γ_1 : from $-X_1$ to $-\delta$ on $\mathbb{R}$
γ_2 : the semicircle in the lower half plane of radius δ and center 0
γ_3 : from δ to X_2 on $\mathbb{R}$
γ_4 : at $x = X_2$ go up to height Y
γ_5 : at height Y travel from X_2 back to $-X_1$
γ_6 : at $x = -X_1$ go down from height Y to the real axis.

We start with

$$\int_{\gamma} \frac{e^{\imath z}}{z}\, dz = 2\pi\imath \operatorname{Res}(f(z)\, dz\, 0),$$

where $f(z) = \frac{e^{\imath z}}{z} = \frac{1}{z} + g(z)$, with g entire. Thus

$$\int_{\gamma} \frac{e^{\imath z}}{z}\, dz = 2\pi\,\imath.$$

We now estimate the integral over γ_4:

$$\left| \int_0^Y \frac{e^{\imath (X_2+\imath y)}}{X_2+\imath y}\, \imath\, dy \right| \le \int_0^Y e^{-y} \frac{1}{|X_2+\imath y|}\, dy \le \frac{1}{X_2} \int_0^Y e^{-y}\, dy < \frac{1}{X_2}.$$

Next we estimate the integral over γ_5:

$$\left| \int_{X_2}^{-X_1} \frac{e^{\imath (x+\imath Y)}}{x+\imath Y}\, dx \right| \le \int_{-X_1}^{X_2} \frac{e^{-Y}}{|x+\imath Y|}\, dx$$

$$\le e^{-Y} \int_{-X_1}^{X_2} \frac{1}{Y}\, dx = \frac{e^{-Y}}{Y}[X_2 + X_1].$$

Also the integral over γ_6:

$$\left| \int_Y^0 \frac{e^{\imath (X_1+\imath y)}}{X_1+\imath y}\, \imath\, dy \right| < \frac{1}{X_1}.$$

We conclude that

$$\int_{\gamma} \frac{e^{\imath z}}{z}\, dz = \lim_{\substack{\delta\to 0+\\ X_1\to+\infty\\ X_2\to+\infty}} \int_{\gamma_1\cup\gamma_2\cup\gamma_3} \frac{e^{\imath z}}{z}\, dz.$$

Finally,

$$\lim_{\delta\to 0+} \left(\int_{\delta}^{X_2} + \int_{-X_1}^{-\delta} \right) \frac{e^{\imath x}}{x}\, dx = \lim_{\delta\to 0+} \left(\int_{\gamma_1\cup\gamma_2\cup\gamma_3} \frac{e^{\imath z}}{z}\, dz + \int_{\gamma_2-} \frac{e^{\imath z}}{z}\, dz \right).$$

But

$$\lim_{\delta\to 0+} \int_{\gamma_2^-} (z^{-1} + g(z))\, dz = \lim_{\delta\to 0+} \int_{\gamma_2^-} z^{-1} dz$$

because g is bounded on a neighborhood of 0 and the length of the path of integration goes to zero (Fig. 6.2). Now

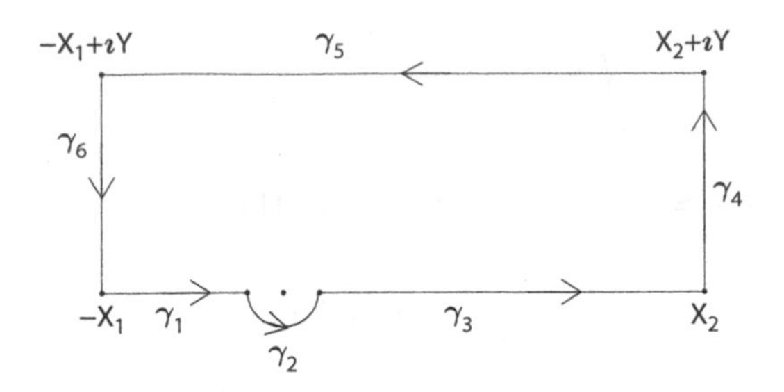

Fig. 6.2 The path of integration for third example

$$\lim_{\delta\to 0+}\int_{\gamma_2^-} z^{-1}\,\mathrm{d}z = \lim_{\delta\to 0+}\int_0^{-\pi}\frac{1}{\delta e^{\imath\,\theta}}\delta e^{\imath\,\theta}\,\imath\,\mathrm{d}\theta = -\pi\imath,$$

and we conclude that

$$\text{pr. v.}\int_{-\infty}^{\infty}\frac{e^{\imath\,x}}{x}\,\mathrm{d}x = \pi\imath.$$

Using the fact that $e^{\imath\,x} = \cos x + \imath\,\sin x$, we see that we have evaluated two real integrals:

$$\text{pr. v.}\int_{-\infty}^{\infty}\frac{\cos x}{x}\,\mathrm{d}x = 0 \quad\text{and}\quad \int_0^{\infty}\frac{\sin x}{x}\,\mathrm{d}x = \frac{\pi}{2}.$$

4. We end with an example that illustrates how multi-valued functions can be used to evaluate definite integrals. We evaluate

$$\int_0^{\infty}\frac{x^{\alpha}}{1+x^2}\mathrm{d}x$$

with $0 < \alpha < 1$.

The method illustrated here will work for integrating on the positive real axis functions of the form

$$x^{\alpha}R(x),$$

where $0 < \alpha < 1$ and R is a rational function without singularities on $\mathbb{R}_{\geq 0}$ and vanishing at infinity to sufficiently high order.

In our example we will use the multi-valued function

$$F(z) = \frac{z^{\alpha}}{1+z^2}$$

to integrate over an appropriate contour where we can use a single-valued branch. We choose the curve γ as follows: start with positive constants ρ (small), θ_0 (small), and R (large). These constants will be restricted in the arguments that

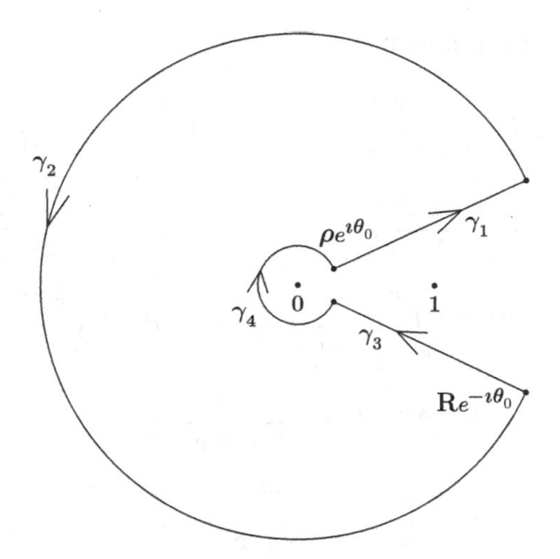

Fig. 6.3 The path of integration for fourth example

follow. The oriented curve $\gamma = \gamma_1 * \gamma_2 * \gamma_3 * \gamma_4$ consists of four components traversed successively and will involve two circles ($|z| = \rho$ and $|z| = R$) and two straight line segments ($\arg z = \pm\theta_0$) (See Fig. 6.3).

- The curve γ_1 consists of the straight line from $\rho e^{\imath\theta_0}$ to $Re^{\imath\theta_0}$. It hence makes sense to restrict $0 < \rho < 1 < R$.
- The curve γ_2 consists of the following portion of the positively oriented circle with center at 0 and radius R:
  $\{Re^{\imath\theta};\ \theta_0 \le \theta \le (2\pi - \theta_0\}$; since we want $\pm\imath$ to be in the interior of the range of γ, we require $0 < \theta_0 < \dfrac{\pi}{2}$.
- The curve γ_3 consists of the straight line segment from $Re^{-\imath\theta_0}$ to $\rho e^{-\imath\theta_0}$.
- Finally, the curve γ_4 consists of the following portion of the negatively oriented circle with center at 0 and radius ρ: $\{\rho e^{\imath(2\pi-\theta)};\ \theta_0 \le \theta \le 2\pi - \theta_0\}$.

We begin evaluating and estimating various integrals. We use the branch of z^α whose arguments lie in $(0, 2\pi\alpha)$; thus a holomorphic function on $\mathbb{C} - \mathbb{R}_{\ge 0}$, with a jump discontinuity at each point on $\mathbb{R}_{>0}$. It is this discontinuity that is the key to the method used here.

- By the residue theorem

$$\frac{1}{2\pi\imath}\int_\gamma F(z)\mathrm{d}z = \text{Res}\,(F(z)\mathrm{d}z, \imath) + \text{Res}\,(F(z)\mathrm{d}z, -\imath)$$

$$= \frac{e^{\imath\frac{\pi}{2}\alpha}}{2\imath} + \frac{e^{\imath\frac{3\pi}{2}\alpha}}{-2\imath} = -\frac{1}{2}\imath\left(e^{\imath\frac{\pi}{2}\alpha} - e^{\imath\frac{3\pi}{2}\alpha}\right).$$

- The above integral is a sum:

$$\int_\gamma F(z)\mathrm{d}z = \int_\rho^R \frac{r^\alpha \mathrm{e}^{\imath\theta_0\alpha}}{1+r^2\mathrm{e}^{2\imath\theta_0}} d(r\mathrm{e}^{\imath\theta_0}) + \int_{\theta_0}^{2\pi-\theta_0} \frac{R^\alpha \mathrm{e}^{\imath\alpha\theta}}{1+R^2\mathrm{e}^{2\imath\theta}} d(R\mathrm{e}^{\imath\theta})$$
$$+ \int_{\mathbb{R}}^{\rho} \frac{r^\alpha \mathrm{e}^{\imath(2\pi-\theta_0)\alpha}}{1+r^2\mathrm{e}^{-2\imath\theta_0}} d(r\mathrm{e}^{-\imath\theta_0}) + \int_{2\pi-\theta_0}^{\theta_0} \frac{\rho^\alpha \mathrm{e}^{\imath\alpha\theta}}{1+\rho^2\mathrm{e}^{2\imath\theta}} d(\rho\mathrm{e}^{\imath\theta})$$
$$= \mathrm{e}^{\imath\theta_0(\alpha+1)} \int_\rho^R \frac{r^\alpha}{1+r^2\mathrm{e}^{2\imath\theta_0}} \mathrm{d}r + \imath R^{1+\alpha} \int_{\theta_0}^{2\pi-\theta_0} \frac{\mathrm{e}^{\imath(\alpha+1)\theta}}{1+R^2\mathrm{e}^{2\imath\theta}} \mathrm{d}\theta$$
$$- \mathrm{e}^{\imath(2\pi-\theta_0)(1+\alpha)} \int_\rho^R \frac{r^\alpha}{1+r^2\mathrm{e}^{-2\imath\theta_0}} \mathrm{d}r$$
$$- \imath\rho^{1+\alpha} \int_{\theta_0}^{2\pi-\theta_0} \frac{\mathrm{e}^{\imath(1+\alpha)\theta}}{1+\rho^2\mathrm{e}^{2\imath\theta}} \mathrm{d}\theta.$$

- We need some estimates:

$$\left| R^{1+\alpha} \int_{\theta_0}^{2\pi-\theta_0} \frac{\mathrm{e}^{\imath(\alpha+1)\theta}}{1+R^2\mathrm{e}^{2\imath\theta}} \mathrm{d}\theta \right| \leq R^{1+\alpha} \frac{1}{R^2-1} 2\pi \to 0 \text{ as } R\to\infty,$$
$$\left| \rho^{1+\alpha} \int_{\theta_0}^{2\pi-\theta_0} \frac{\mathrm{e}^{\imath(\alpha+1)\theta}}{1+\rho^2\mathrm{e}^{2\imath\theta}} \mathrm{d}\theta \right| \leq \rho^{1+\alpha} \frac{1}{1-\rho^2} 2\pi \to 0 \text{ as } \rho\to 0.$$

- The value of $\int_\gamma F(z)\mathrm{d}z$ is independent of the choices of ρ, θ_0, and R. We can hence let the first two of these constants approach zero and the last approach infinity to obtain

$$2\pi\imath\left(-\frac{1}{2}\imath\left(\mathrm{e}^{\imath\frac{\pi}{2}\alpha} - \mathrm{e}^{\imath\frac{3\pi}{2}\alpha}\right)\right) = \int_0^\infty \frac{r^\alpha}{1+r^2}\mathrm{d}r - \mathrm{e}^{2\pi\imath(1+\alpha)} \int_0^\infty \frac{r^\alpha}{1+r^2}\mathrm{d}r.$$

- The last identity yields at once that

$$\int_0^\infty \frac{x^\alpha}{1+x^2}\mathrm{d}x = \pi\frac{\mathrm{e}^{\imath\frac{\pi}{2}\alpha}}{1+\mathrm{e}^{\imath\pi\alpha}},$$

a real number, as expected.

6.7 Appendix: Cauchy Principal Value

We have studied integration of holomorphic forms ω along paths γ. For some applications it is useful to allow ω to have some singularities, simple poles in our case, on γ. We describe a path toward this goal of evaluating integrals of differential forms with some singularities.

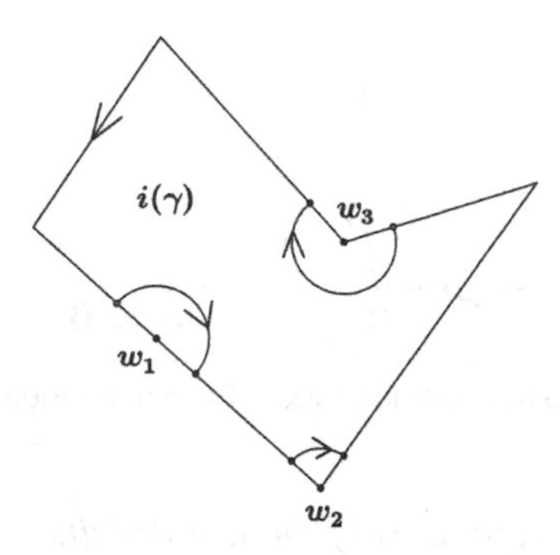

Fig. 6.4 A modification of a Jordan path

Definition 6.31. We consider an oriented pdp $\gamma \subset \mathbb{C}$, and a differential form $\omega = f(z)\,dz$ that is holomorphic in a neighborhood D of γ except for isolated singularities on the range of γ that are simple poles. Because $\int_\gamma \omega$ is not defined, we want to introduce two paths γ_δ and $\gamma_{d,\delta}$, the second is disconnected, for which $\int_{\gamma_\delta} \omega$ and $\int_{\gamma_{d,\delta}} \omega$ are defined and finite. Let $w_1, w_2, \ldots, w_m$ be the set of singularities of ω on γ. Let $\epsilon \in \mathbb{R}_{>0}$ be the minimum of the finite set consisting of one half the distances between the various w_k and the radii of the largest discs about these points contained in D. Let $0 < \delta < \epsilon$. For each k, let C_k be the positively oriented circle of radius δ and center w_k. This circle intersects γ in a finite set of points. Let $a_{\delta,k}$ be the last point before w_k in this intersection and $b_{\delta,k}$ the first point after w_k. These two points define two arcs on C_k. Choose one of these arcs; call it $\widetilde{\alpha_{\delta,k}}$, the *arc subtended at* w_k, and give it the orientation consistent with that of γ. Let $-2\pi < \alpha_{\delta,k} < 2\pi$ be the angle of this arc measured from $a_{\delta,k}$. The *δ-modification* γ_δ of the curve γ is obtained by replacing for each k the segment of γ between $a_{\delta,k}$ and $b_{\delta,k}$ by the arc $\widetilde{\alpha_{\delta,k}}$. The *disconnected δ-modification* $\gamma_{d,\delta}$ of the curve γ is obtained by removing for each k the segment of γ between $a_{\delta,k}$ and $b_{\delta,k}$. See Figs. 6.4 and 6.5. We define the *principal value*

$$\text{pr. v.} \int_\gamma \omega = \lim_{\delta \to 0} \int_{\gamma_{d,\delta}} \omega,$$

provided the limit exists.

Remark 6.32. Each of the sets $\{a_{\delta,k}\}$, $\{b_{\delta,k}\}$, and $\{\alpha_{\delta,k}\}$ is bounded. Hence we can construct a sequence $\{\delta_n\}$ that converges to zero with the property that all three sequences $\{a_{\delta_n,k}\}$, $\{b_{\delta_n,k}\}$, and $\{\alpha_{\delta_n,k}\}$ have limits denoted by a_k, b_k α_k, respectively. Once these limits are known to exist, it is easy to see that the three nets[4] $\{a_{\delta,k}\}$, $\{b_{\delta,k}\}$, and $\{\alpha_{\delta,k}\}$ also converge to the appropriate limits.

[4]This topological concept and its properties are not discussed here.

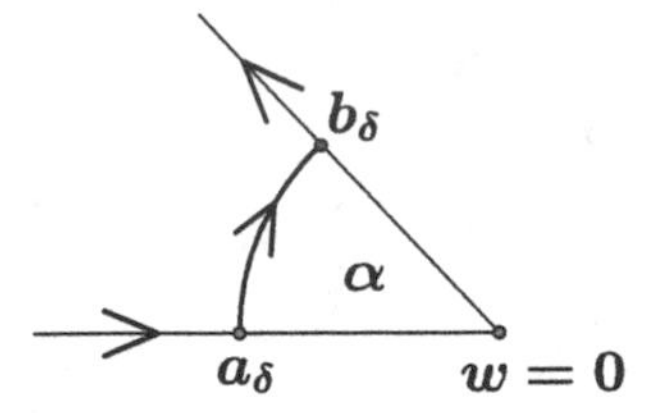

Fig. 6.5 The modification localized (after a translation and rotation)

Lemma 6.33. *Under the hypothesis of the last definition, there exists a constant $M > 0$ independent of δ such that*

$$\int_{\gamma_\delta} \omega - \int_{\gamma_{d,\delta}} \omega = -\imath \sum_{k=1}^{m} \alpha_{\delta,k} \operatorname{Res}(\omega, w_k) + \delta\epsilon(\delta),$$

where $|\epsilon(\delta)| < M$. Hence

$$\lim_{\delta\to 0}\left(\int_{\gamma_\delta} \omega - \int_{\gamma_{d,\delta}} \omega\right) = -\imath \sum_{k=1}^{m} \alpha_k \operatorname{Res}(\omega, w_k). \tag{6.4}$$

Proof. Only the first identity needs verification. Fix k. By a translation and rotation, we may assume that $w_k = 0$ and $a_k \in \mathbb{R}_{<0}$. Write

$$f(z) = \frac{\rho}{z} + g(z) \text{ for } |z| < \delta,$$

with M_0 a bound for $|g|$ in $U_{0,\delta}$. Hence

$$\begin{aligned}\int_{\widetilde{\alpha_{\delta,k}}} \omega &= \int_{\pi}^{\pi-\alpha_k} \left(\frac{\rho}{\delta e^{\imath\theta}} + g(\delta e^{\imath\theta})\right) d(\delta e^{\imath\theta}) \\ &= \imath\rho \int_{\pi}^{\pi-\alpha_{\delta,k}} d\theta + \imath\delta \int_{\pi}^{\pi-\alpha_{\delta,k}} e^{\imath\theta} g(\delta e^{\imath\theta}) d\theta.\end{aligned}$$

The first of the integrals in the last line is easily evaluated to yield $-\alpha_{\delta,k}$; the absolute value of the second is bounded by $M_0|\alpha_{\delta,k}| < 2\pi M_0$. From this the lemma follows. □

Theorem 6.34 (Residue Theorem, Version 2). *Let γ be a closed positively oriented Jordan curve in a domain $D \subset \mathbb{C}$ with $i(\gamma) \subset D$. Let f be a holomorphic function in the domain D except for isolated singularities at $z_1, \ldots, z_n$ in $i(\gamma)$ and simple poles $w_1, \ldots, w_m$ on* range (γ). *Then*

$$\text{pr. v.} \int_\gamma f(z)\,\mathrm{d}z = 2\pi\imath \sum_{j=1}^{n} \operatorname{Res}(f(z)\,\mathrm{d}z, z_j) + \imath \sum_{k=1}^{m} \alpha_k \operatorname{Res}(f(z)\,\mathrm{d}z, w_k),$$

where the α_k are defined by Remark 6.32.

Proof. As in previous arguments, δ is a small positive number, but in this case at most the minimum of the distances from any of the w_k to the z_j. We will use much of the notation previously introduced in this section. By introducing new line segments and thus integrating over a finite number of paths, rather than just a single path, it suffices to assume that $m = 1$ and $n = 0$. Thus we may take γ to be a path in the closure of $i(\gamma)$ from $a_{\delta,1}$ to w_1 (where $f(z)\,\mathrm{d}z$ has a simple pole) followed by a second one from w_1 to $b_{\delta,1}$ and a third from $b_{\delta,1}$ back to $a_{\delta,1}$. As before by translating and rotating the picture we may assume that $w_1 = 0$ and $a_{\delta,1} \in \mathbb{R}_{<0}$. Note that

$$\int_{\gamma_\delta} f(z)\,\mathrm{d}z = 0.$$

From the lemma (equation (6.4))

$$\text{pr. v.} \int_\gamma f(z)\,\mathrm{d}z = \imath\alpha_1 \operatorname{Res}(f(z)\,\mathrm{d}z, w_1),$$

and this suffices to establish the theorem. □

Exercises

6.1. Use Rouché's theorem to prove the fundamental theorem of algebra.

6.2. Let g be a holomorphic function on $|z| < R$, $R > 1$, with $|g(z)| \leq 1$ for all $|z| < R$.

1. Show that for all $t \in \mathbb{C}$ with $|t| < 1$, the equation

$$z = tg(z)$$

   has a unique solution $z = s(t)$ in the disc $|z| < 1$.
2. Show that $t \mapsto s(t)$ is a holomorphic function on the disc $|t| < 1$ with $s(0) = 0$.
3. When is s a constant function?
4. When is s injective?

6.3. Verify (6.3) (in the proof of the argument principle) using Laurent series expansions for f and F.

6.4. If f is a holomorphic function on $0 < |z| < 1$ and f does not assume any value w with $|w - 1| < 2$, what can you conclude?

6.5. Evaluate the following integrals:

(a) $\int_0^{\pi} \frac{d\theta}{\sqrt{5} + \cos\theta}$

(b) $\int_{|z|=1} \frac{z^6 dz}{7z^7 - 1}$

(c) $\int_{|z-100\pi|=\frac{199}{2}\pi} z \cot z dz$

(d) $\int_{|z-\frac{\pi}{2}|=3.15} z \tan z \, dz$

6.6. Compute:

(a) $\int_{-\infty}^{\infty} \frac{dx}{1+x^6}$

(b) $\int_{-\infty}^{\infty} \frac{(1+x)}{1+x^4} dx$

(c) $\int_{-\infty}^{\infty} \frac{x}{4+x^4} dx$

(d) $\int_0^{\infty} \frac{dx}{1+x^3}$

(e) $\int_0^{\infty} \frac{x^{\alpha}}{1+x^3} dx$, where $0 < \alpha < 1$

6.7. Let f be an entire function such that $|f(z)| = 1$ for $|z| = 1$. What are the possible values for $f(0)$ and for $f(17)$?

6.8. Find all functions f that are meromorphic in a neighborhood of $\{|z| \leq 1\}$ and satisfy $|f(z)| = 1$ for $|z| = 1$, f has a double pole at $z = \frac{1}{2}$, a triple zero at $z = -\frac{1}{3}$, and no other zeros or poles in $\{|z| < 1\}$.

6.9. Suppose f is an entire function satisfying $f(n) = n^5$ and $f\left(-\frac{n}{2}\right) = n^7$ for all $n \in \mathbb{Z}_{>0}$. How many zeros does the function $g(z) = [f(z) - e][f(z) - \pi]$ have?

6.10. Evaluate

$$\int_{|z|=3} \frac{f'(z)}{f(z)-1} dz,$$

where $f(z) = 2 - 2z + z^2 + \frac{z^3}{81}$.

6.11. Suppose f is holomorphic for $|z| < 1$ and $f\left(\frac{1}{n}\right) = \frac{7}{n^3}$ for $n = 2, 3, \ldots$ What can be said about $f'''(0)$?

6.12. Let f be an entire function such that $|f(z)| \le |z|^{\frac{23}{3}}$ for all $|z| > 10$. Compute $f^{(8)}(10.001)$.

6.13. Evaluate the following real integrals using residues:

$$\int_{-\infty}^{\infty} \frac{\cos x}{1+x^2} dx\,, \qquad \int_{-\infty}^{\infty} \frac{\sin x}{1+x^2} dx.$$

6.14. Find **all** Laurent series of the form $\sum_{-\infty}^{\infty} a_n z^n$ for the functions

•
$$f(z) = \frac{z^2}{(1-z)^2(1+z)},$$

•
$$f(z) = \frac{1}{(z-1)(z-2)(z-3)},$$

and

•
$$f(z) = \frac{2-z^2}{z(1-z)(2-z)}.$$

6.15. If f is entire and satisfies $|f''(z) - 3| \ge 0.001$ for all $z \in \mathbb{C}$, $f(0) = 0$, $f(1) = 2$, $f(-1) = 4$, what is $f(\imath)$?

6.16. If f is an entire function such that $\Re f(z) > -2$ for all $z \in \mathbb{C}$ and $f(\imath) = \imath + 2$, what is $f(-\imath)$?

6.17. If f is holomorphic on $0 < |z| < 2$ and satisfies $f(\frac{1}{n}) = n^2$ and $f(-\frac{1}{n}) = n^3$ for all $n \in \mathbb{Z}_{>0}$, what kind of singularity does f have at 0?

6.18. Let D be a bounded domain in $\mathbb{C}$ with smooth boundary. Assume f is a nonconstant holomorphic function in a neighborhood of the closure of D such that $|f|$ is constant on ∂D, say, $|f| = c$ on ∂D. Show that f takes on each value d such that $|d| < c$ at least once in D.

6.19. Suppose f is holomorphic in a neighborhood of the closure of the unit disc. Show that for $|z| \le 1$

$$f(z)(1 - |z|^2) = \frac{1}{2\pi\imath} \int_{|\tau|=1} \frac{1 - \bar{z}\tau}{\tau - z} f(\tau)\, d\tau,$$

and conclude that the following inequality holds:

$$|f(z)|\,(1 - |z|^2) \le \frac{1}{2\pi} \int_0^{2\pi} \left|f(\exp^{\imath\,\theta})\right| d\theta.$$

6.20. Let f be an entire function. Suppose there exist positive constants A and B such that $|f(z)| \le A + B\,|z|^{10}$ for all $z \in \mathbb{C}$. Show that f is a polynomial. What is its degree?

6.21. Suppose f is meromorphic in a neighborhood of the closed unit disc and that $|f(z)| = 1$ for $|z| = 1$. Find the most general such function.

6.22. Let C denote the positively oriented unit circle. Consider the function

$$f(z) = \frac{2\,z^{26}}{81} + \exp\left(z^{21}\right)\left(z - \frac{1}{2}\right)^2\left(z - \frac{1}{3}\right)^3.$$

Evaluate the following integrals:

$$\int_C f(z)\,\mathrm{d}z\,, \quad \int_C f'(z)\,\mathrm{d}z\,, \quad \int_C \frac{f'(z)}{f(z)}\,\mathrm{d}z.$$

6.23. If f is holomorphic for $0 < |z| < 1$ and satisfies $f\left(\frac{1}{n}\right) = n^2$, $f\left(-\frac{1}{n}\right) = 2n^2$ for $n = 2, 3, 4, \ldots$, what can you say about f?

6.24. Suppose f is entire and $f(z) \ne t^2$ for all $z \in \mathbb{C}$ and for all $t \in \mathbb{R}$. Show that f is constant.

6.25. If f is holomorphic for $0 < |z| < 1$ and satisfies $f\left(\frac{1}{n}\right) = n^{-2}$ and $f\left(-\frac{1}{2}\right) = 2n^{-2}, n = 2, 3, 4, \ldots$, find $\liminf_{z\to 0} |f(z) - 2|$.

6.26. Find $\displaystyle\int_0^\infty \frac{\sin^2 2x}{x^2}\,\mathrm{d}x$ using residues.

6.27. Prove the following extension of the maximum modulus principle. Let f be holomorphic and bounded on $|z| < 1$, and continuous on $|z| \le 1$ except maybe at $z = 1$. If $\left|f(\mathrm{e}^{\imath\,\theta})\right| \le A$ for $0 < \theta < 2\pi$ then $|f(z)| \le A$ for all $|z| < 1$.

6.28. Let $\mathbb{D}$ denote the unit disc and let $\{f_n\}$ be a sequence of holomorphic functions in $\mathbb{D}$ such that $\lim_{n\to\infty} f_n = f$ uniformly on compact subsets of $\mathbb{D}$. Suppose that each f_n takes on the value 0 at most seven times on $\mathbb{D}$ (counted with multiplicity). Prove that either $f \equiv 0$ or f takes on the value 0 at most seven times on $\mathbb{D}$ (counted with multiplicity).

6.29. Show that the function $f(z) = z + 2z^2 + 3z^3 + 4z^4 + \cdots$ is injective in the unit disc $\mathbb{D} = \{z \in \mathbb{C}; |z| < 1\}$. Find $f(\mathbb{D})$.

6.30. Suppose f is a nonconstant function holomorphic on the unit disc $\{z \in \mathbb{C};\ |z| < 1\}$ and continuous on $\{z \in \mathbb{C};\ |z| \le 1\}$ such that for all $\theta \in \mathbb{R}$, the value $f(\mathrm{e}^{\imath\theta})$ is on the boundary of the triangle with vertices 0, 1, and $\imath$. Is there a z_0 with $|z_0| < 1$ such that $f(z_0) = \frac{1}{10}(1+\imath)$? Is there a z_0 with $|z_0| < 1$ such that $f(z_0) = \frac{1}{2}(1+\imath)$?

6.31. Is there a function f holomorphic for $|z| < 1$ and continuous for $|z| \leq 1$ that satisfies

$$f(\mathrm{e}^{\imath\,\theta}) = \cos\theta + 2\imath\,\sin\theta, \quad \text{for all } \theta \in \mathbb{R}?$$

6.32. The integral $I = \int_0^\infty \mathrm{e}^{-x^2}\mathrm{d}x$ can be evaluated using both real and complex analytic methods.

1. By working with $\int_0^\infty \mathrm{e}^{-x^2}\mathrm{d}x \int_0^\infty \mathrm{e}^{-y^2}\mathrm{d}y$ and Fubini's theorem, evaluate I.
2. Evaluate I using residues by considering a change of variables $t = x^2$ and evaluating $\int_0^\infty \mathrm{e}^{-x^2}\cos 2mx\,\mathrm{d}x$ by integrating along a proper closed curve.

6.33. Formulate and prove forms of the residue theorem and the argument principle for domains in the sphere $\mathbb{C} \cup \{\infty\}$.

6.34. Describe the type of singularity at ∞ for the following functions $f : \mathbb{C}_{\neq 0} \to \mathbb{C}$:

(a) $f(z) = \exp\left(\frac{1}{z}\right)$

(b) $f(z) = z^n,\, n \in \mathbb{Z}$

(c) $f(z) = \dfrac{\sin(z)}{z}$

6.35. Prove that the inverse function of an injective entire function f cannot be entire unless f is a polynomial of degree one.

Chapter 7
Sequences and Series of Holomorphic Functions

We now turn from the study of a single holomorphic function to the study of collections of holomorphic functions. In the first section we will see that under the appropriate notion of convergence of a sequence of holomorphic functions, the limit function inherits several properties that the approximating functions have, such as being holomorphic. In the second section we show that the space of holomorphic functions on a domain can be given the structure of a complete metric space. We then apply these ideas and results to obtain, as an illustrative example, a series expansion for the cotangent function. In the fourth section we characterize the compact subsets of the space of holomorphic functions on a domain. This powerful characterization is used in Sect. 7.5, to study approximations of holomorphic functions and, in particular, to prove Runge's theorem, which describes conditions under which a holomorphic function can be approximated by rational functions with prescribed poles. The characterization will also be used in Chap. 8 to prove the Riemann mapping theorem.

7.1 Consequences of Uniform Convergence on Compact Sets

We begin by recalling some notation and introducing some new symbols. Let D be a domain in $\mathbb{C}$; we denote by $\mathbf{C}(D)$ the vector space of continuous complex-valued functions on D, and recall that $\mathbf{H}(D) \subset \mathbf{C}(D)$ is the vector space of holomorphic functions on D (see Definition 3.57). We say that a compact disc $\operatorname{cl} U_z(r)$ has rational center if $z = x + \imath\, y$ with x and y in $\mathbb{Q}$.

Proposition 7.1. *A necessary and sufficient condition for a sequence of functions $\{f_n\} \subset \mathbf{C}(D)$ to converge uniformly on all compact subsets of D is for the sequence to converge uniformly on all compact discs with rational centers and rational radii contained in D.*

R.E. Rodríguez et al., *Complex Analysis: In the Spirit of Lipman Bers*,
Graduate Texts in Mathematics 245, DOI 10.1007/978-1-4419-7323-8_7,

Proof. Every compact set contained in D can be covered by finitely many such discs. □

It is clear that if a sequence of functions $\{f_n\} \subset \mathbf{C}(D)$ converges uniformly to a function f on all compact subsets of D, then for all z in D we have pointwise convergence: $\lim_{n\to\infty} f_n(z) = f(z)$. The converse is not true: uniform convergence on all compact subsets of D is stronger than pointwise convergence. To see this observe that we know from Theorem 2.23 that if a sequence of functions $\{f_n\} \subset \mathbf{C}(D)$ converges uniformly to a function f on all compact subsets of D, then $f \in \mathbf{C}(D)$. On the other hand, it is easy to construct an example of a sequence of continuous functions converging at every point of the domain to a discontinuous function (see Exercise 7.1).

We proceed to describe some consequences of uniform convergence on all compact subsets of D, also called *locally uniform convergence*, for $\mathbf{H}(D)$. The first of these is that $\mathbf{H}(D)$ is closed under locally uniform convergence.

Theorem 7.2. *If $\{f_n\} \subset \mathbf{H}(D)$ and $\{f_n\}$ converges uniformly on all compact subsets of D, then the limit function f is holomorphic on D.*

Proof. We already know that $f \in \mathbf{C}(D)$. Let γ be any closed curve homotopic to a point in D. Then, by Cauchy's theorem,

$$\int_\gamma f_n(z)\, \mathrm{d}z = 0.$$

By uniform convergence it follows that

$$\int_\gamma f(z)\, \mathrm{d}z = \lim_{N\to\infty} \int_\gamma f_n(z)\, \mathrm{d}z = 0,$$

and then, by Morera's theorem, f is holomorphic on D. □

Corollary 7.3. *If $\{f_n\} \subset \mathbf{H}(D)$ and $\sum_{n=1}^{\infty} f_n$ converges uniformly on all compact subsets of D, then the limit function (also denoted by $\sum_{n=1}^{\infty} f_n$) is holomorphic on D.*

Theorem 7.2 has no analog in real variables: it is easy to see (at least pictorially) that the absolute value function on $\mathbb{R}$, which has no derivative at 0, can be uniformly approximated by differentiable functions (see Exercise 7.1). A more extreme example was constructed by Weierstrass, that of a continuous function defined on $[0, 1]$ which is nowhere differentiable and uniformly approximated by polynomials.

We will shortly see that uniform convergence of a sequence of holomorphic functions on all compact subsets of their common domain of definition implies uniform convergence of the derivatives on the same sets. This is another feature of holomorphic functions not shared by real differentiable functions: it is easy to construct a sequence of differentiable functions converging uniformly on a closed interval with the property that the sequence of derivatives does not converge uniformly there. We leave this construction to the reader as Exercise 7.2.

Theorem 7.4. *If $\{f_n\} \subset \mathbf{H}(D)$ and $f_n \to f$ uniformly on all compact subsets of D, then $f_n' \to f'$ uniformly on all compact subsets of D.*

Proof. Since $f \in \mathbf{H}(D)$, it is enough to check uniform convergence of the derivatives on all compact subdiscs $R \subset D$, with $\partial R = \gamma$ positively oriented. For $z \in i(\gamma)$, we have

$$f'(z) = \frac{1}{2\pi\imath}\int_\gamma \frac{f(w)}{(w-z)^2}\,\mathrm{d}w = \lim_{n\to\infty}\frac{1}{2\pi\imath}\int_\gamma \frac{f_n(w)}{(w-z)^2}\,\mathrm{d}w = \lim_{n\to\infty} f_n'(z);$$

this convergence is uniform in any smaller compact subdisc, such as

$$\widetilde{R} = \{z \in i(\gamma); \inf\{|z-w|\,; w \in \partial R\} \geq \delta > 0\},$$

with δ sufficiently small. □

Theorem 7.5. *Let $\{f_n\}$ be a sequence of holomorphic functions on D such that $f_n \to f$ uniformly on all compact subsets of D. If $f_n(z) \neq 0$ for all $z \in D$ and all $n \in \mathbb{Z}_{>0}$, then either*

(a) *f is identically zero, or*
(b) *$f(z) \neq 0$ for all $z \in D$.*

Proof. Assume that there is $c \in D$ with $f(c) = 0$ and that f is not identically zero. Then there exists a circle γ with center c such that $\operatorname{cl} i(\gamma) \subset D$ and $f(z) \neq 0$ for all $z \in \operatorname{cl} i(\gamma) - \{c\}$. By the argument principle, the number N of zeros of f in $i(\gamma)$ is given by

$$N = \frac{1}{2\pi\imath}\int_\gamma \frac{f'(z)}{f(z)}\,\mathrm{d}z \geq 1.$$

But

$$\int_\gamma \frac{f'(z)}{f(z)}\,\mathrm{d}z = \lim_{n\to\infty}\int_\gamma \frac{f_n'(z)}{f_n(z)}\,\mathrm{d}z = 0.$$ □

Remark 7.6. An equivalent formulation for this theorem is the following, sometimes referred to as Hurwitz's theorem.

Theorem 7.7. *Let $\{f_n\}$ be a sequence of holomorphic functions on D such that $f_n \to f$ uniformly on all compact subsets of D, and assume f is not identically zero on D. For every disc U such that $\operatorname{cl} U \subset D$ with the property that $f \neq 0$ on ∂U, there exists $N \in \mathbb{Z}_{>0}$ such that f and f_n have the same number of zeros in U for all $n \geq N$.*

Definition 7.8. Let $f \in \mathbf{H}(D)$. We call f *simple, univalent,* or *schlicht* if it is one-to-one (injective) on D; thus a homeomorphism onto $f(D)$.

Theorem 7.9 (Hurwitz). *Assume D is a domain in $\mathbb{C}$. If $\{f_n\}$ is a sequence in $\mathbf{H}(D)$ with $f_n \to f$ uniformly on all compact subsets of D and f_n is schlicht for each n, then either f is constant or schlicht.*

Proof. Assume that f is neither constant nor schlicht; thus, in particular, there exist z_1 and z_2 in D with $z_1 \neq z_2$ and $f(z_1) = f(z_2)$. For each $n \in \mathbb{Z}_{>0}$, set $g_n(z) = f_n(z) - f_n(z_2)$ on the domain $D' = D - \{z_2\}$. Then $g_n \in \mathbf{H}(D')$, g_n never vanishes in D' and $g_n \to g = f - f(z_2)$ uniformly on all compact subsets of D'. But g is not identically zero and vanishes at z_1; we have thus obtained a contradiction to Theorem 7.5. □

7.2 A Metric on $\mathbf{C}(D)$

We introduce, for use in the proof of the compactness theorem of this chapter and in the proof of the Riemann mapping theorem in Chap. 8, a *metric* on $\mathbf{C}(D)$ for any domain D in $\mathbb{C}$. The metric ρ on $\mathbf{C}(D)$ will have the property that convergence in the ρ-metric is equivalent to uniform convergence on all compact subsets of D.

For K compact in D and $f \in \mathbf{C}(D)$, set

$$||f||_K = \max\{|f(z)|\,;\, z \in K\}.$$

Consider the set of compact (closed) discs contained in D with rational radii and rational centers. There are countably many such discs and they cover D. Call this collection of discs $\{D_i\}_{i \in \mathbb{Z}_{>0}}$.

For $n \in \mathbb{Z}_{>0}$, let

(1)
$$K_n = \bigcup_{i \leq n} D_i,$$

then $\{K_n\}$ is an *exhaustion* of D; that is,

(2)
$$\text{Each } K_n \text{ is compact,}$$

(3)
$$K_n \subset K_{n+1} \text{ for all } n \in \mathbb{Z}_{>0}, \text{ and}$$

(4)
$$\bigcup_{n \in \mathbb{Z}_{>0}} \operatorname{int} K_n = D,$$

where int K denotes the interior of the set K.

From now on, we shall use only properties (2), (3), and (4) of our exhaustion and *not* how these sets were constructed.

Remark 7.10. A crucial consequence of these properties that we will use often is that given an exhaustion $\{K_n\}$ of D, *each compact subset K of D is contained in K_n for some n.*

For $f \in \mathbf{C}(D)$ and $i \in \mathbb{Z}_{>0}$, we set

$$M_i(f) = ||f||_{K_i},$$

and note that

$$M_{i+1} \geq M_i.$$

We define

$$d(f) = \sum_{i=1}^{\infty} 2^{-i} \min(1, M_i(f)) \leq \sum_{i=1}^{\infty} 2^{-i} = 1. \tag{7.1}$$

7.2.1 Properties of d

For all f and $g \in \mathbf{C}(D)$:

(1) $d(f) \geq 0$, and $d(f) = 0$ if and only if $f \equiv 0$
(2) $d(f+g) \leq d(f) + d(g)$
(3) For each i, $2^{-i} \min(1, M_i(f)) \leq d(f)$
(4) For each i, $d(f) \leq M_i(f) + 2^{-i}$

Proof. Properties (1) and (3) are immediate from the definition of d. To prove (2), observe that

$$\begin{aligned} d(f+g) &= \sum_{i=1}^{\infty} 2^{-i} \min(1, M_i(f+g)) \\ &\leq \sum_{i=1}^{\infty} 2^{-i} \min(1, M_i(f) + M_i(g)) \\ &\leq \sum_{i=1}^{\infty} 2^{-i} [\min(1, M_i(f)) + \min(1, M_i(g))] \\ &= d(f) + d(g). \end{aligned}$$

For property (4)

$$d(f) = \sum_{j \leq i} 2^{-j} \min(1, M_j(f)) + \sum_{j > i} 2^{-j} \min(1, M_j(f))$$

$$\leq \sum_{j\leq i} 2^{-j} M_j(f) + \sum_{j>i} 2^{-j}$$
$$\leq \left(\sum_{j\leq i} 2^{-j} \right) M_i(f) + 2^{-i}$$
$$\leq M_i(f) + 2^{-i}.$$

□

Finally, we define the *metric* on $\mathbf{C}(D)$ we have been seeking:

$$\rho(f, g) = d(f - g). \tag{7.2}$$

We list some immediate properties of ρ; only one of them requires proof.

7.2.2 *Properties of ρ*

For all f, g, and h in $\mathbf{C}(D)$, the following hold:

(1) $\rho(f, g) \geq 0$ and $\rho(f, g) = 0$ if and only if $f = g$
(2) $\rho(f, g) = \rho(g, f)$
(3) $\rho(f, g) \leq \rho(f, h) + \rho(h, g)$
(4) $\rho(f + h, g + h) = \rho(f, g)$; that is, ρ is translation invariant
(5) $\rho(f, g) \leq 1$; that is, ρ is a bounded metric

Proof of property (3)

$$\rho(f, g) = d(f - g) = d(f - h + h - g) \leq d(f - h) + d(h - g).$$

□

Note that properties (1)–(3) say that ρ is a metric on $\mathbf{C}(D)$.

Theorem 7.11. *Convergence in the ρ-metric in $\mathbf{C}(D)$ is equivalent to uniform convergence on all compact subsets of D.*

Proof. Let $\{f_n\} \subset \mathbf{C}(D)$ and assume that $\{f_n\}$ is ρ-convergent. Since for every compact set $K \subset D$ there is an i in $\mathbb{Z}_{>0}$ such that $K \subset K_i$, it suffices to show uniform convergence on K_i for each i.

Given $0 < \epsilon < 1$, we can choose $N \in \mathbb{Z}_{>0}$ large so that

$$\rho(f_m, f_n) = d(f_m - f_n) \leq d(f_m - f) + d(f - f_n) < 2^{-i}\epsilon$$

for all $m, n \geq N$.

Now

$$2^{-i}\min(1, M_i(f_m - f_n)) \leq d(f_m - f_n) < 2^{-i}\epsilon,$$

and thus

$$M_i(f_m - f_n) < \epsilon < 1;$$

that is,

$$||f_m - f_n||_{K_i} < \epsilon.$$

The above inequality implies that the sequence $\{f_n\}$ converges uniformly on K_i. If $\epsilon \geq 1$, then use $\epsilon_0 = \frac{3}{4}$ and proceed as above.

We have actually shown more than claimed: if $\{f_n\}$ is a ρ-Cauchy sequence in $\mathbf{C}(D)$, then there exists an $f \in \mathbf{C}(D)$ such that $f_n \to f$ uniformly on all compact subsets of D.

Conversely, assume that $f_n \to f$ uniformly on K_i for all i. Thus

$$\lim_{n\to\infty} M_i(f - f_n) = 0 \text{ for all } i.$$

Given $\epsilon > 0$, first choose i such that $2^{-i} < \dfrac{\epsilon}{2}$ and next choose N such that $M_i(f - f_n) < \dfrac{\epsilon}{2}$ for all $n \geq N$. Then

$$d(f - f_n) \leq M_i(f - f_n) + 2^{-i} < \frac{\epsilon}{2} + \frac{\epsilon}{2} = \epsilon.$$

□

Corollary 7.12. *The topology of the metric space* $(\mathbf{C}(D), \rho)$ *is independent of the choice of exhaustion* $\{K_n\}_{n\in\mathbb{Z}_{>0}}$ *of* D.

Corollary 7.13. ρ *is a complete metric on* $\mathbf{C}(D)$.

Because of Theorem 7.11, we can reformulate the results of the previous section in terms of the metric ρ. In particular, Theorems 7.2 and 7.4 can now be phrased as in the following corollary. We already remarked that $\mathbf{H}(D) \subset \mathbf{C}(D)$. We let $\rho|_{\mathbf{H}(D)}$ denote the restriction of the metric ρ to $\mathbf{H}(D)$.

Corollary 7.14. $\mathbf{H}(D)$ *is a closed subspace of* $(\mathbf{C}(D), \rho)$. *As such,* $(\mathbf{H}(D), \rho|_{\mathbf{H}(D)})$ *is a complete metric space. Furthermore,* $f \mapsto f'$ *is a continuous linear operator from* $\mathbf{H}(D)$ *to itself.*

There is an alternate description of the topology induced by ρ, that is, the topology of the metric space $(\mathbf{C}(D), \rho)$. Namely,

Definition 7.15. Given $f \in \mathbf{C}(D)$, K compact $\subset D$ and $\epsilon > 0$, we define

$$N_f(\epsilon) = \{g \in \mathbf{C}(D);\ \rho(g, f) < \epsilon\}$$

and

$$V_f(K,\epsilon) = \{g \in \mathbf{C}(D);\ ||g - f||_K < \epsilon\}.$$

Remark 7.16. For any $f \in \mathbf{C}(D)$, a basis for the neighborhood system at f (with respect to the topology induced by ρ on $\mathbf{C}(D)$) is given by the sets $N_f(\epsilon)$, with $\epsilon > 0$. That is, the open sets $U \subseteq \mathbf{C}(D)$ that contain f are precisely those for which there exists $\epsilon > 0$ such that $N_f(\epsilon) \subseteq U$.

We will now show that the same is true for the collection

$$\{V_f(K,\epsilon);\ K \text{ compact } \subset D \text{ and } \epsilon > 0\}.$$

Theorem 7.17. *For any $f \in \mathbf{C}(D)$, a basis for the neighborhood system at f (with respect to the topology induced by ρ on $\mathbf{C}(D)$) is given by the sets $V_f(K,\epsilon)$.*

Proof. It is enough to show that

(1) Given $V_f(K,\epsilon)$, there exists an $N_f(\delta) \subseteq V_f(K,\epsilon)$,

and

(2) Given $N_f(\delta)$, there exists a $V_f(K,\epsilon) \subseteq N_f(\delta)$.

To show (1), we assume without loss of generality that $0 < \epsilon < 1$. Choose i such that $K \subset K_i$ and set $\delta = 2^{-i}\epsilon$. If $g \in N_f(\delta)$, then $d(g-f) < 2^{-i}\epsilon$. Thus

$$2^{-i}\min(1, M_i(g-f)) < 2^{-i}\epsilon$$

and then

$$M_i(g-f) = ||g-f||_{K_i} < \epsilon.$$

But

$$||g-f||_K \le ||g-f||_{K_i};$$

that is, $g \in V_f(K,\epsilon)$.

To show (2), choose i such that $2^{-i} < \frac{\delta}{2}$. For $g \in V_f\left(K_i, \frac{\delta}{2}\right)$, we have

$$M_i(g-f) < \frac{\delta}{2}.$$

Hence

$$\rho(g,f) = d(g-f) < M_i(g-f) + 2^{-i} < \delta;$$

that is, $g \in N_f(\delta)$. □

Remark 7.18. The theorem proves that $f_n \to f$ in the ρ-metric if and only if for all compact $K \subset D$ and all $\epsilon > 0$, there exists $N = N(K,\epsilon)$ in $\mathbb{Z}_{>0}$ such that $||f - f_n||_K < \epsilon$ for all $n > N$.

We can apply these concepts to convergence of series of meromorphic functions.

Definition 7.19. Let $\{f_n\}$ be a sequence in $\mathbf{M}(D)$, the meromorphic functions on D. We say that $\sum f_n$ *converges uniformly (absolutely)* on a subset A of D if there exists an integer N such that f_n is holomorphic on A for all $n > N$ and $\sum\limits_{N+1}^{\infty} f_n$ converges uniformly (absolutely) on A.

Theorem 7.20. *Let $\{f_n\} \subset \mathbf{M}(D)$. If $\sum f_n$ converges uniformly on all compact subsets of D, then the series $f = \sum f_n$ is a meromorphic function on D, and $\sum f_n'$ converges uniformly on all compact subsets to f'.*

Proof. The proof is trivial. □

7.3 The Cotangent Function

As an application of the ideas developed in the last two sections, we establish a series expansion formula for the cotangent function.

Theorem 7.21. *For all z in $\mathbb{C} - \mathbb{Z}$ the following equalities hold:*

$$\pi \cot \pi z = \frac{\pi \cos \pi z}{\sin \pi z} = \frac{1}{z} + \sum_{n \in \mathbb{Z},\, n \neq 0} \left[\frac{1}{z-n} + \frac{1}{n} \right]$$

$$= \frac{1}{z} + 2z \sum_{n=1}^{\infty} \left[\frac{1}{z^2 - n^2} \right]. \tag{7.3}$$

We first observe that the meromorphic function

$$F(z) = \frac{\cos \pi z}{\sin \pi z}$$

has its poles at the integers, and that each of these poles is simple with residue equal to 1. It would seem more natural to sum the series $\sum\limits_{n=-\infty}^{\infty} \frac{1}{z-n}$, but this one does not converge (Exercise 7.5).

We claim that $\sum\limits_{n \in \mathbb{Z},\, n \neq 0} \left[\frac{1}{z-n} + \frac{1}{n} \right] = \sum\limits_{n \neq 0} \frac{z}{n(z-n)}$ converges absolutely and uniformly on all compact subsets of $\mathbb{C}$. To verify this claim, assume that $|z| \leq R$ with $R > 0$. Then

$$\sum_{|n| \geq 2R} \frac{|z|}{|n|\,|n-z|} \leq \sum_{|n| \geq 2R} \frac{R}{|n|\,(|n| - R)}$$

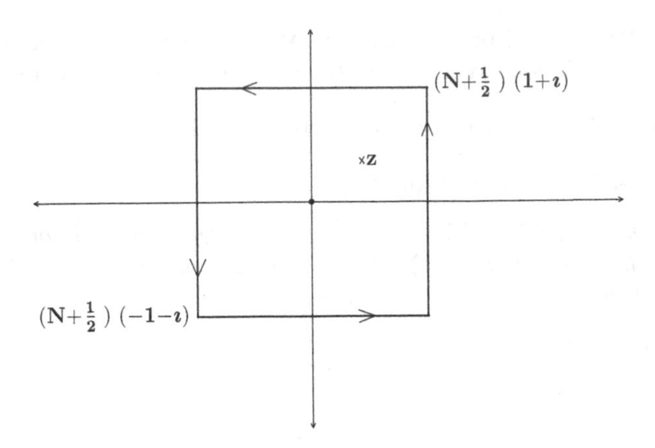

Fig. 7.1 The square C_N

$$\leq \sum_{|n|\geq 2R} \frac{R}{|n|\left|\frac{n}{2}\right|} \leq 2R\sum_{n\neq 0}\frac{1}{|n|^2} < +\infty.$$

We can now verify the expansion (7.3) for $\pi \cot \pi z$.

Proof of Theorem 7.21. For $N \in \mathbb{Z}_{>0}$, let C_N be the positively oriented boundary of the square with vertices $\left(N + \frac{1}{2}\right)(\pm 1 \pm \imath)$; see Fig. 7.1.

Then

$$\frac{1}{2\pi\imath}\int_{C_N} \frac{\cot \pi t}{t-z}\,\mathrm{d}t = \sum_{t\in i(C_N)} \operatorname{Res}\left(\frac{\cot \pi t}{t-z}, t\right).$$

Here $z \in \mathbb{C}$ is fixed: we take $z \in i(C_N)$ and $z \notin \mathbb{Z}$. The poles of the function $H(t) = \frac{\cot \pi t}{t-z}$ occur at $t = z$ and at $t = n \in \mathbb{Z}$, and they are all simple. Furthermore, we see that

$$\operatorname{Res}\left(\frac{\cot \pi t}{t-z}, z\right) = \cot \pi z$$

and

$$\operatorname{Res}\left(\frac{\cot \pi t}{t-z}, n\right) = \lim_{t\to n}(t-n)\frac{\cos \pi t}{\sin \pi t}\frac{1}{t-z} = \frac{1}{\pi(n-z)}.$$

Thus we have

$$\begin{aligned}\frac{1}{2\pi\imath}\int_{C_N}\frac{\cot \pi t}{t-z}\,\mathrm{d}t &= \cot \pi z + \frac{1}{\pi}\sum_{n=-N}^{N}\frac{1}{n-z}\\ &= \cot \pi z + \frac{1}{\pi}\sum_{\substack{n=-N\\ n\neq 0}}^{N}\left[\frac{1}{n-z}-\frac{1}{n}\right] - \frac{1}{\pi z},\end{aligned}$$

where the last equality holds because

$$\sum_{\substack{n=-N \\ n\neq 0}}^{N} \frac{1}{n} = 0.$$

Hence it suffices to prove

Lemma 7.22. *We have*

$$\lim_{N\to\infty} \int_{C_N} \frac{\cot \pi t}{t-z}\, dt = 0.$$

Remark 7.23. Once the lemma is verified, we will also have obtained

$$\pi \cot \pi z = \frac{1}{z} - \sum_{n=1}^{\infty} \left[\frac{1}{n-z} - \frac{1}{n+z}\right] = \frac{1}{z} - \sum_{n=1}^{\infty} \frac{2z}{n^2 - z^2},$$

where the last series converges uniformly and absolutely on all compact subsets of $\mathbb{C} - \mathbb{Z}$.

Proof of Lemma 7.22. We proceed in stages:

(1)
$$\frac{1}{2\pi\imath} \int_{C_N} \frac{\cot \pi t}{t}\, dt = 0.$$

As usual, for $G(t) = \dfrac{\cot \pi t}{t}$, we have

$$\begin{aligned}\frac{1}{2\pi\imath} \int_{C_N} \frac{\cot \pi t}{t}\, dt &= \sum_{t \in i(C_N)} \operatorname{Res}(G(t)dt, t) \\ &= \operatorname{Res}(G(t)dt, 0) + \sum_{\substack{n=-N \\ n\neq 0}}^{N} \frac{1}{\pi n}.\end{aligned}$$

The last sum is clearly zero, and the residue of $G(t)\,dt$ at zero is 0 because G is an even function.

(2) $$\int_{C_N} \frac{\cot \pi t}{t-z}\, dt = \int_{C_N} \cot \pi t \left[\frac{1}{t-z} - \frac{1}{t}\right] dt = \int_{C_N} \cot \pi t \left[\frac{z}{t(t-z)}\right] dt,$$

where the first equality holds by (1).

(3) There exists an $M > 0$ (independent of N) such that

$$|\cot \pi t| \leq M \quad \text{for all } t \in C_N.$$

For $t = u + \imath v$,

$$|\cos \pi t|^2 = \cos^2 \pi u + \sinh^2 \pi v,$$
$$|\sin \pi t|^2 = \sin^2 \pi u + \sinh^2 \pi v,$$

and thus

$$|\cot \pi t|^2 = \frac{\cos^2 \pi u + \sinh^2 \pi v}{\sin^2 \pi u + \sinh^2 \pi v}.$$

On the vertical sides of C_N we have $u = \pm N \pm \frac{1}{2}$ and hence

$$\cos^2 \pi u = \cos^2 \left(\pi \left(\pm N \pm \frac{1}{2}\right)\right) = 0,$$
$$\sin^2 \pi u = \sin^2 \left(\pi \left(\pm N \pm \frac{1}{2}\right)\right) = 1, \text{ and}$$
$$|\cot \pi t|^2 = \frac{\sinh^2 \pi v}{1 + \sinh^2 \pi v} \leq 1.$$

On the horizontal sides of C_N, $v = \pm N \pm \frac{1}{2}$ and hence

$$|\cot \pi t|^2 \leq \frac{1 + \sinh^2 \pi \left(\pm N \pm \frac{1}{2}\right)}{\sinh^2 \pi \left(\pm N \pm \frac{1}{2}\right)} \to 1 \text{ as } N \to \infty.$$

Thus there exists an $M > 0$ such that $|\cot \pi t| \leq M$ for t on the horizontal sides of C_N, and the claim is proved.

(4) If we denote by $L(C_N)$ the length of C_N, then

$$\begin{aligned}\left|\int_{C_N} \frac{\cot \pi t}{t - z}\, \mathrm{d}t\right| &= \left|\int_{C_N} \cot \pi t \left[\frac{z}{t(t-z)}\right] \mathrm{d}t\right| \\ &\leq \int_{C_N} \frac{M\,|z|}{|t|\,|t-z|}\, |\mathrm{d}t| \\ &\leq \frac{M\,|z|}{\left|N + \frac{1}{2}\right| \left(\left|N + \frac{1}{2}\right| - |z|\right)} L(C_N) \\ &= \frac{M\,|z|}{\left|N + \frac{1}{2}\right| \left(\left|N + \frac{1}{2}\right| - |z|\right)} 4\,(2N + 1) \to 0 \text{ as } N \to \infty.\end{aligned}$$

□

We have thus established the formula (7.3) for the cotangent function. Differentiating the series (7.3) term by term, we obtain the following expansion.

Corollary 7.24. *For all $z \in \mathbb{C} - \mathbb{Z}$,*

$$\frac{\pi^2}{\sin^2 \pi z} = \sum_{n=-\infty}^{\infty} \frac{1}{(z-n)^2}.$$

In particular, setting $z = \frac{1}{2}$,

$$\frac{\pi^2}{4} = \sum_{n=-\infty}^{\infty} \frac{1}{(2n-1)^2}.$$

7.4 Compact Sets in $\mathbf{H}(D)$

We return to the study of $\mathbf{C}(D)$ with the ρ-metric.

A metric space X is compact if and only if every sequence in X has a subsequence which converges to a point in X, and a subset X of $\mathbb{R}^n$ is compact if and only if it is closed and bounded, a result we generalize to $\mathbf{H}(D)$.

Definition 7.25. Let $A \subseteq \mathbf{C}(D)$. We say that A is *bounded in the strong sense* or *strongly bounded* if for all compact $K \subset D$ and all $\epsilon > 0$ there exists a $\lambda > 0$ such that

$$A \subseteq \lambda V_0(K, \epsilon) = \{g \in \mathbf{C}(D);\ g = \lambda f \text{ with } ||f||_K < \epsilon\}.$$

Remark 7.26. For a subset A in a metric space (X, ρ), one defines

$$\text{diam } A = \sup\{\rho(f, g);\ f \text{ and } g \in A\}.$$

Usually one says that A is bounded if diam $A < +\infty$. For a bounded metric (as in our case), this concept is not very useful. Hence we have introduced the concept of "strongly bounded sets."

Lemma 7.27. *A set $A \subseteq \mathbf{C}(D)$ is strongly bounded if and only if for each compact $K \subset D$, there exists an $M(K) > 0$ such that $||f||_K \leq M(K)$ for all $f \in A$; that is, A is strongly bounded if and only if the functions in A are uniformly bounded on each compact subset of D.*

Proof. We leave the proof as Exercise 7.6. □

Theorem 7.28. *A compact subset $A \subseteq \mathbf{C}(D)$ is closed and strongly bounded.*

Proof. Since A is a compact subset of the metric space $\mathbf{C}(D)$, it is closed. If $K \subset D$ is compact, then the function

$$f \mapsto ||f||_K$$

is continuous on the compact set A. Thus it is a bounded function, and hence A is strongly bounded. □

Lemma 7.29. *Let $c \in \mathbb{C}$ and $D = U(c, R)$ for some $R > 0$. Assume $A \subseteq \mathbf{H}(D)$ is strongly bounded, and let $\{f_k\}_{k\in\mathbb{Z}_{>0}} \subseteq A$. The sequence $\{f_k\}$ converges uniformly on all compact subsets of D if and only if $\lim_{k\to\infty} f_k^{(n)}(c)$ exists (in $\mathbb{C}$) for all integers $n \geq 0$.*

Proof. If $f_k \to f$ uniformly on all compact subsets of D, then for every nonnegative integer n, $f_k^{(n)} \to f^{(n)}$ uniformly on all compact subsets of D; in particular, $f_k^{(n)}(c) \to f^{(n)}(c)$, as a set consisting of one point is certainly compact.

Conversely, it suffices to show that $\{f_k\}$ converges uniformly on cl $U(c, r)$ with $0 < r < R$. Choose r_0 such that $r < r_0 < R$. Since A is strongly bounded, there exists an $M = M(r_0)$ such that

$$|f_k(z)| \leq M \text{ for } |z - c| \leq r_0 \text{ and all } k > 0.$$

Write

$$f_k(z) = \sum_{n\geq 0} a_{n,k}(z-c)^n \text{ for } |z - c| < R;$$

then Cauchy's inequalities (5.3) tell us that

$$|a_{n,k}| \leq \frac{M}{r_0^n} \text{ for all } n \text{ and } k.$$

Assume $|z - c| \leq r$. Then

$$\begin{aligned} |f_k(z) - f_m(z)| &\leq \left| \sum_{n=0}^{\infty} a_{n,k}(z-c)^n - \sum_{n=0}^{\infty} a_{n,m}(z-c)^n \right| \\ &\leq \left| \sum_{n=0}^{N} (a_{n,k} - a_{n,m})(z-c)^n \right| + 2M \sum_{n=N+1}^{\infty} \left(\frac{r}{r_0}\right)^n \end{aligned}$$

for all N, k, m.

Let $\epsilon > 0$ and choose $N_0 \in \mathbb{Z}_{>0}$ such that

$$2M \sum_{n=N+1}^{\infty} \left(\frac{r}{r_0}\right)^n < \frac{\epsilon}{2} \text{ for } N > N_0.$$

Finally, choose N_1 such that $k, m \geq N_1$ implies

$$\left| \sum_{n=0}^{N_0} (a_{n,k} - a_{n,m})(z-c)^n \right| < \frac{\epsilon}{2}.$$

This will be achieved by requiring, for example, that

$$|a_{n,k} - a_{n,m}| < \frac{\epsilon}{2N_0 r_0^n};$$

this last finite set of inequalities can be satisfied because

$$\lim_{k,m\to\infty} |a_{n,k} - a_{n,m}| = 0 \text{ for each } n,$$

since for each n

$$\left\{ a_{n,k} = \frac{f_k^{(n)}(c)}{n!} \right\}_k$$

is a Cauchy sequence of complex numbers, because we are assuming that $\lim_{k\to\infty} f_k^{(n)}(c)$ exists for each n. □

Theorem 7.30 (Compactness Theorem). *Let D be a domain in $\mathbb{C}$. Then every closed subset A of $\mathbf{H}(D)$ that is bounded in the strong sense is compact.*

Proof. Cover D by countably many open discs $\{U(z_i, r_i)\}_{i\in\mathbb{Z}_{>0}}$ whose closures are contained in D. For each $i \in \mathbb{Z}_{>0}$ and each $n \in \mathbb{Z}_{\geq 0}$, consider the mapping

$$\lambda_i^n : \mathbf{H}(D) \to \mathbb{C},\ \lambda_i^n(f) = f^{(n)}(z_i);$$

the maps $\{\lambda_i^n\}$ are $\mathbb{C}$-linear and continuous.

Given a sequence $\{f_k\}_{k\in\mathbb{Z}_{>0}}$ in A, we consider the set of numbers

$$\lambda_i^n(f_k) = f_k^{(n)}(z_i).$$

We show that there exists $B \subseteq \mathbb{Z}_{>0}$, $|B| = \infty$, such that

$$\lim_{\substack{k\in B\\ k\to+\infty}} f_k^{(n)}(z_i) \text{ exists for all } n \text{ and } i. \tag{7.4}$$

Assertion (7.4) suffices to prove the theorem; for then, by Lemma 7.29, the sequence $\{f_k\}_{k\in B}$ converges uniformly on the closed disc cl $U(z_i, r_i)$ for each i, which implies that the same sequence converges uniformly on all compact subsets of D. Since A is closed, $\lim_{\substack{k\in B\\ k\to+\infty}} f_k \in A$. Thus every sequence in A has a subsequence converging to a point of A, and A is hence compact. To establish (7.4), we use the "Cantor diagonalization" method.[1]

[1] This method is often used in analysis.

Since A is strongly bounded, for each i, there exists an $M(i) \in \mathbb{R}_{>0}$ such that $|f(z)| \leq M(i)$ for all z in cl $U(z_i, r_i)$ and all f in A. Thus

$$\left| f_k^{(n)}(z_i) \right| \leq \frac{M(i)}{r_i^n}\, n!.$$

Now (as i and n vary)

$$f \mapsto \lambda_i^n(f)$$

form a countable set of mappings. Renumber these mappings as

$$\{\mu_1, \mu_2, \ldots, \mu_m, \ldots\}.$$

For $m = 1$,

$$\{\mu_1(f_k)\}_{k \in \mathbb{Z}_{>0}}$$

is a bounded sequence of complex numbers, and therefore there exists a subsequence B_1 of $\mathbb{Z}_{>0}$ such that

$$\lim_{\substack{k \in B_1 \\ k \to +\infty}} \mu_1(f_k) \text{ exists.}$$

For $m = 2$,

$$\{\mu_2(f_k)\}_{k \in B_1} \subseteq \{\mu_2(f_k)\}_{k \in \mathbb{Z}_{>0}}$$

is a bounded sequence of complex numbers. Therefore there exists a subsequence B_2 of B_1 such that

$$\lim_{\substack{k \in B_2 \\ k \to +\infty}} \mu_2(f_k) \text{ exists.}$$

Continue to obtain a nested sequence of sets

$$B_1 \supseteq B_2 \supseteq \cdots \supseteq B_m \supseteq \cdots$$

with the property that

$$\lim_{\substack{k \in B_m \\ k \to +\infty}} \mu_m(f_k) \text{ exists.}$$

At last, *diagonalize* (justifying the name of the procedure); that is, let

$$B = \{n_1, n_2, \ldots, n_m, \ldots\},$$

where n_m is the m-th term of B_m. Then

$$\lim_{\substack{k \in B \\ k \to +\infty}} \mu_m(f_k) \text{ exists for all } m \in \mathbb{Z}_{>0}$$

because

$$\{n_m, n_{m+1}, \ldots\} \subseteq B_m.$$

□

Corollary 7.31. *A set $A \subseteq \mathbf{H}(D)$ is compact if and only if it is closed and bounded in the strong sense.*

Definition 7.32. A set $A \subseteq \mathbf{H}(D)$ is *relatively compact* if cl A is compact. This definition clearly makes sense in much more general settings.

Corollary 7.33 (Montel's Theorem). *Every strongly bounded subset of $\mathbf{H}(D)$ is relatively compact.*

Note that the converse to Montel's theorem also holds.

Definition 7.34. Let A be a strongly bounded set in $\mathbf{H}(D)$ and let $\{f_k\}_{k\in\mathbb{Z}_{>0}}$ be a sequence in A. We say that $f \in \mathbf{H}(D)$ is *adherent* to $\{f_k\}_{k\in\mathbb{Z}_{>0}}$ if it is a limit point of this sequence; that is, if for every $\varepsilon > 0$ there exists $k \in \mathbb{Z}_{>0}$ such that $0 < \rho(f, f_k) < \varepsilon$.

Remark 7.35. If $\{x_k\}_{k\in\mathbb{Z}_{>0}}$ is a sequence in a compact metric space X, then there exists a convergent subsequence of $\{x_k\}_{k\in\mathbb{Z}_{>0}}$; furthermore, if every subsequence of $\{x_k\}_{k\in\mathbb{Z}_{>0}}$ that converges has the same limit, then the sequence $\{x_k\}_{k\in\mathbb{Z}_{>0}}$ converges.

Theorem 7.36 (Vitali's Theorem). *Let D be a domain in $\mathbb{C}$, and assume that the elements in a sequence $\{f_k\}_{k\in\mathbb{Z}_{>0}} \subset \mathbf{H}(D)$ are uniformly bounded on compact subsets of D. Let $S \subseteq D$ and assume that S has a limit point in D. If $\lim_{k\to\infty} f_k(z)$ exists (pointwise) for all $z \in S$, then the sequence $\{f_k\}$ converges uniformly on compact subsets of D.*

Proof. The assumptions imply that the set $A = \{f_k\}_{k\in\mathbb{Z}_{>0}} \subset \mathbf{H}(D)$ is strongly bounded. It follows from Montel's theorem that its closure is compact, and therefore every subsequence of the sequence $\{f_k\}$ has a converging sub-subsequence. Thus there exists an f in $\mathbf{H}(D)$ adherent to A. Say that f and g are both adherent to $\{f_k\}$; then

$$f(z) = \lim_{k\to\infty} f_k(z) = g(z) \text{ for } z \in S$$

and thus $f = g$ on D. By our previous remark, $\{f_k\}$ converges to f in the ρ-metric.

□

7.5 Runge's Theorem

We consider the problem of approximating holomorphic functions by rational functions. We regard a nonconstant polynomial as a rational function whose only pole is at infinity. The ability to uniformly approximate a holomorphic function depends on the region where the function is being approximated, as well as upon the function itself. The strongest statement about uniform approximation of holomorphic functions that we prove is Runge's approximation theorem. A number of proofs appear in the literature; ours is a variant of these.[2]

We have already proved a form of Runge's theorem for an open disc Ω: a holomorphic function on Ω has a power series expansion at the center of the disc; for every positive integer n, we obtain a polynomial of degree n by discarding all the higher order terms in the series. These polynomials converge to the function uniformly on any compact subset of the disc.

On the other hand, we also know that uniform polynomial approximation does not hold in general. For instance, consider a punctured disc $\Omega = \{z \in \mathbb{C};\ 0 < |z-c| < R\}$, with $R > 0$ and c arbitrary, and the analytic function on Ω defined by $f(z) = \dfrac{1}{z-c}$ (we take advantage of the fact that it is a rational function whose only pole is at c — not in Ω, of course); if f were uniformly approximated by a sequence of polynomials $\{p_n\}$ in the closed annulus $K = \{0 < r \le |z-c| \le \rho < R\}$, then by taking $\gamma(t) = \dfrac{r+\rho}{2}\exp(2\pi\imath t)$ for $0 \le t \le 2\pi$ we would obtain the contradiction that

$$0 = \lim_{n\to\infty} \int_\gamma p_n(z)\,\mathrm{d}z = \int_\gamma f(z)\,\mathrm{d}z = 2\pi\imath.$$

However, truncation of the Laurent series expansion for f on Ω shows that f is indeed uniformly approximated on K by rational functions whose poles lie outside Ω. This fact is generalized to arbitrary open sets Ω by the next.

Theorem 7.37 (Runge). *Let K be a compact subset of $\mathbb{C}$ and let S be a subset of $\widehat{\mathbb{C}} - K$ that intersects nontrivially each connected component of $\widehat{\mathbb{C}} - K$. If f is a holomorphic function on an open set Ω containing K, then it can be uniformly approximated on K by rational functions with simple poles lying on S; that is, for every $\epsilon > 0$ there exists a rational function R with possibly simple poles only in S such that*

$$|f(z) - R(z)| < \epsilon \text{ for all } z \in K.$$

Runge's theorem is the implication (1) $\Longrightarrow$ (8) of the fundamental theorem. The converse follows from Theorem 7.2.

We can always choose for S a smallest set consisting of one point from each connected component of $\widehat{\mathbb{C}} - K$. For the important special case where $\widehat{\mathbb{C}} - K$ is connected and S is chosen as $S = \{\infty\}$, Runge's theorem asserts that each function that is analytic in an open neighborhood of K can be uniformly approximated in K by a sequence of polynomials.

---

[2] We follow a course outlined by S. Grabiner, *A short proof of Runge's theorem*, Am. Math. Monthly **83** (1976), 807–808, and rely on arguments appearing in Conway's book listed in our bibliography.

An outline for the proof of Runge's theorem is given in Sect. 7.5.2, after some needed preliminaries from real analysis and topology given in the next subsection. The proof of the theorem depends on three major lemmas that are stated and proved in the subsequent subsections; the first two are given in Sect. 7.5.3 and the third in Sect. 7.5.4.

7.5.1 Preliminaries for the Proof of Runge's Theorem

We rely on some new terminology and notation. We also call the reader's attention to some elementary topological concepts that we will need. If F and K are subsets of $\mathbb{C}$ with F closed and K compact, then the *distance* between these two sets is the nonnegative real number

$$d(F, K) = \inf\{|z - w|\,; z \in F \text{ and } w \in K\},$$

that is easily seen to satisfy

$$d(F, K) = 0 \text{ if and only if } F \cap K \neq \emptyset.$$

In particular, if F consists of only one complex number c, we set

$$d(c, K) = d(\{c\}, K) = \inf\{|c - w|\,; w \in K\}.$$

If A and B are connected subsets of $\mathbb{C}$ that are not disjoint, then $A \cup B$ is connected. If C is a connected component of the set A, then C is an open subset of A.

Let K and S be the sets described in the hypothesis of Runge's theorem, and define $B(K, S)$ to be the set of continuous complex-valued functions on K that are uniform limits of sequences of rational functions with poles only in S. Sums and products of elements of $B(K, S)$ are obviously also elements of $B(K, S)$, as are products of constants by elements of $B(K, S)$. We summarize the (additional) properties of this algebra in the following lemma. The proof is left as an exercise.

Lemma 7.38. *The algebra $B(K, S)$ contains the rational functions with poles in S, and is closed under uniform limits in K.*

Runge's theorem asserts that every holomorphic function on a neighborhood of K belongs to $B(K, S)$. To establish this, we will also need the following topological result.

Lemma 7.39. *Let U and V be open subsets of $\mathbb{C}$ with $V \subseteq U$ and $\partial V \cap U = \emptyset$. If H is a connected component of U and $H \cap V \neq \emptyset$, then $H \subseteq V$.*

Proof. Let c be a point in $H \cap V$ and let G be the connected component of V containing c. It is enough to show that $G = H$. Now $H \cup G$ is connected, contained in U and contains c. Since H is the component of U, containing c, $G \subseteq H$. Furthermore, $\partial G \subseteq \partial V$ and so $\partial G \cap H = \emptyset$. This implies that $H - G = H - \mathrm{cl}\, G$ and, therefore, that $H - G$ is open in H. Since G is also open in H, the conclusion that $G = H$ follows. □

7.5.2 *Proof of Runge's Theorem*

We outline the steps in the proof of Runge's theorem. The details needed to fill in the outline will be completed in the next two subsections. Let Ω, K, S, and f be as in the statement of Runge's theorem. The proof of the theorem (that $f \in B(K, S)$) consists of four steps. We establish:

(1) (Lemma 7.40 of the next section) There exists a finite collection of oriented line segments $\gamma_1, \gamma_2, \ldots, \gamma_n$ in $\Omega - K$ such that

$$f(z) = \frac{1}{2\pi\imath} \sum_{j=1}^{n} \int_{\gamma_j} \frac{f(w)}{w - z}\, dw \quad \text{for all } z \in K.$$

(2) It suffices to prove that each of the above integrals $\int_{\gamma_j} \frac{f(w)}{w - z}\, dw$ can be approximated uniformly on K by finite sums of rational functions $z \mapsto P\left(\frac{1}{z-c}\right)$, where P is a polynomial and c is a point in S. We hence drop the subscript j from the notation. It will be convenient to regard $P\left(\frac{1}{z-\infty}\right)$ as a polynomial in z. The observation that every uniformly convergent series in h on K (in particular $P(h)$) belongs to $B(K, S)$ whenever $h \in B(K, S)$ will be used repeatedly in our arguments that follow.

(3) (Lemma 7.41) The line integral $\int_{\gamma} \frac{f(w)}{w - z}\, dw$ can be uniformly approximated on K by Riemann sums of the form

$$\sum_k \frac{a_k}{b_k - z}, \quad \text{with } a_k \in \mathbb{C} \text{ and } b_k \in \text{range } \gamma.$$

(4) (Lemma 7.42) Each summand $\frac{a_k}{b_k - z}$ can be approximated uniformly on K by appropriate finite sums, and thus belongs to $B(K, S)$.

7.5.3 Two Major Lemmas

We establish two results: the first gives an extension of Cauchy's integral formula, and the second provides an approximation by a rational function to a function defined by an integral.

Lemma 7.40. *Let K be a compact subset of $\mathbb{C}$ and let Ω be an open set containing K. Then there exists a finite collection of oriented line segments $\gamma_1, \gamma_2, \ldots, \gamma_n$ in $\Omega - K$ such that for every holomorphic function f on Ω,*

$$f(z) = \frac{1}{2\pi \imath} \sum_{i=1}^{n} \int_{\gamma_i} \frac{f(w)}{w - z} \mathrm{d}w \tag{7.5}$$

for all $z \in K$.

Proof. After enlarging K if necessary, we may assume that

$$K = \mathrm{cl}(\mathrm{int}(K)).$$

For example, we enlarge K if it consists of a single point.

To simplify statements we adopt the standard convention that a curve $\gamma : [a, b] \to \widehat{\mathbb{C}}$ and its range $\{\gamma(t);\ t \in [a, b]\}$ are both called γ. For any positive real number δ we consider a rectangular grid of horizontal and vertical lines in the plane $\mathbb{C}$ so that consecutive parallel lines are at distance δ apart. We let $R_1, R_2, \ldots, R_m$ be the rectangles in the grid that have nonempty intersection with K. Since K is compact, there are only a finite number of such rectangles. We can (and from now on do) choose δ such that $R_j \subset \Omega$ for all j; if $\Omega = \mathbb{C}$ any $\delta > 0$ suffices, and otherwise it is enough to consider any $0 < \delta < \frac{1}{2} d(K, \mathbb{C} - \Omega)$, since $z \in R_j$ implies that $d(z, K) < \sqrt{2}\delta$. As usual, the boundary of R_j is denoted by ∂R_j and is oriented in the counterclockwise direction. The integrals of a continuous form along the common boundaries of any pair of contiguous R_i and R_j cancel out (as in the proof of Goursat's Theorem 4.61). This last observation implies that we can choose a set $\mathcal{S}$ of curves whose ranges are a subset of the sides in $\bigcup_{j=1}^{m} \partial R_j$, and such that the set $\mathcal{S} = \{\gamma_i; 1 \le i \le n\}$ satisfies

(1) If γ_i is in $\mathcal{S}$, it lies on a side of only one R_j.
(2) If γ_i is in $\mathcal{S}$, then it is disjoint from K.
(3) For any continuous function g on $\bigcup_{j=1}^{m} \partial R_j$, we have

$$\sum_{j=1}^{m} \int_{\partial R_j} g(z) \mathrm{d}z = \sum_{i=1}^{n} \int_{\gamma_i} g(z) \mathrm{d}z. \tag{7.6}$$

Each γ_i is an oriented line segment in $\Omega - K$. It remains to prove that equation (7.5) holds with these γ_i. If $z \in K$ and z is not on the boundary of any of the rectangles, then the function

$$w \mapsto g(w) = \frac{1}{2\pi\iota} \frac{f(w)}{w-z}, \quad w \in \bigcup_{j=1}^{m} \partial R_j,$$

is continuous. Thus, we have, by (7.6),

$$\frac{1}{2\pi\iota} \sum_{j=1}^{m} \int_{\partial R_j} \frac{f(w)}{w-z} \mathrm{d}w = \frac{1}{2\pi\iota} \sum_{i=1}^{n} \int_{\gamma_i} \frac{f(w)}{w-z} \mathrm{d}w.$$

Assume that z belongs to the interior of exactly one of the R_j, call it R_t. If $j \neq t$, then $z \notin R_j$ and

$$\frac{1}{2\pi\iota} \int_{\partial R_j} \frac{f(w)}{w-z} \mathrm{d}w = 0;$$

also, since $z \in R_t$, by Cauchy's integral formula, we have

$$\frac{1}{2\pi\iota} \int_{\partial R_t} \frac{f(w)}{w-z} \mathrm{d}w = f(z).$$

Thus (7.5) holds for all $z \in R_t$. Since range γ_i does not intersect K, both sides of this equation are continuous functions of z on K, and they agree on the set of points z in K that are not on the boundary of any rectangle R_j, a dense subset of K. Thus they agree for all $z \in K$. □

Lemma 7.41. *Let γ be a pdp and let K be a compact set disjoint from the range of γ. If f is a continuous function on γ and ϵ is any positive real number, then there exists a rational function R, with only simple poles, all lying on* range γ, *such that*

$$\left| \int_{\gamma} \frac{f(w)}{w-z} \mathrm{d}w - R(z) \right| < \epsilon \text{ for all } z \in K.$$

Proof. We assume that γ is not a constant, and thus that its length is positive. Since K and the image of γ are disjoint, $d(K, \text{range } \gamma) > 0$, and we can choose a number r with

$$0 < r < d(K, \text{range } \gamma).$$

If γ is parameterized by $[0, 1]$, then for all $0 \leq s, t \leq 1$ and all $z \in K$ we have

$$\begin{aligned}
&\left| \frac{f(\gamma(t))}{\gamma(t) - z} - \frac{f(\gamma(s))}{\gamma(s) - z} \right| \\
&\quad \leq \frac{1}{r^2} \left| f(\gamma(t))\gamma(s) - f(\gamma(s))\gamma(t) - z\left(f(\gamma(t)) - f(\gamma(s))\right) \right| \\
&\quad \leq \frac{1}{r^2} \Big[|f(\gamma(t))|\,|\gamma(s) - \gamma(t)| + |\gamma(t)|\,|f(\gamma(s)) - f(\gamma(t))| \\
&\qquad + |z|\,|f(\gamma(s)) - f(\gamma(t))| \Big].
\end{aligned}$$

Since γ and f are continuous functions and K is a compact set, there is a constant $C > 0$ such that $|z| < \frac{C}{2}$ for all $z \in K$, $|\gamma(t)| \leq \frac{C}{2}$ and $|f(\gamma(t))| \leq C$ for all $t \in [0,1]$. Thus for all s and t in $[0,1]$ and all $z \in K$ we have

$$\left|\frac{f(\gamma(t))}{\gamma(t)-z} - \frac{f(\gamma(s))}{\gamma(s)-z}\right| \leq \frac{C}{r^2}\left[\,|\gamma(s)-\gamma(t)| + |f(\gamma(s)) - f(\gamma(t))|\,\right].$$

Since both γ and $f \circ \gamma$ are uniformly continuous on $[0,1]$, there is a partition of $[0,1]$ with $0 = t_0 < t_1 < \cdots < t_n = 1$ such that

$$\left|\frac{f(\gamma(t))}{\gamma(t)-z} - \frac{f(\gamma(t_j))}{\gamma(t_j)-z}\right| < \frac{\epsilon}{L(\gamma)} \tag{7.7}$$

for $t_{j-1} \leq t \leq t_j$, $1 \leq j \leq n$, and all $z \in K$, where $L(\gamma)$ denotes the length of γ (recall Definition 4.62).

Define the function R as follows. For $z \neq \gamma(t_j)$, $j = 1, \ldots, n$,

$$R(z) = \sum_{j=1}^{n} f(\gamma(t_j))\frac{\gamma(t_j) - \gamma(t_{j-1})}{\gamma(t_j) - z}.$$

Then R is a rational function whose poles are simple and contained in the set

$$\{\gamma(t_1), \gamma(t_2), \ldots, \gamma(1)\};$$

in particular, they are contained in range γ. Now inequality (7.7) gives

$$\begin{aligned}\left|\int_\gamma \frac{f(w)}{w-z}\,\mathrm{d}w - R(z)\right| &= \left|\sum_{j=1}^{n}\int_{t_{j-1}}^{t_j}\left(\frac{f(\gamma(t))}{\gamma(t)-z} - \frac{f(\gamma(t_j))}{\gamma(t_j)-z}\right)\gamma'(t)\right|\,\mathrm{d}t \\ &\leq \frac{\epsilon}{L(\gamma)}\sum_{j=1}^{n}\int_{t_{j-1}}^{t_j}|\gamma'(t)|\,\mathrm{d}t = \epsilon\end{aligned}$$

for all $z \in K$. □

7.5.4 *Approximating* $\dfrac{1}{z-c}$

Of central importance in the proof of Runge's theorem is

Lemma 7.42. *For every $c \in \mathbb{C} - K$, the rational function*

$$g : z \mapsto (z-c)^{-1} \in B(K, S).$$

Proof. The proof consists of several claims that will need verification.

- Let us choose $R \in \mathbb{R}_{>0}$ so that $K \subset U(0, R)$. Let $z_0 \in \mathbb{C}$ be arbitrary with $|z_0| > R$ and let h be a rational function with a pole only at z_0 (thus if $z_0 \in \mathbb{C}$, h is finite at infinity). Then h can be approximated uniformly on K by polynomials; in particular, $h \in B(K, \{\infty\})$. This assertion is obvious if $z_0 = \infty$. In general,[3] the Taylor series for h at 0 is of the form $\sum_{n=0}^{\infty} a_n z^n$. This series converges uniformly on every compact sub-disc of $\{|z| < |z_0|\}$ centered at 0, hence certainly on K. Since its individual terms $a_n z^n$ are in $B(K, \{\infty\})$ and this space is closed under uniform convergence on K, $h \in B(K, \{\infty\})$.
- Assume that $\infty \in S$. We claim that

$$B(K, (S - \{\infty\}) \cup \{z_0\}) \subseteq B(K, S).$$

Indeed, if $f \in B(K, (S - \{\infty\}) \cup \{z_0\})$ and $\{R_j\}$ is a sequence of rational functions with poles in $(S - \{\infty\}) \cup \{z_0\}$ uniformly approximating f on K, we can write

$$f = \lim_{j \to \infty} R_j = \lim_{j \to \infty} (R_{1,j} + R_{2,j}),$$

where $R_{1,j}$ has all its poles (if any) in $S - \{\infty\}$ and $R_{2,j}$ has a unique pole (if any) at z_0. Here and below limits are understood in the sense of "uniformly on K", and estimates on absolute values of functions are on the set K.

We have shown in the previous assertion that $R_{2,j} \in B(K, \{\infty\})$ and that there exist polynomials $P_{i,j}$ (they belong to $B(K, \{\infty\})$) such that

$$\lim_{i \to \infty} P_{i,j} = R_{2,j}.$$

For each j, choose i sufficiently large so that

$$|R_{2,j} - P_{i,j}| < \frac{1}{j}.$$

Now the rational function $R_{1,j} + P_{i,j}$ has poles only in S. We claim that

$$f = \lim_{j \to \infty} (R_{1,j} + P_{i,j}).$$

To verify this claim, let $\epsilon > 0$. Choose j sufficiently large so that

$$j > \frac{2}{\epsilon} \text{ and } \left|f - (R_{1,j} + R_{2,j})\right| < \frac{\epsilon}{2}.$$

[3]The argument that follows also applies for $z_0 = \infty$.

Then

$$\begin{aligned}\left|f-(R_{1,j}+P_{i,j})\right| &= \left|f-(R_{1,j}+R_{2,j})+(R_{2,j}-P_{i,j})\right| \\ &\leq \left|f-(R_{1,j}+R_{2,j})\right|+\left|R_{2,j}-P_{i,j}\right| \\ &< \frac{\epsilon}{2}+\frac{1}{j}<\epsilon.\end{aligned}$$

Hence $f \in B(K, S)$.

- Thus it is sufficient to prove the lemma for $S \subset \mathbb{C}$; for this we will rely on Lemma 7.39.
- Let $U = \mathbb{C} - K$ and let

$$V = \{w \in \mathbb{C} - K;\ z \mapsto (z-w)^{-1} \in B(K, S)\}.$$

Then $S \subseteq V \subseteq U$. We want to show that $U = V$. We show first that

$$\text{if } a \in V \text{ and } |b-a| < d(a, K), \text{ then } b \in V. \tag{7.8}$$

Assume $a \in V$ and $|b-a| < d(a, K)$. Then there is a real number r, $0 < r < 1$, such that $|b-a| < r\,|z-a|$ for all $z \in K$. Note that

$$(z-b)^{-1} = (z-a)^{-1}\left(1-\frac{b-a}{z-a}\right)^{-1}; \tag{7.9}$$

since $\dfrac{|b-a|}{|z-a|} < r < 1$ for all $z \in K$, we use the Weierstrass M-test to conclude that the series (in the variable z)

$$\left(1-\frac{b-a}{z-a}\right)^{-1} = \sum_{n=0}^{\infty}\left(\frac{b-a}{z-a}\right)^{n} \tag{7.10}$$

converges uniformly on K. Lemma 7.38 and equation (7.9) imply that (7.8) holds; thus V is an open subset of $\mathbb{C}$.

We show next that $\partial V \cap U = \emptyset$. Indeed, if $b \in \partial V$, let $\{a_n\}$ be a sequence of elements of V converging to b. Since $b \notin V$, it follows that $|b-a_n| \geq d(a_n, K)$, and letting $n \to \infty$ we obtain $0 = d(b, K)$; that is, $b \in K$, and therefore $b \notin U$.

We now apply Lemma 7.39. If H is any connected component of $U = \mathbb{C} - K$, then by the definition of S there exists $s \in H \cap S$. But then $s \in H \cap V \neq \emptyset$ and the lemma implies $H \subseteq V$. Therefore every connected component of U lies in V; consequently $U \subseteq V$, and thus $U = V$. □

As pointed out earlier, we have completed the proof of Runge's theorem. An important special case is the following

Corollary 7.43. *If D is any simply connected domain in the plane and f is a holomorphic function in D, then f can be approximated uniformly on compact subsets of D by polynomials.*

Exercises

7.1. (a) Show that Theorem 7.2 has no analogue for real variables in that the absolute value function on $\mathbb{R}$, which has no derivative at 0, can be uniformly approximated by differentiable functions.
(b) Use (a) to construct a sequence of continuous functions on a domain $D \subseteq \mathbb{C}$, converging at every point of D and such that the limit function is not continuous.

7.2. Construct an example of a sequence of real differentiable functions converging uniformly to a real differentiable function on a closed interval such that the sequence given by the derivatives does not converge uniformly there. (Hint: The sequence $f_n(x) = x^n$ does not converge uniformly on $[0, 1]$.)

7.3. Show that both possibilities in Theorem 7.5 do occur.

7.4. Show that the series $\displaystyle\sum_{n=1}^{\infty} \frac{1}{n^2} = \frac{\pi^2}{6}$.

7.5. Prove that $\displaystyle\sum_{n=1}^{\infty} \frac{1}{z-n}$ does not converge.

7.6. Prove Lemma 7.27.

7.7. This exercise requires some familiarity with standard topics in functional analysis.

(1) Prove Lemma 7.38; that is, that $B(K, S)$ is an algebra closed under uniform limits on K.
(2) Introduce a norm on $B(K, S)$ so that it becomes a Banach algebra.
(3) There are several other (more function theoretic) proofs of Runge's theorem that rely on tools not presented in this book (for example, the Hahn–Banach theorem and/or the Riesz representation theorem). After consulting the literature, outline an alternate proof.

7.8. Show that the family

$$\mathcal{S} = \{f \in \mathbf{H}(\mathbb{D});\ f \text{ is injective},\ f(0) = 0,\ f'(0) = 1\}$$

is closed in $\mathbf{H}(\mathbb{D})$.

The members of $\mathcal{S}$ are usually called "schlicht functions."It is true that $\mathcal{S}$ is compact, but proving it is strongly bounded requires Koebe's distortion theorem, beyond the scope of this book.

Chapter 8
Conformal Equivalence and Hyperbolic Geometry

In this chapter, we study *conformal maps* between domains in the extended complex plane $\widehat{\mathbb{C}}$; these are one-to-one meromorphic functions. Our goal here is to characterize all simply connected domains in the extended complex plane. The first two sections of this chapter study the action of a quotient of the group of two-by-two nonsingular complex matrices on the extended complex plane, namely, the group $\mathrm{PSL}(2,\mathbb{C})$ and the projective special linear group. This group is also known as the Möbius group. In the third section we characterize simply connected proper domains in the complex plane by establishing the Riemann mapping theorem (RMT). This extraordinary theorem tells us that there are conformal maps between any two such domains.

The study of the Möbius group is intimately connected with hyperbolic geometry, a subject that has increasingly become an essential part of complex variable theory. In the next to last section of this chapter we study this geometry: we define the non-Euclidean metric, also known as the hyperbolic or Poincaré metric, study the disc and half-plane models for the hyperbolic plane, find their geodesics, and show that their sense-preserving isometries are subgroups of the Möbius group. We end this section by using Schwarz's lemma to establish the deep connection between complex variables and geometry given in Theorem 8.41, which states that a holomorphic self-map of a proper simply connected domain in the plane is either an isometry or a contraction in the hyperbolic metric.

As a further application based on Möbius transformations, the last section is devoted to a study of certain bounded analytic functions on the unit disc known as *finite Blaschke products*.

Definition 8.1. An injective meromorphic function is called a *conformal map*.[1] A map f is *anti-conformal* if its conjugate $\bar{f}$ is conformal.

---

[1]In geometry, $\mathbf{C}^1$-maps are called *conformal* if they preserve angles. We have seen in Proposition 6.27 that in the orientation-preserving case these are precisely the holomorphic functions with nowhere vanishing derivatives. Thus the two definitions agree locally for sense-preserving transformations. In our definition we also require global injectivity.

R.E. Rodríguez et al., *Complex Analysis: In the Spirit of Lipman Bers*, Graduate Texts in Mathematics 245, DOI 10.1007/978-1-4419-7323-8_8,

Our definition of conformality is the correct notion of isomorphism in the category of meromorphic mappings, since the inverse of a conformal map is also conformal. Thus the concept introduces a natural equivalence relation on the family of domains on the sphere, called *conformal equivalence*.

Definition 8.2. Let D be a domain in $\widehat{\mathbb{C}}$. Aut D is defined as the group (under composition) of *conformal automorphisms* (or conformal bijections) of D; that is, it consists of the conformal maps from D onto itself.

There are two naturally related problems:

Problem I. Describe Aut D for a given D.

Problem II. Given two domains D and D', determine when they are conformally equivalent.

We solve Problem Ifor $D = \widehat{\mathbb{C}}$, $D = \mathbb{C}$, and $D = \mathbb{D}$ (the unit disc $\{z \in \mathbb{C};\ |z| < 1\}$ in $\widehat{\mathbb{C}}$), and Problem IIfor D and D' any pair of simply connected domains in $\widehat{\mathbb{C}}$.

8.1 Fractional Linear (Möbius) Transformations

We describe the (orientation preserving) Möbius group, and show that for the domains $D = \widehat{\mathbb{C}}$, $\mathbb{C}$, a disc or a half plane, the group Aut D is a subgroup of this group.

Definition 8.3. A *fractional linear transformation* (or *Möbius transformation*) is a meromorphic function $A : \widehat{\mathbb{C}} \to \widehat{\mathbb{C}}$ of the form

$$z \mapsto A(z) = \frac{az+b}{cz+d}, \tag{8.1}$$

where a, b, c, and d are complex numbers such that $ad - bc \neq 0$. Specifically,

$$A(z) = \begin{cases} \dfrac{az+b}{cz+d} & \text{if } c \neq 0, z \neq \infty \text{ and } z \neq -\dfrac{d}{c}, \\ \dfrac{a}{c} & \text{if } c \neq 0 \text{ and } z = \infty, \\ \infty & \text{if } c \neq 0 \text{ and } z = -\dfrac{d}{c}, \\ \dfrac{a}{d}z + \dfrac{b}{d} & \text{if } c = 0 \text{ and } z \neq \infty, \\ \infty & \text{if } c = 0 \text{ and } z = \infty. \end{cases} \tag{8.2}$$

From now on, the abbreviated notation (8.1) will be interpreted as the expanded version (8.2). Without loss of generality we assume subsequently that $ad - bc = 1$

(*the reader should prove* that there is really no loss of generality in this assumption; that is, establish Exercise 8.1). Also, whenever convenient we will multiply each of the four constants a, b, c, and d by -1, since this does not alter the Möbius transformation's action on $\widehat{\mathbb{C}}$ nor the condition $ad - bc = 1$.

Remark 8.4. A Möbius transformation is an element of the group $\text{Aut}(\widehat{\mathbb{C}})$, and the set of all Möbius transformations is a group under composition, the *Möbius group*. We will soon see that these two groups coincide.

Remark 8.5. Other related groups are the matrix group

$$\text{SL}(2,\mathbb{C}) = \left\{ \begin{bmatrix} a & b \\ c & d \end{bmatrix};\ a,b,c,d \in \mathbb{C}, ad - bc = 1 \right\},$$

the corresponding quotient group

$$\text{PSL}(2,\mathbb{C}) = \text{SL}(2,\mathbb{C})/\{\pm I\},$$

where $I = \begin{bmatrix} 1 & 0 \\ 0 & 1 \end{bmatrix}$ is the identity matrix, and the *extended Möbius group* of orientation preserving and reversing transformations, consisting of the maps

$$z \mapsto \frac{az+b}{cz+d} \text{ and } z \mapsto \frac{a\bar{z}+b}{c\bar{z}+d}, \text{ with } ad - bc = 1.$$

Here orientation reversing means that angles are preserved in magnitude but reversed in sense (as the map $z \to \bar{z}$ does). It is clear that

$$1 \to \{\pm I\} \to \text{SL}(2,\mathbb{C}) \to \text{Aut}(\widehat{\mathbb{C}}) \tag{8.3}$$

is an exact sequence, where the first two arrows denote inclusion, and by the last arrow, a matrix $\begin{bmatrix} a & b \\ c & d \end{bmatrix}$ in $\text{SL}(2,\mathbb{C})$ is sent to the element of $\text{Aut}(\widehat{\mathbb{C}})$ given by (8.1). An exact sequence is, of course, one where for any pair of consecutive maps in the sequence, the kernel of the second map coincides with the image of the first one.

It is also clear that the image of the last arrow in the sequence (8.3) is precisely the Möbius group, and, therefore, that it is isomorphic to $\text{PSL}(2,\mathbb{C})$, the quotient of $\text{SL}(2,\mathbb{C})$ by $\pm I$ as defined above. It is natural to ask whether the last arrow is surjective; that is, whether the Möbius group coincides with $\text{Aut}(\widehat{\mathbb{C}})$. We will see that this is the case in Theorem 8.17.

Let A be an element of $\text{PSL}(2,\mathbb{C})$. The square of the trace of a preimage of A in $\text{SL}(2,\mathbb{C})$ is the same for both of the two preimages of A. Thus even though the trace of an element in the Möbius group is not well defined, *the trace squared* of an element in $\text{PSL}(2,\mathbb{C})$ is. Thus it makes sense to have the following:

Definition 8.6. For A in the Möbius group, given by (8.1) with $ad - bc = 1$, we define $\text{tr}^2 A = (a + d)^2$.

8.1.1 *Fixed Points of Möbius Transformations*

Let A be any element of the Möbius group different from the identity map. We are interested in the *fixed points* of A in $\widehat{\mathbb{C}}$; that is, those $z \in \widehat{\mathbb{C}}$ with $A(z) = z$.

If $A(z) = \dfrac{az+b}{cz+d}$ with $ad - bc = 1$, then for a fixed point z of A we have either $z = \infty$, or $z \in \mathbb{C}$ and $cz^2 + (d - a)z - b = 0$. We consider two cases:

Case 1: $c = 0$. In this case ∞ is a fixed point of A and $ad = 1$. If $d = a$ then $A(z) = z + \dfrac{b}{a}$ with $ab \neq 0$ ($b \neq 0$ because A is not the identity map), and A has no other fixed point. If $d \neq a$, then $A(z) = \dfrac{a}{d}z + \dfrac{b}{d}$, and A has one more fixed point, at $\dfrac{b}{d-a}$ in $\mathbb{C}$.

We note that in this case A has precisely one fixed point if and only if $\text{tr}^2 A = 4$.

Case 2: $c \neq 0$. In this case ∞ is not fixed by A, and the fixed points of A are given by

$$\frac{a - d \pm \sqrt{(a-d)^2 + 4bc}}{2c} = \frac{(a-d) \pm \sqrt{\text{tr}^2 A - 4}}{2c}.$$

We have thus proved.

Proposition 8.7. *If A is a Möbius transformation different from the identity map, then A has either one or two fixed points in $\widehat{\mathbb{C}}$. It has exactly one if and only if* $\text{tr}^2 A = 4$.

8.1.2 *Cross Ratios*

Proposition 8.8. *Given three distinct points z_2, z_3, z_4 in $\widehat{\mathbb{C}}$, there exists a unique Möbius transformation S with $S(z_2) = 1$, $S(z_3) = 0$, and $S(z_4) = \infty$.*

Proof. The proof has two parts.

Uniqueness: If S_1 and S_2 are Möbius transformations that solve our problem, then $S_1 \circ S_2^{-1}$ is a Möbius transformation that fixes 1, 0 and ∞ and hence, by Proposition 8.7, it is the identity map.

Existence: If the z_i are complex numbers, then

$$S(z) = \frac{z - z_3}{z - z_4} \frac{z_2 - z_4}{z_2 - z_3}$$

is the required map. If one of the z_i equals ∞, use a limiting procedure to obtain

$$S(z) = \begin{cases} \dfrac{z - z_3}{z - z_4}, & \text{if } z_2 = \infty, \\[2ex] \dfrac{z_2 - z_4}{z - z_4}, & \text{if } z_3 = \infty, \\[2ex] \dfrac{z - z_3}{z_2 - z_3}, & \text{if } z_4 = \infty, \end{cases}$$

respectively. □

Corollary 8.9. *If $\{z_i\}$ and $\{w_i\}$ ($i = 2, 3, 4$) are two triples of distinct points in $\widehat{\mathbb{C}}$, then there exists a unique Möbius transformation S with $S(z_i) = w_i$; thus the Möbius group is uniquely triply transitive on $\widehat{\mathbb{C}}$.*

Definition 8.10. The *cross ratio* (z_1, z_2, z_3, z_4) of four distinct points in $\widehat{\mathbb{C}}$ is the image of z_1 under the Möbius transformation taking z_2 to 1, z_3 to 0, and z_4 to ∞; that is,

$$(z_1, z_2, z_3, z_4) = \frac{z_1 - z_3}{z_1 - z_4} \frac{z_2 - z_4}{z_2 - z_3}$$

if the four points are finite, with the corresponding limiting values if one of the z_i equals ∞.

As we will see in the next proposition, it is useful to view the cross ratio as a Möbius transformation (a function of z_1) $S = S_{z_2,z_3,z_4}$ that takes the four distinct ordered points z_1, z_2, z_3, z_4 to the four distinct ordered points $w_1 = S(z_1) = (z_1, z_2, z_3, z_4), w_2 = 1, w_3 = 0, and w_4 = \infty$. It hence makes sense to allow one repetition among the four points z_j and hence have S defined on $\widehat{\mathbb{C}}$ and conclude that $(z_2, z_2, z_3, z_4) = 1$, for example. This point of view will be used from now on when needed.

Proposition 8.11. *If z_1, z_2, z_3, z_4 are four distinct points in $\widehat{\mathbb{C}}$, and T is any Möbius transformation, then*

$$(T(z_1), T(z_2), T(z_3), T(z_4)) = (z_1, z_2, z_3, z_4).$$

Proof. If we define $S(z) = (z, z_2, z_3, z_4)$ for $z \in \mathbb{C} - \{0, 1\}$, then $S \circ T^{-1}$ is a Möbius transformation taking $T(z_2)$ to 1, $T(z_3)$ to 0 and $T(z_4)$ to ∞. Therefore

$$(T(z_1), T(z_2), T(z_3), T(z_4)) = S \circ T^{-1}(T(z_1)) = S(z_1) = (z_1, z_2, z_3, z_4).$$

□

Definition 8.12. A *circle* in $\widehat{\mathbb{C}}$ is either an Euclidean (ordinary) circle in $\mathbb{C}$, or a straight line in $\mathbb{C}$ together with ∞ (this is a circle passing through ∞). See Exercise 3.21 for a justification for the name.

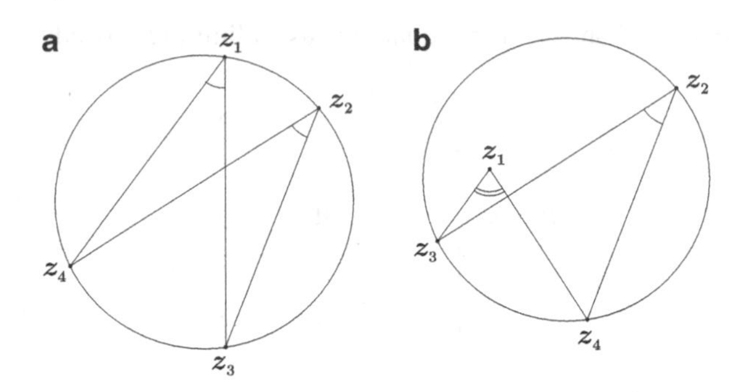

Fig. 8.1 The cross ratio arguments. (**a**) On a circle. (**b**) Not on a circle

Proposition 8.13. *The cross ratio of four distinct points in* $\widehat{\mathbb{C}}$ *is a real number if and only if the four points lie on a circle in* $\widehat{\mathbb{C}}$.

Proof. This is an elementary geometric argument that goes as follows. It is clear that

$$\arg(z_1, z_2, z_3, z_4) = \arg \frac{z_1 - z_3}{z_1 - z_4} - \arg \frac{z_2 - z_3}{z_2 - z_4}.$$

It is also clear from the geometry of the situation (see Fig. 8.1 and Exercise 8.3) that the two quantities on the right-hand side differ by πn, with $n \in \mathbb{Z}$, if and only if the four points lie on a circle in $\widehat{\mathbb{C}}$. □

Theorem 8.14. *A Möbius transformation maps circles in* $\widehat{\mathbb{C}}$ *to circles in* $\widehat{\mathbb{C}}$.

Proof. This follows immediately from Propositions 8.11 and 8.13. □

We use the following *standard notation* in the rest of this chapter: $\mathbb{D}$ denotes the unit disc $\{z \in \mathbb{C};\ |z| < 1\}$ and $\mathbb{H}^2$ the upper half plane $\{z \in \mathbb{C};\ \Im z > 0\}$. Note that both $\mathbb{D}$ and $\mathbb{H}^2$ should be regarded as discs in $\widetilde{\mathbb{C}}$, since they are bounded by circles in $\widehat{\mathbb{C}}$: the unit circle S^1 and the extended real line $\widehat{\mathbb{R}} = \mathbb{R} \cup \{\infty\}$, respectively.

The next result shows that these two discs in $\widehat{\mathbb{C}}$ are conformally equivalent.

Corollary 8.15. *If* $w(z) = \dfrac{z - \imath}{z + \imath}$ *for* $z \in \mathbb{H}^2$, *then* w *is a conformal map of* $\mathbb{H}^2$ *onto* $\mathbb{D}$.

Proof. All Möbius transformations, in particular w, are conformal. A calculation shows that w maps $\widehat{\mathbb{R}} = \mathbb{R} \cup \{\infty\}$ onto S^1 (the unit circle centered at 0) and $w(\imath) = 0$. By connectivity considerations, it follows that $w(\mathbb{H}^2) = \mathbb{D}$. □

8.2 Aut(D) for $D = \widehat{\mathbb{C}}, \mathbb{C}, \mathbb{D}$, and $\mathbb{H}^2$

Theorem 8.16. *A function $f : \mathbb{C} \to \mathbb{C}$ belongs to* Aut($\mathbb{C}$) *if and only if there exist a and b in $\mathbb{C}$, $a \neq 0$, such that $f(z) = az + b$ for all $z \in \mathbb{C}$.*

Proof. The if part is trivial. For the only if part, note that f is an entire function, and we can use its Taylor series at zero to conclude that

$$f(z) = \sum_{n=0}^{\infty} a_n z^n \text{ for all } z \in \mathbb{C}.$$

If ∞ were an essential singularity of f, then $f(|z| > 1)$ would be dense in $\mathbb{C}$. But

$$f(|z| > 1) \cap f(|z| < 1) \text{ is empty}$$

since f is injective. Thus ∞ is either a removable singularity or a pole of f; in any case, there is a nonnegative integer N such that $a_n = 0$ for all $n > N$ and $a_N \neq 0$; that is, f is a polynomial of degree N. If N were bigger than one or equal to zero, then f would not be injective. □

Theorem 8.17. Aut($\widehat{\mathbb{C}}$) $\cong$ PSL($2, \mathbb{C}$). *Thus the last arrow in the exact sequence (8.3) corresponds to a surjective map.*

Proof. We need only show that Aut($\widehat{\mathbb{C}}$) is contained in the Möbius group. Let f be an element of Aut($\widehat{\mathbb{C}}$). If $f(\infty) = \infty$, then f is a Möbius transformation by Theorem 8.16. If $f(\infty) = c \neq \infty$, then consider the Möbius transformation $A(z) = \frac{1}{z-c}$ and conclude that $B = A \circ f$ in Aut($\widehat{\mathbb{C}}$) and fixes ∞; therefore B is a Möbius transformation. But then so is $f = A^{-1} \circ B$. □

We now provide a characterization of the elements of Aut($\mathbb{D}$); it shows that they form a subgroup of Aut($\widehat{\mathbb{C}}$). Another useful characterization is given in Exercise 8.5.

Theorem 8.18. *A function B defined on $\mathbb{D}$ is in* Aut($\mathbb{D}$) *if and only if there exist a and b in $\mathbb{C}$ such that $|a|^2 - |b|^2 = 1$ and*

$$B(z) = \frac{az + b}{\overline{b}z + \overline{a}}$$

for all $z \in \mathbb{D}$.

Proof. **The if part:** Assume that B is of the above form and observe that a is different from zero. We show that $B \in \mathrm{Aut}(\mathbb{D})$. This follows from the following easy to prove facts:

(1) Mappings B of the given form constitute a group under composition. In particular, if $B(z) = \dfrac{az+b}{\overline{b}z+\overline{a}}$ with $|a|^2 - |b|^2 = 1$, then $B^{-1}(z) = \dfrac{\overline{a}z - b}{-\overline{b}z + a}$ has the same form as B.

(2) $|z| = 1$ if and only if $|B(z)| = 1$.

(3) $|B(0)| = \left|\dfrac{b}{a}\right| < 1$.

(4) $B(\mathbb{D})$ is connected. Thus, from (2), either $B(\mathbb{D})$ is contained in $\mathbb{D}$ or $B(\mathbb{D}) \cap \mathbb{D}$ is empty. From (3) we see that $B(\mathbb{D}) \subseteq \mathbb{D}$. It follows from (1) that $B^{-1}(\mathbb{D}) \subseteq \mathbb{D}$.

(5) Obviously $\mathbb{D} = B \circ B^{-1}(\mathbb{D})$, which implies that $B(\mathbb{D}) = \mathbb{D}$.

The only if part: Let $f \in \mathrm{Aut}(\mathbb{D})$ and $w = f(z)$. Then $f^{-1} \in \mathrm{Aut}(\mathbb{D})$ and $z = f^{-1}(w)$.

(6) If $f(0) = 0$, then it follows by the Schwarz' lemma applied first to f^{-1} and then to f that

$$|z| = \left|f^{-1}(w)\right| \le |w| = |f(z)| \le |z| \text{ for all } z \in \mathbb{D}.$$

The same lemma implies that there exists a $\theta \in \mathbb{R}$ such that $f(z) = \mathrm{e}^{\imath\theta} z$ for all $z \in \mathbb{D}$. So we can take $a = \mathrm{e}^{\imath\frac{\theta}{2}}$ and $b = 0$ to conclude that f has the required form.

(7) If $f(0) = c \neq 0$, then $0 < |c| < 1$ and we set $C(z) = \dfrac{z-c}{1-\overline{c}z}$.

It follows from Exercise 8.5 (see also Exercise 2.2) that the Möbius transformation C belongs to $\mathrm{Aut}(\mathbb{D})$. Since $C \circ f$ fixes the origin, it follows from (6) that $C \circ f$ is of the required form, and therefore so is $f = C^{-1} \circ (C \circ f)$ by (1).

□

Just as in Sect. 8.1 we defined $\mathrm{PSL}(2,\mathbb{C})$ as the quotient of $\mathrm{SL}(2,\mathbb{C})$ by $\pm I$ and then proved that it is isomorphic to the group $\mathrm{Aut}(\widehat{\mathbb{C}})$, we can define the group $\mathrm{PSL}(2,\mathbb{R}) = \mathrm{SL}(2,\mathbb{R})/\{\pm I\}$ of appropriate matrices with real coefficients modulo plus or minus the identity matrix and obtain the following description:

Theorem 8.19. $\mathrm{Aut}(\mathbb{H}^2) \cong \mathrm{PSL}(2,\mathbb{R})$.

Proof. Consider the conformal map $w : \mathbb{H}^2 \to \mathbb{D}$ given in Corollary 8.15. Then

$$\mathrm{Aut}(\mathbb{H}^2) = w^{-1}\,\mathrm{Aut}(\mathbb{D})\,w.$$

By the preceding theorem, any element f of $\mathrm{Aut}(\mathbb{D})$ may be written as

$$f(z) = \frac{az+b}{\overline{b}z+\overline{a}}$$

with $|a|^2 - |b|^2 = 1$. Denote $a = a_1 + \imath\, a_2$, $b = b_1 + \imath\, b_2$. Then

$$(w^{-1} \circ f \circ w)(z) = \frac{-(a_1+b_1)z + b_2 - a_2}{(a_2+b_2)z + b_1 - a_1}$$

with $-(a_1+b_1)(b_1-a_1) + (a_2+b_2)(a_2-b_2) = |a|^2 - |b|^2 = 1$; thus we have associated to any element $w^{-1} \circ f \circ w$ of $\mathrm{Aut}(\mathbb{H}^2)$ the image in $\mathrm{PSL}(2,\mathbb{R})$ of the matrix $\begin{bmatrix} -(a_1+b_1) & b_2-a_2 \\ a_2+b_2 & b_1-a_1 \end{bmatrix}$ in $\mathrm{SL}(2,\mathbb{R})$.

Conversely, every matrix $\begin{bmatrix} a & b \\ c & d \end{bmatrix}$ in $\mathrm{SL}(2,\mathbb{R})$ induces a Möbius transformation given by $S(z) = \dfrac{az+b}{cz+d}$. Since S preserves the circle $\widehat{\mathbb{R}}$ and

$$\Im S(\imath) = \frac{ad-bc}{c^2+d^2} = \frac{1}{c^2+d^2}$$

is positive, we conclude that S belongs to $\mathrm{Aut}(\mathbb{H}^2)$. □

8.3 The Riemann Mapping Theorem

We now combine the results about Möbius transformations of the previous two sections with results from Chap. 7 about compact and bounded families of holomorphic functions to show that every simply connected domain D in $\mathbb{C}$, other than $\mathbb{C}$ itself, is conformally equivalent to the unit disc; any conformal map from D onto the unit disc $\mathbb{D}$ will be called a *Riemann map*.

Recall that a set A is a *proper subset* of a set B if $A \subset B$ (thus, in particular, $A \neq B$).

Theorem 8.20 (Riemann Mapping Theorem). *Let D be a nonempty proper simply connected open subset of $\mathbb{C}$, and let $c \in D$. Then there exists a unique conformal map $f : D \to \mathbb{D}$ with $f(c) = 0$, $f'(c) > 0$, and $f(D) = \mathbb{D}$.*

Proof. We are looking for a map for the pair (D, c). The argument has two parts, existence and uniqueness.

Existence. We first reduce to a special case.

First Reduction: It suffices to assume that D is bounded.

Proof. Since $D \neq \mathbb{C}$, we can choose $b \in \mathbb{C} - D$; since D is simply connected there is a branch $g(z)$ of $\log(z - b)$ on D. Thus

$$e^{g(z)} = z - b \quad \text{for all} \quad z \in D.$$

The function g is injective: for if $g(w) = g(z)$, then $w - b = z - b$. Furthermore, if $d \in D$, then $g(z) - g(d) \neq 2\pi\imath$ for all $z \in D$. Otherwise

$$z - b = e^{g(z)} = e^{g(d)+2\pi\imath} = e^{g(d)} = d - b.$$

Since $g(D)$ is an open set, we can choose $d \in D$ and $\delta > 0$ such that

$$|w - g(d)| < \delta \Rightarrow w \in g(D).$$

Thus

$$|w - g(d) - 2\pi\imath| < \delta \Rightarrow w \notin g(D),$$

since otherwise $w = g(z_0)$ for some z_0 in D, and

$$|w - g(d) - 2\pi\imath| = |[g(z_0) - 2\pi\imath] - g(d)| < \delta$$

implies that $g(z_0) - 2\pi\imath = g(z_1)$ for some z_1 in D, a contradiction.

Now

$$F(z) = \frac{1}{g(z) - g(d) - 2\pi\imath}$$

is a conformal map from D onto $F(D)$, and the simply connected domain $F(D)$ is contained in the bounded set $\operatorname{cl} U\left(0, \frac{1}{\delta}\right)$. □

We are reduced to solving the mapping problem for $(F(D), F(c))$, and may thus assume that D is bounded.

Second Reduction: We may also assume that $c = 0$.

Proof. The map $G : z \mapsto z - c$ takes the bounded domain D onto its (bounded) translate D'. If F is a solution of our problem for $(D', 0)$, then $F \circ G$ solves the problem for $(D, 0)$. □

Thus we now assume that D is bounded, simply connected and $c = 0 \in D$.

Proof of the Theorem Under These Assumptions. We define

$$\begin{aligned} \mathcal{F} = \{ & f \in \mathbf{H}(D); f \text{ is either conformal or identically zero,} \\ & f(0) = 0, f'(0) \in \mathbb{R}, f'(0) \geq 0 \text{ and } |f(z)| < 1 \text{ for all } z \in D\}. \end{aligned}$$

Our first observation is that $\mathcal{F}$ is nonempty. Of course, $f \equiv 0$ is in $\mathcal{F}$. This is not good enough for much. Since D is bounded, there exists an $M > 0$ such that

$$|z| \leq M \text{ for all } z \in D. \tag{8.4}$$

Hence, for every $a \in \mathbb{R}$ such that $a > M$, the function $f(z) = \dfrac{z}{a}$ for z in D belongs to $\mathcal{F}$.

Next we show that $\mathcal{F}$ is compact using Corollary 7.31.Each element f of $\mathcal{F}$ satisfies $||f||_K \leq 1$ for all compact sets $K \subset D$; hence $\mathcal{F}$ is bounded in the strong sense. To show that $\mathcal{F}$ is closed, let $\{f_n\} \subset \mathcal{F}$ be a sequence such that $f_n \to f$ uniformly on all compact subsets of D. Then $f \in \mathbf{H}(D)$, by Theorem 7.2, and since each f_n vanishes at 0, so does f. It is now convenient to consider two cases:

1. $f_n \equiv 0$ for infinitely many distinct n. In this case $f \equiv 0$ and hence certainly $f \in \mathcal{F}$.
2. $f_n \equiv 0$ for only finitely many n. In this case we may assume that each f_n is a conformal map; then $f_n'(0) > 0$ for all n, and thus $f'(0) \geq 0$. Hurwitz's Theorem 7.9 says that f is either constant (hence identically zero) or univalent (that is, one-to-one). Since $|f_n(z)| < 1$ for all $z \in D$, we conclude that $|f(z)| \leq 1$ for all $z \in D$. If $|f(z_0)| = 1$ for some $z_0 \in D$, then $|f| \equiv 1$ by the maximum modulus principle; this is a contradiction to $f(0) = 0$. Thus $|f(z)| < 1$ for all $z \in D$, and we conclude that $f \in \mathcal{F}$; thus $\mathcal{F}$ is closed, and therefore compact.

We now complete the proof of the existence part. If

$$S = \{f'(0); f \in \mathcal{F}\},$$

then $S \subseteq \mathbb{R}_{\geq 0}$. We claim that S is bounded from above. Indeed, choose $\epsilon > 0$ so that cl $U(0, \epsilon) \subset D$. If $\gamma(\theta) = \epsilon\, e^{\imath\theta}$ for $0 \leq \theta \leq 2\pi$ is the circle centered at 0 with radius ϵ, then for any $f \in \mathcal{F}$ we have

$$f'(0) = \frac{1}{2\pi\imath} \int_\gamma \frac{f(z)}{z^2}\, dz$$

and thus

$$|f'(0)| \leq \frac{1}{2\pi} \frac{2\pi\epsilon}{\epsilon^2} = \frac{1}{\epsilon}.$$

If $\mu = \sup S$, then $\mu \geq \frac{1}{M} > 0$, with M as in (8.4), because $\dfrac{1}{a} \in S$ for all $a > M$, as we saw above. Also, there exists a sequence $\{f_n\} \subset \mathcal{F}$ such that

$$\lim_{n\to\infty} f_n'(0) = \mu.$$

Since $\mathcal{F}$ is compact, there exists a convergent subsequence $\{f_{n_k}\}$ with $\lim\limits_{k\to\infty} f_{n_k} = f \in \mathcal{F}$. Since $f'(0) = \mu$, f is a conformal map.

Since $f(D) \subseteq \mathbb{D}$, to show that $f(D) = \mathbb{D}$, we assume for contradiction that $f(D) \neq \mathbb{D}$ and construct an $h \in \mathcal{F}$ with $h'(0) > \mu$, thus contradicting the fact that $\mu = \sup S$. Namely, if $f(D) \neq \mathbb{D}$, then there exists $w_0 = r e^{\imath\theta}$ with $0 < r < 1$ such that $w_0 \in \partial f(D)$. We now construct h as follows:

1. Let $g_1(z) = e^{-\imath\theta} f(z)$. The map g_1 is the map f followed by a rotation through the angle $-\theta$; g_1 sends w_0 to r.
2. Let $p(z) = \dfrac{r - g_1(z)}{1 - r g_1(z)}$. The map p is g_1 followed by an automorphism of $\mathbb{D}$ that sends r to 0 (see Exercise 8.5).

   Note that $p(z) \neq 0$ for all $z \in D$, since $p(z) = 0$ if and only if $g_1(z) = r$ if and only if $f(z) = w_0$.
3. Let $q(z) = p(z)^{\frac{1}{2}}$, where we choose the branch of the square root[2] with $q(0) = r^{\frac{1}{2}} > 0$.

   The map q is injective because $q(z_1) = q(z_2)$ if and only if $p(z_1) = p(z_2)$. Furthermore, $|q(z)| < 1$ for all $z \in D$.
4. Let $g_2(z) = \dfrac{r^{\frac{1}{2}} - q(z)}{1 - r^{\frac{1}{2}} q(z)}$. The map g_2 is q followed by an automorphism of $\mathbb{D}$ that sends $r^{\frac{1}{2}}$ to 0.
5. Let $h(z) = e^{\imath\theta} g_2(z)$. The map h is g_2 followed by a rotation through the angle θ.

Conclusion: h is a univalent mapping of D into $\mathbb{D}$.

We calculate $h(0)$ and $h'(0)$. In order to use the chain rule we need to see what happens to zero under all the maps used to construct h. It is easily checked that

$$g_1(0) = 0,\ p(0) = r,\ q(0) = r^{\frac{1}{2}},\ g_2(0) = 0, \text{ and } h(0) = 0.$$

Aside: Let $\alpha, \beta, \gamma, and \delta$ be complex numbers. If $A(z) = \dfrac{\alpha z + \beta}{\gamma z + \delta}$ for all $z \in \mathbb{C} - \left\{\dfrac{-\delta}{\gamma}\right\}$, then $A'(z) = \dfrac{\alpha\delta - \beta\gamma}{(\gamma z + \delta)^2}$.

The calculation of the derivative of h at zero proceeds as follows:

$$\begin{aligned}
h'(0) &= e^{\imath\theta} g_2'(0) = e^{\imath\theta} \frac{r-1}{(1 - r^{\frac{1}{2}} q(0))^2} q'(0) = e^{\imath\theta} \frac{r-1}{(1-r)^2} q'(0) \\
&= \frac{e^{\imath\theta}}{r-1} \left(\frac{1}{2}\right) p(0)^{-\frac{1}{2}} p'(0) = \frac{e^{\imath\theta}}{r-1} \left(\frac{1}{2}\right) r^{-\frac{1}{2}} \frac{-1 + r^2}{(1 - r g_1(0))^2} g_1'(0) \\
&= \frac{1}{2} e^{\imath\theta} \frac{1}{r-1} \frac{1}{r^{\frac{1}{2}}} \frac{r^2 - 1}{1} e^{-\imath\theta} \mu = \frac{r+1}{2 r^{\frac{1}{2}}} \mu.
\end{aligned}$$

---

[2]By Exercise 5.1 there certainly exists a holomorphic function q whose square is p. Hence $-q$ is also such a function. These are the two *branches* of the square root of p.

Finally

$$\frac{r+1}{2r^{\frac{1}{2}}} > 1 \text{ if and only if } 0 < r < 1,$$

arriving at the contradiction $h'(0) > \mu$ that finishes the existence proof.

Uniqueness. This is a straightforward argument using the Schwarz's lemma. □

Corollary 8.21. *If D is a nonempty simply connected domain in $\widehat{\mathbb{C}}$, then D is conformally equivalent to one and only one of the following domains:* (i) $\widehat{\mathbb{C}}$, (ii) $\mathbb{C}$, *or* (iii) $\mathbb{D}$.

The case (i), (ii), *or* (iii) *occurs when the boundary of D consists of no points, one point, or more than one point, respectively. In the last case the boundary of D contains a continuum (a homeomorphic image of a closed interval containing more than one point).*

Proof. **Existence**: If $D \subsetneq \widehat{\mathbb{C}}$, we may first reduce to the case $D \subseteq \mathbb{C}$ by observing that if D contains ∞, we can choose $c \in \mathbb{C} - D$, and setting $F(z) = \frac{1}{z-c}$ we have that $F(D) \subseteq \mathbb{C}$ is a nonempty simply connected domain not containing ∞ and conformally equivalent to D. If the result holds for $F(D)$, then it also holds for D. If D is a proper subset of $\mathbb{C}$, the result follows from the RMT.

Uniqueness: We need to prove that no two of the simply connected domains $\widehat{\mathbb{C}}$, $\mathbb{C}$, and $\mathbb{D}$ are conformally equivalent. But $\widehat{\mathbb{C}}$ is compact, and hence cannot be conformally equivalent to either $\mathbb{C}$ or $\mathbb{D}$. On the other hand, a conformal map from $\mathbb{C}$ onto $\mathbb{D}$ would be a nonconstant entire bounded function, a contradiction to Liouville's theorem. □

8.4 Hyperbolic Geometry

Let D be a simply connected domain in the extended complex plane with two or more boundary points. In this section we establish that such a domain[3] carries a conformally invariant metric, known as the *Poincaré* or *hyperbolic* metric. These domains are called *hyperbolic*; they are all conformally equivalent to the unit disc, by the RMT.

We show that conformal equivalences between these domains preserve the hyperbolic metric; that is, they are isometries (distance preserving maps) with respect to the hyperbolic metrics on the respective domains. Endowed with these equivalent metrics, the upper half-plane $\mathbb{H}^2$ and the unit disc $\mathbb{D}$ become models for non-Euclidean (also known as hyperbolic or Lobachevsky) geometry. As we

[3]The metric may be defined on all domains in $\widehat{\mathbb{C}}$ with two or more boundary points, not necessarily simply connected; to prove this takes us beyond the scope of this book.

have shown, the groups Aut($\mathbb{H}^2$) and Aut($\mathbb{D}$) of conformal automorphisms of these domains consist of Möbius transformations, a class of maps much easier to study than the group of conformal automorphisms of an arbitrary D. It is a remarkable fact that these Möbius functions constitute the full group of orientation-preserving isometries of $\mathbb{H}^2$ and $\mathbb{D}$ with their respective hyperbolic metrics. We conclude this section using Schwarz's lemma and the hyperbolic metric to establish a deep connection between complex analysis and geometry. Namely, holomorphic maps between hyperbolic domains are either isometries or contractions with respect to their hyperbolic metrics.

We first define the Poincaré metric in a general setting; that is, on an arbitrary simply connected domain D with two or more boundary points (Sect. 8.4.1). We subsequently study it in more detail on $\mathbb{H}^2$ and $\mathbb{D}$, where specific computations are most easily carried out (Sects. 8.4.2 and 8.4.3). The results that follow from these computations transfer to the general setting because of the conformal equivalence established in the RMT. Finally in Sect. 8.4.4 we establish the result about contractions.

8.4.1 The Poincaré Metric

Definition 8.22. Let D be a simply connected domain in the extended complex plane with two or more boundary points. We define the *(infinitesimal form of the) Poincaré metric* on D

$$\lambda_D(z)\,|dz|$$

as follows. First, in the unit disc, set

$$\lambda_{\mathbb{D}}(z) = \frac{2}{1-|z|^2}, \quad z \in \mathbb{D}. \tag{8.5}$$

Next, for arbitrary D, choose a Riemann map $\pi : D \to \mathbb{D}$ and define λ_D by

$$\lambda_D(w) = \lambda_{\mathbb{D}}(\pi(w))\,\left|\pi'(w)\right|, \quad w \in D. \tag{8.6}$$

Our first task is to show that $\lambda_D(w)$ is well defined for all simply connected domains[4] D and all $w \in D$. Toward this end, let A be a conformal automorphism of $\mathbb{D}$. Recall that there exist complex numbers a and b with $|a|^2 - |b|^2 = 1$ such that

$$A(z) = \frac{az+b}{\overline{b}z+\overline{a}}, \quad z \in \mathbb{D}.$$

[4]With two or more boundary points.

An easy calculation now shows that

$$\lambda_{\mathbb{D}}(A(z))\,|A'(z)| = \lambda_{\mathbb{D}}(z), \quad z \in \mathbb{D}. \tag{8.7}$$

Let $w_0 \in D$ be arbitrary and suppose that π and ρ are two Riemann maps of D onto $\mathbb{D}$, with

$$\pi(w_0) = z_0 \text{ and } \rho(w_0) = t_0$$

for two points z_0 and $t_0 \in \mathbb{D}$. We need to show that

$$\lambda_{\mathbb{D}}(z_0)\,|\pi'(w_0)| = \lambda_{\mathbb{D}}(t_0)\,|\rho'(w_0)|.$$

But $A = \rho \circ \pi^{-1}$ is in Aut($\mathbb{D}$) and $A(z_0) = t_0$; it now follows from (8.7) that

$$\begin{aligned}\lambda_{\mathbb{D}}(z_0)\,|\pi'(w_0)| &= \lambda_{\mathbb{D}}(A(z_0))\,|A'(z_0)|\,|\pi'(w_0)| \\ &= \lambda_{\mathbb{D}}(A(z_0))\,|\rho'(\pi^{-1}(z_0))(\pi^{-1})'(z_0)|\,|\pi'(w_0)| \\ &= \lambda_{\mathbb{D}}(t_0)\,|\rho'(w_0)|.\end{aligned}$$

Remark 8.23. (1) If $w \in D$ is arbitrary and we choose the Riemann map π to satisfy $\pi(w) = 0$, then

$$\lambda_D(w) = 2\,|\pi'(w)|.$$

(2) It is easy to see, using (8.6) and Corollary 8.15, that

$$\lambda_{\mathbb{H}^2}(z) = \frac{1}{\Im z} \quad \text{for all } z \in \mathbb{H}^2. \tag{8.8}$$

The important invariance property of our metric is described in our next result.

Proposition 8.24. *For every conformal map f defined on D,*

$$\lambda_{f(D)}(f(z))\,|f'(z)| = \lambda_D(z) \text{ for all } z \in D.$$

Proof. If π is a Riemann map (for D), so is $\pi \circ f^{-1}$ (for $f(D)$). □

Any infinitesimal metric on D allows us to define lengths of paths in D, and hence a distance function on the domain. We work, of course, with the length element

$$ds = \lambda_D(z)\,|dz|.$$

Definition 8.25. We define *the hyperbolic length of a piecewise differentiable curve* γ in D by

$$l_D(\gamma) = \int_\gamma \lambda_D(z)\,|dz|;$$

and if z_1 and z_2 are any two points in D, *the hyperbolic (or Poincaré) distance* between them by

$$\rho_D(z_1, z_2) = \inf\{l_D(\gamma); \gamma \text{ is a pdp in } D \text{ from } z_1 \text{ to } z_2\}. \tag{8.9}$$

We leave to the reader (Exercise 8.13) the verification that ρ_D defines a metric on D.

An *isometry* from one metric space to another is a distance preserving map between them. It follows from Proposition 8.24 that for every conformal map f defined on D and every pdp γ in D,

$$l_{f(D)}(f \circ \gamma) = l_D(\gamma)$$

and

$$\rho_{f(D)}(f(z_1), f(z_2)) = \rho_D(z_1, z_2) \text{ for all } z_1 \text{ and } z_2 \in D;$$

that is, ρ is conformally invariant and f is an isometry between D and $f(D)$ with respect to the appropriate hyperbolic metrics. In particular, every element of $\text{Aut}(D)$ is an isometry for the hyperbolic metric on D.

8.4.2 Upper Half-plane Model

We know from Remark 8.23 that in $\mathbb{H}^2$ we have $ds = \frac{|dz|}{\Im(z)}$. The hyperbolic length of an arbitrary curve γ in $\mathbb{H}^2$ and the hyperbolic distance between two points in $\mathbb{H}^2$ may be hard to calculate directly from their definitions; an indirect approach is technically less complicated. We show that given any two distinct points in $\mathbb{H}^2$, they lie on either a unique Euclidean circle centered on the real axis or on a unique straight line perpendicular to the real axis. The corresponding portion of the circle or straight line lying in $\mathbb{H}^2$ is called a *hyperbolic line* or *geodesic*; the unique portion of the geodesic between the two points is called a *geodesic path* or *geodesic segment.* The name is justified by showing that the hyperbolic length of a geodesic segment realizes the hyperbolic distance between its two end points.

A straight line in $\mathbb{C}$ is a circle in $\mathbb{C} \cup \{\infty\}$ passing through infinity (see Exercise 3.21). It is not useful, in general, to assign centers to these circles. However, if such a line intersects $\mathbb{R}$ in one point and is perpendicular to $\mathbb{R}$ at that point, we consider that point to be the *center* of the circle. In the current context, we shall be interested only in lines perpendicular to $\mathbb{R}$ and use the related fact that a Euclidean circle with center on the real axis is perpendicular to the real axis.

Definition 8.26. For a circle C in $\mathbb{C} \cup \{\infty\}$ centered on the real axis, the part of C lying in the upper half plane is called a *hyperbolic line* or a *geodesic* in $\mathbb{H}^2$. The reason for the terminology will shortly become clear.

The following lemma establishes the existence of a geodesic path between two points; the proof of its uniqueness follows.

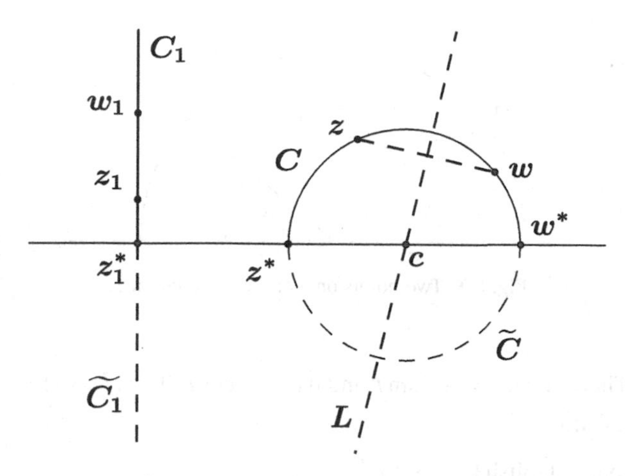

Fig. 8.2 Unique circles (in $\hat{\mathbb{C}}$) perpendicular to $\mathbb{R}$ through two pairs of points in the upper half plane

Lemma 8.27. *For every pair z and w of distinct points in $\mathbb{H}^2$, there exists a unique circle centered at the real axis passing through them, and a unique geodesic in $\mathbb{H}^2$ passing through them.*

Proof. If $\Re(z) = \Re(w)$, take $\widetilde{C}$ to be the Euclidean line through z and w. Otherwise, let L be the perpendicular bisector of the Euclidean line segment connecting z and w. If c is the point where L intersects the real line, we take $\widetilde{C}$ to be the circle with center c passing through z and w. See Fig. 8.2. The portion C of $\widetilde{C}$ in $\mathbb{H}^2$ gives the sought geodesic. □

Definition 8.28. Let z and w be two distinct points in $\mathbb{H}^2$. The arc of the unique geodesic determined by z and w between them is the *geodesic segment* or *geodesic path* joining z and w.

The next two lemmas compute the hyperbolic length of the geodesic segment between two points in $\mathbb{H}^2$.

Lemma 8.29. *Let P and Q be two points in $\mathbb{H}^2$ lying on an Euclidean circle C centered on the real axis, and let γ be the arc of C in $\mathbb{H}^2$ between P and Q. Assume further that the radii from the center of C to P and Q make respective angles α and β with the positive real axis. Then*

$$l_{\mathbb{H}^2}(\gamma) = \left| \log \frac{\csc(\beta) - \cot(\beta)}{\csc(\alpha) - \cot(\alpha)} \right|.$$

Proof. Assume the circle C has radius r and is centered at c (see Fig. 8.3). Let $z = (x, y)$ be an arbitrary point on γ and let t be the angle that the radius from z to the center of C makes with the positive real axis; then $x = c + r \cos t$ and

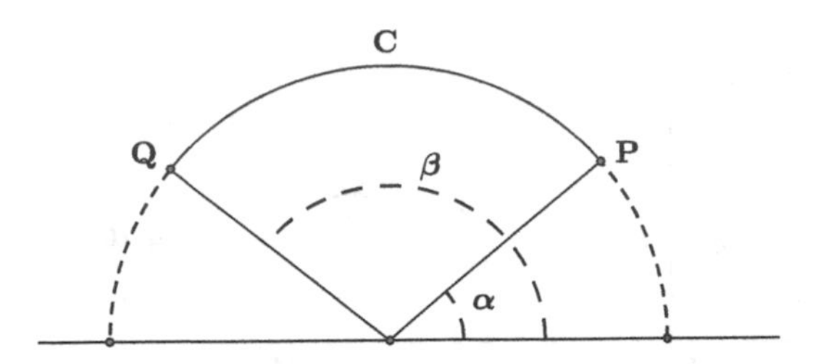

Fig. 8.3 Two points on a circle centered on $\mathbb{R}$

$y = r\sin t$. Therefore $dx = -r\sin t$ and $dy = r\cos t$. Thus $l_{\mathbb{H}^2}(\gamma) = \left|\int_\alpha^\beta \csc t \, dt\right|$ and the result follows. □

Similarly one establishes

Lemma 8.30. *Let $P = x_P + \imath\, y_P$ and $Q = x_P + \imath\, y_Q$ be two points in $\mathbb{H}^2$ lying on a straight line C perpendicular to the real axis, and let γ be the segment of C in $\mathbb{H}^2$ between P and Q. Then*

$$l_{\mathbb{H}^2}(\gamma) = \left|\log \frac{y_P}{y_Q}\right|.$$

The next definition and the following two lemmas allow us to prove that the hyperbolic length of a geodesic segment minimizes the hyperbolic lengths of all pdp's joining two distinct points in $\mathbb{H}^2$; they will also provide an explicit formula for the hyperbolic distance in $\mathbb{H}^2$.

Definition 8.31. We have shown in Lemma 8.27 that any two distinct points z and w in $\mathbb{H}^2$ lie on a unique circle $\widetilde{C}$ centered on the real axis, and on a unique geodesic C. If $\Re z \neq \Re w$, C is the portion in $\mathbb{H}^2$ of an Euclidean circle $\widetilde{C}$ centered on the real axis; we let z^* and w^* denote the points on $\widetilde{C} \cap \mathbb{R}$ closest to z and w, respectively. If $\Re(z) = \Re(w)$, C is the portion in $\mathbb{H}^2$ of a straight line perpendicular to $\mathbb{R}$; if $\Im z < \Im w$ we let $z^* = \Re(z)$ and $w^* = \infty$; if $\Im z > \Im w$ we let $z^* = \infty$ and $w^* = \Re(w)$. See Fig. 8.2.

Lemma 8.32. *Let z and w distinct points in $\mathbb{H}^2$. There exists a unique $T \in \operatorname{Aut}(\mathbb{H}^2)$ such that $T(z^*) = 0$, $T(z) = \imath$, $T(w) = \imath\, y$ with $y > 1$, and $T(w^*) = \infty$, where z^* and w^* are as in Definition 8.31.*

Proof. Consider the unique circle $\widetilde{C}$ centered on the real axis and passing through z and w. Since the Möbius group is triply transitive, there exists a unique Möbius transformation T that maps z^*, z, w^* to $0, \imath, \infty$, respectively. Since Möbius transformations map circles to circles, T maps $\widetilde{C}$ onto the imaginary axis union $\{\infty\}$, and hence $T(w) = \imath\, y$ for some real y. Since Möbius transformations preserve orthogonality of curves, T maps $\mathbb{R} \cup \{\infty\}$ onto itself. Since T maps $z \in \mathbb{H}^2$ to $\imath \in \mathbb{H}^2$, T is in $\operatorname{Aut}(\mathbb{H}^2)$, and because it is orientation preserving, $y > 1$. □

Lemma 8.33. *If z and w are two distinct points in $\mathbb{H}^2$, then the hyperbolic length of the geodesic segment $\widetilde{\gamma}$ joining them is shorter than the hyperbolic length of any other pdp γ in $\mathbb{H}^2$ joining them.*

Proof. Write $z = x_z + \imath\, y_z$ and $w = x_w + \imath\, y_w$.

First consider the case $x_z = x_w$. Assume the curve γ is parameterized by the closed interval $[a, b] \subset \mathbb{R}$ and $\gamma(t) = x(t) + \imath\, y(t)$. Then x and y are differentiable functions except at finitely many points, and

$$l_{\mathbb{H}^2}(\gamma) = \left| \int_a^b \frac{\sqrt{x'(t)^2 + y'(t)^2}}{y(t)}\, \mathrm{d}t \right| \ge \int_a^b \frac{|y'(t)|}{y(t)} \mathrm{d}t \ge \left| \log \frac{y_w}{y_z} \right| = l_{\mathbb{H}^2}(\widetilde{\gamma}),$$

where the last equality follows from Lemma 8.30. Furthermore, equality of lengths is attained if and only if x is constant and y' does not change sign where it exists; that is, if and only if γ is a reparametrization of $\widetilde{\gamma}$.

For the case $x_z \ne x_w$, by Lemma 8.32 we can find a T in $\text{Aut}(\mathbb{H}^2)$ such that $T(z^*) = 0$ and $T(w^*) = \infty$. Furthermore, the image under T of the geodesic segment between z and w is the segment on the imaginary axis between $T(z)$ and $T(w)$, and both segments have the same hyperbolic length. Similarly, $T \circ \gamma$ is a pdp in $\mathbb{H}^2$ joining $T(z)$ and $T(w)$, with the same hyperbolic length as γ, and we are reduced to the previous case. □

We have established

Theorem 8.34. *For any two distinct points z and w in $\mathbb{H}^2$, the geodesic segment joining z to w is the unique curve that achieves the infimum defined by (8.9).*

Using cross ratios to simplify notation, a routine computation establishes

Proposition 8.35. *For any two distinct points z and w in $\mathbb{H}^2$, the hyperbolic distance between z and w is equal to the length of the geodesic segment γ joining z and w, and is given by*

$$\rho_{\mathbb{H}^2}(z, w) = l_{\mathbb{H}^2}(\gamma) = \log |(z^*, w^*, w, z)| = \log |(0, \infty, \imath\, y, \imath)| \tag{8.10}$$

$$= \log y = \log \frac{|z - \overline{w}| + |z - w|}{|z - \overline{w}| - |z - w|}, \tag{8.11}$$

where y is the real number (greater than one) given in Lemma 8.32.

Proof. A fractional linear transformation preserves the cross ratio of any four points, and fractional linear transformations mapping $\mathbb{H}^2$ to itself are isometries for the hyperbolic metric. This yields all the equalities except for the last one, which is a calculation (see Exercise 8.15). □

Remark 8.36. It is useful to record the following simplified form for the last identity for $\imath\, y$ and $\imath\, v$ in $\imath\, \mathbb{R}_{>0}$

$$\rho_{\mathbb{H}^2}(\imath\, y, \imath\, v) = \left| \log \frac{y}{v} \right|.$$

As mentioned at the end of the previous subsection, the set $\mathrm{PSL}(2,\mathbb{R})$ of conformal automorphisms of $\mathbb{H}^2$ acts as a group of hyperbolic isometries of $\mathbb{H}^2$. This also follows from two facts: fractional linear transformations preserve the cross ratios and map circles to circles. We now proceed to establish the converse.

Proposition 8.37. *An orientation preserving isometry f of $(\mathbb{H}^2, \rho_{\mathbb{H}^2})$ that fixes the imaginary axis pointwise is the identity map.*

Proof. Let $z = x + \imath\, y$ and $f = u + \imath\, v$. For all positive real numbers t we have

$$\rho_{\mathbb{H}^2}(z, \imath t) = \rho_{\mathbb{H}^2}(f(z), f(\imath t)) = \rho_{\mathbb{H}^2}(u(z) + \imath v(z), \imath t).$$

We calculate using (8.10) or Exercise 8.15 that this is equivalent to

$$[x^2 + (y - t)^2]v(z) = [u(z)^2 + (v(z) - t)^2]y;$$

hence also equivalent to

$$(x^2 + y^2)v(z) - (u^2(z) + v^2(z))y = (y - v(z))t^2$$

for all positive t. Since the LHS of the last equation is independent of t, the RHS must vanish identically. Hence $v(z) = y$, $u(z)^2 = x^2$, and $f(z) = \pm x + \imath\, y$. Because f is continuous, the same sign holds for all z; thus either $f(z) = z$ or $f(z) = -\bar{z}$ for all z; since f is orientation preserving, we conclude that f is the identity map. □

Theorem 8.38. *The set of orientation-preserving isometries of $\mathbb{H}^2$ with respect to the hyperbolic metric is precisely the set of fractional linear transformations mapping $\mathbb{H}^2$ to itself; that is,* $\mathrm{PSL}(2,\mathbb{R})$.

Proof. If g is such an isometry of $\mathbb{H}^2$, it preserves geodesics. Thus there is a fractional linear transformation f that preserves $\mathbb{H}^2$ and such that $f \circ g$ leaves invariant the imaginary axis. Following this map by an isometry of the form $z \mapsto kz$ with $k \in \mathbb{R}_{>0}$ and then (if necessary) by $z \mapsto \dfrac{-1}{z}$, we may assume that $f \circ g$ fixes $\imath$ and leaves invariant the intervals $(\imath, \infty)$ and $(0, \imath)$ on the imaginary axis. Using the fact that $\rho_{\mathbb{H}^2}(\imath\, y, \imath v) = \left|\log\left(\frac{y}{w}\right)\right|$ for $y,\ v \in \mathbb{R}_{>0}$, we see that $f \circ g$ is the identity on the imaginary axis and hence also on $\mathbb{H}^2$, by the previous proposition. We conclude that g is a Möbius transformation. □

8.4.3 Unit Disc Model

Statements about the hyperbolic metric on the upper half plane can be translated to the unit disc model, where the *length* differential is $\mathrm{d}s = \lambda_{\mathbb{D}}(z)|\mathrm{d}z| = \dfrac{2\,|\mathrm{d}z|}{1 - |z|^2}$.

Most of these translations are routine, some are left as exercises, and other, do not require translation. We emphasize the following two results.

Theorem 8.39. *The set of orientation-preserving isometries of $\mathbb{D}$ consists of the fractional linear transformations mapping $\mathbb{D}$ to itself, that is,* Aut($\mathbb{D}$).

Proposition 8.40. *For all z and $w \in \mathbb{D}$,*

$$\rho_{\mathbb{D}}(w,z) = \log \frac{|1-w\bar{z}| + |w-z|}{|1-w\bar{z}| - |w-z|}. \tag{8.12}$$

In particular,

$$\rho_{\mathbb{D}}(0,z) = \log \frac{1+|z|}{1-|z|}. \tag{8.13}$$

Proof. See Exercise 8.16. □

8.4.4 Contractions and the Schwarz's Lemma

A deep connection between function theory and geometry is established through Schwarz's lemma. Recall that not every holomorphic self-map of $\mathbb{D}$ is a Möbius transformation (for instance, $z \mapsto z^2$), only conformal automorphisms are, and, as we have seen, these are isometries in the hyperbolic metric. However, the following result holds.

Theorem 8.41. *Holomorphic self-maps of the unit disc do not increase distances with respect to the hyperbolic metric; that is, for all holomorphic self-maps F of $\mathbb{D}$ and all z and $w \in \mathbb{D}$,*

$$\rho_{\mathbb{D}}(F(z), F(w)) \le \rho_{\mathbb{D}}(z,w)$$

and

$$\lambda_{\mathbb{D}}(F(z))|F'(z)| \le \lambda_{\mathbb{D}}(z).$$

Furthermore, if for some distinct z and $w \in \mathbb{D}$

$$\rho_{\mathbb{D}}(F(z), F(w)) = \rho_{\mathbb{D}}(z,w),$$

or for some $z \in \mathbb{D}$

$$\left|F'(z)\right| = \frac{\lambda_{\mathbb{D}}(z)}{\lambda_{\mathbb{D}}(F(z))},$$

then F is a conformal self-map of $\mathbb{D}$.

Proof. The two inequalities certainly hold for constant maps F. So assume that $F : \mathbb{D} \to \mathbb{D}$ is holomorphic and nonconstant.

Assume first that $F(0) = 0$. By Schwarz's lemma (Theorem 5.34),

$$|F(w)| \le |w| \text{ for } |w| < 1 \text{ and } \left|F'(0)\right| \le 1.$$

These are the Euclidean analogues of the two inequalities in the theorem for the points $z = 0$ and arbitrary $w \in \mathbb{D}$. We intend to use equation (8.13) to translate these to the non-Euclidean setting. We start with

$$\frac{1+|F(w)|}{1-|F(w)|} = (1+|F(w)|)(1+|F(w)|+|F(w)|^2+\cdots)$$
$$\leq (1+|w|)(1+|w|+|w|^2+\cdots) = \frac{1+|w|}{1-|w|}$$

and then apply (8.13) to obtain

$$\rho_{\mathbb{D}}(0, F(w)) \leq \rho_{\mathbb{D}}(0, w).$$

Let c and d denote two different points in $\mathbb{D}$; choose conformal automorphisms A and B of the unit disc such that $B(0) = c$ and $A(F(c)) = 0$. Then $A \circ F \circ B$ is a holomorphic self-map of the unit disc that fixes 0. Hence, by what was already established,

$$\rho_{\mathbb{D}}(A(F(B(0))), A(F(B(z)))) \leq \rho_{\mathbb{D}}(0, z)$$

for all z in $\mathbb{D}$. Since B and A are isometries,

$$\rho_{\mathbb{D}}(0, z) = \rho_{\mathbb{D}}(B(0), B(z))$$

and

$$\rho_{\mathbb{D}}(A(F(B(0))), A(F(B(z)))) = \rho_{\mathbb{D}}(F(B(0)), F(B(z)))$$

for all z in $\mathbb{D}$, and we conclude that

$$\rho_{\mathbb{D}}(F(B(0)), F(B(z))) \leq \rho_{\mathbb{D}}(B(0), B(z))$$

for all z in $\mathbb{D}$. Taking $z = B^{-1}(d)$, we obtain the first required inequality

$$\rho_{\mathbb{D}}(F(c), F(d)) \leq \rho_{\mathbb{D}}(c, d). \tag{8.14}$$

If we multiply each side of the last equation by

$$\left|\frac{1}{c-d}\right| = \left|\frac{1}{F(c)-F(d)} \cdot \frac{F(c)-F(d)}{c-d}\right|$$

and take the limit as c approaches d, we get the infinitesimal form of our required formula (the second inequality)

$$\lambda_{\mathbb{D}}(F(d))\,|F'(d)| \leq \lambda_{\mathbb{D}}(d). \tag{8.15}$$

We leave it to the reader to verify that equality in either (8.14) or (8.15) implies that F is conformal (thus also an isometry). □

Definition 8.42. Let M be a metric space with distance d. A self-map f of M is a *contraction* if $d(f(x), f(y)) < d(x, y)$ for all x and y in M with $x \neq y$.

Theorem 8.41 can be restated as

Theorem 8.43. *A holomorphic self-map of the unit disc is either an isometry or a contraction with respect to the hyperbolic metric.*

8.5 Finite Blaschke Products

For $a \in \mathbb{D}$, set

$$B_a(z) = -\frac{|a|}{a}\frac{z-a}{1-\bar{a}z}, \; z \in \mathbb{C},$$

with the understanding that $\frac{|a|}{a} = 1$ if $a = 0$.

Of course, B_a is the Riemann map f for $\mathbb{D}$ normalized (as before) by sending a to 0 but with (the change that) $\arg f'(a) = \pi - \arg a$; it is an automorphism of $\mathbb{D}$.

Let $\mathbb{A} = \{a_0, a_1, \ldots\}$ be a nonempty finite or countable sequence of complex numbers lying in the unit disc $\mathbb{D}$. Define

$$B_{\mathbb{A}} = B = \prod_i B_{a_i}.$$

Definition 8.44. The function $B_{\mathbb{A}}$ is the (finite or infinite) *Blaschke product associated to* $\mathbb{A}$. In the infinite case there are, of course, convergence issues (see Sect. 10.6).

For the rest of this section we will study *finite* Blaschke products. It follows immediately from the definition that we have

Proposition 8.45. *If $\mathbb{A} = \{a_0, a_1, \ldots, a_n\}$ is a nonempty finite sequence of points in $\mathbb{D}$, then*

(a) $B = B_{\mathbb{A}}$ *is a meromorphic function on $\mathbb{C} \cup \{\infty\}$, with zeros precisely*[5] *at the $n+1$ points $\{a_i\}$ and poles at the $n+1$ points $\left\{\frac{1}{\bar{a}_i}\right\}$, and*

---

[5]If a_i appears ν times in our list $\mathbb{A}$, then

$$\nu_{a_i}(B) = \nu \text{ and } \nu_{\frac{1}{\bar{a}_i}}(B) = -\nu.$$

(b) *B is a self-map of the closed unit disc, that maps the open unit disc holomorphically onto itself, and the unit circle onto itself, with $B(0) = \prod_i |a_i|$.*

Blaschke products transform beautifully under automorphisms of $\mathbb{D}$, as shown next.

Proposition 8.46. *Let $\mathbb{A} = \{a_0, a_1, \ldots, a_n\}$ be a nonempty finite sequence of points in $\mathbb{D}$, and let T be any element of $\operatorname{Aut}(\mathbb{D})$. Then*

$$B_{\mathbb{A}} \circ T = \lambda\, B_{T^{-1}(\mathbb{A})},$$

where λ is a constant of absolute value 1, and

$$T^{-1}(\mathbb{A}) = \{T^{-1}(a_0),\ T^{-1}(a_1), \ldots\}.$$

Proof. Since T belongs to $\operatorname{Aut}(\mathbb{D})$, there exist complex numbers a and b, with $|a|^2 - |b|^2 = 1$, such that $T(z) = \dfrac{a\,z + b}{\overline{b}\,z + \overline{a}}$ for all z in $\mathbb{D}$. It suffices to compute the action of T on the function $z \mapsto \dfrac{z - c}{1 - \overline{c}z}$, where $c \in \mathbb{D}$. A calculation shows that

$$\frac{T(z) - c}{1 - \overline{c}\,T(z)} = \frac{z - \dfrac{\overline{a}c - b}{-\overline{b}c + a}}{\dfrac{\overline{a} - b\overline{c}}{a - \overline{b}c}\left(1 - \dfrac{a\overline{c} - \overline{b}}{-b\overline{c} + \overline{a}}z\right)}.$$

The proof is completed by observing that $\left|\dfrac{\overline{a} - b\overline{c}}{a - \overline{b}c}\right| = 1$, and recalling that $T^{-1}(w) = \dfrac{\overline{a}w - b}{-\overline{b}w + a}$. □

Theorem 8.47. *Let f be a holomorphic self-map of the open unit disc $\mathbb{D}$, $\mathbb{A} = \{a_0, a_1, \ldots, a_n\}$ a nonempty finite collection of points in $\mathbb{D}$ and $B = B_{\mathbb{A}}$. Assume that $f(a_i) = 0$ for each a_i in $\mathbb{A}$, with multiplicities; that is, $\nu_{a_i}(f) \geq \nu_{a_i}(B)$ for all i. Then*

(a) *$|f(z)| \leq |B(z)|$ for all $z \in \mathbb{D}$, and $|f'(a_i)| \leq |B'(a_i)|$ for all i in $\{0, 1, \ldots, n\}$*
(b) *If $|f(a)| = |B(a)|$ for some $a \in \mathbb{D}$ with $a \neq a_i$ for all i, then there is $\lambda \in \mathbb{C}$ with $|\lambda| = 1$ such that*

$$f(z) = \lambda\, B(z) \text{ for all } z \in \mathbb{D} \tag{8.16}$$

(c) *If a_i appears ν times in the sequence $\mathbb{A}$ and if*

$$0 = f(a_i) = f'(a_i) = \cdots = f^{(\nu-1)}(a_i)$$

and

$$\left|f^{(\nu)}(a_i)\right| = \left|B^{(\nu)}(a_i)\right|,$$

then there is $\lambda \in \mathbb{C}$ *with* $|\lambda| = 1$ *such that* (8.16) *holds.*

Proof. To prove (a), it suffices to consider the special case with $a_0 = 0$. To verify this claim, let T be an automorphism of $\mathbb{D}$ that sends 0 to a_0; note that if we have the first inequality in (a) for $f \circ T$ and $B_{T^{-1}(\mathbb{A})}$, then, as a result of the last proposition, for all $z \in \mathbb{D}$,

$$|f(T(z))| \leq \left|B_{T^{-1}(\mathbb{A})}(z)\right| = |B(T(z))|,$$

and since $T(z)$ is an arbitrary point in $\mathbb{D}$, we thus also have the inequality for f and B.

We now assume that $a_0 = 0$ (thus $f(0) = 0$) and let $\mathcal{A}^1$ denote the sequence obtained from $\mathcal{A}$ after removing all occurrences of 0, and let B^1 be the Blaschke product associated to $\mathcal{A}^1$, if this sequence is nonempty. If $\mathcal{A}^1$ is empty, set $B^1 \equiv 1$.

The function $F(z) = \dfrac{f(z)}{z^{\nu_0(B)-1}B^1(z)}$ is certainly holomorphic on $\mathbb{D}$ and $F(0) = 0$. We claim that

$$|F(z)| \leq 1$$

for all $z \in \mathbb{D}$. Fix such a point z. The restrictions of $\left|B^1\right|$ to circles of radius r, with $0 \leq r \leq 1$, yield a family of functions that uniformly approach the constant function 1 as r approaches 1. Hence, for all $\epsilon > 0$, we can choose an r such that $\left|B^1(w)\right| \geq 1 - \epsilon$ for all w of absolute value r. Without loss of generality we can choose $r \geq |z|$. Hence, by the maximum principle for $|w| \leq r$, it follows that

$$|F(w)| = \frac{|f(w)|}{r^{\nu_0(B)-1}\left|B^1(w)\right|} \leq \frac{1}{r^{\nu_0(B)-1}(1-\epsilon)}.$$

Since ϵ is arbitrary and the inequality holds for all $|z| \leq r < 1$, the claim is proved. By Schwarz's lemma we obtain $|F(z)| \leq |z|$, or, equivalently,

$$|f(z)| \leq |z| \left|z^{\nu_0(B)-1}\right| \left|B^1(z)\right| = |B(z)|$$

for all $z \in \mathbb{D}$, proving the first inequality in (a).

For the inequality about derivatives, note that by the first inequality,

$$\left|\frac{f(z) - f(a_i)}{z - a_i}\right| = \left|\frac{f(z)}{z - a_i}\right| \leq \left|\frac{B(z)}{z - a_i}\right| = \left|\frac{B(z) - B(a_i)}{z - a_i}\right|.$$

Taking limits as z approaches a_i yields the second inequality.

To prove (b), note that it follows from (a) that the function $\dfrac{f}{B_{\mathbb{A}}}$ is analytic on $\mathbb{D}$, and its modulus is at most 1 on $\mathbb{D}$. If $|f(a)| = |B(a)|$ for some $a \in \mathbb{D} - \mathcal{A}$, then the modulus of $\dfrac{f}{B_{\mathbb{A}}}$ at a is 1; therefore this function is constant and (b) follows.

To prove (c), we may assume that $a_i = 0$. By l'Hopital's rule,

$$\left|\frac{f}{B_{\mathbb{A}}}(0)\right| = \left|\frac{f^{(\nu)}(0)}{B_{\mathbb{A}}^{(\nu)}(0)}\right| = 1,$$

and (8.16) follows. □

The particular case when $\mathbb{A}$ consists of one point leads to the following interesting result, a generalization of Schwarz's lemma that the reader may have encountered before as an exercise.

Corollary 8.48 (Schwarz-Pick Lemma). *Let f be a holomorphic self-map of the open unit disc $\mathbb{D}$. If $a \in \mathbb{D}$, then*

$$\left|\frac{f(z) - f(a)}{1 - \overline{f(a)}f(z)}\right| \le \left|\frac{z-a}{1-\bar{a}z}\right| \quad \text{for all } z \in \mathbb{D} \text{ and } \left|f'(a)\right| \le \frac{1 - |f(a)|^2}{1-|a|^2}.$$

Furthermore, equality for one $z \neq a$ in the first inequality or equality in the second inequality implies that f is an automorphism of $\mathbb{D}$.

Proof. Apply the previous theorem to the function $B_{f(a)} \circ f$. □

This corollary may also be deduced by applying Schwarz's lemma to the function $B_{f(a)} \circ f \circ B_a^{-1}$ (see Exercise 5.4) or, more interesting, is equivalent to Theorem 8.43 (Exercise 8.22).

Exercises

8.1. A matrix $A = \begin{bmatrix} a & b \\ c & d \end{bmatrix}$ with $a, b, c, d \in \mathbb{C}$, $ad - bc \neq 0$, acts on $\widehat{\mathbb{C}}$ by $z \mapsto \frac{az+b}{cz+d}$. Show that for each $t \in \mathbb{C}_{\neq 0}$, A and tA induce the same action.

8.2. Suppose the four distinct points z_1, z_2, z_3, z_4 in $\widehat{\mathbb{C}}$ are permuted. What effect will this have on the cross ratio (z_1, z_2, z_3, z_4)?

8.3. Prove in detail that the angles at z_1 and z_2 in Fig. 8.1 are equal precisely when the four points lie on a circle in $\widehat{\mathbb{C}}$.

8.4. In the text we established, using a geometric argument, that the cross ratio of four distinct points $\{z_1, z_2, z_3, z_4\}$ on the Riemann sphere $\mathbb{C} \cup \{\infty\}$ is real if and only if the four points lie on a circle. This exercise presents an alternate, purely analytic, way of establishing this result.

1. Show that every Möbius transformation maps the extended real line $\mathbb{R} \cup \{\infty\}$ onto a circle in $\mathbb{C} \cup \{\infty\}$.

2. Use the last assertion to show that (z_1, z_2, z_3, z_4) is real if and only if the four distinct points $\{z_1, z_2, z_3, z_4\}$ lie on a circle in $\mathbb{C} \cup \{\infty\}$.

8.5. Show that $f \in \mathrm{Aut}(\mathbb{D})$ if and only if there exist θ in $\mathbb{R}$ and $a \in \mathbb{D}$ such that

$$f(z) = \mathrm{e}^{\imath\theta} \frac{z-a}{1-\bar{a}\,z} \quad \text{for all } z \in \mathbb{D}.$$

8.6. This exercise explores the transitivity properties of groups of Möbius transformations and provides alternate formulations and proofs for many of the properties of Möbius transformations developed in this chapter by other methods.

1. We begin with the study of $\mathrm{PSL}(2, \mathbb{C})$ viewed as the group of conformal automorphisms of $\widehat{\mathbb{C}}$, and let $a,\ b,\ c$ be three distinct points in $\widehat{\mathbb{C}}$.

   - Assume that all the points are finite. Show that the Möbius transformation

   $$A(z) = \frac{c(a-b)z + a(b-c)}{(a-b)z + (b-c)}, \quad z \in \widehat{\mathbb{C}},$$

   maps the ordered triple $0,\ 1,\ \infty$ in $\widehat{\mathbb{C}}$ to the ordered triple $a,\ b,\ c$ in $\mathbb{C}$.
   - Show that by appropriate limiting processes applied to the formula for A, we can extend the above result to ordered triples of distinct points $a,\ b,\ c$ in $\widehat{\mathbb{C}}$. What are the resulting formulae?
   - Show by direct computation that the identity is the only Möbius transformation that fixes 0, 1, and ∞.
   - Determine the analogues of the above formulae for mappings of an arbitrary triple in $\widehat{\mathbb{C}}$ to another such triple. Thus establishing once again the unique triple transitivity of $\mathrm{PSL}(2, \mathbb{C})$ which can be summarized as: given two ordered triples of distinct points (a_0, b_0, c_0) and (a, b, c) in $\widehat{\mathbb{C}}$, there exists a unique $A \in \mathrm{PSL}(2, \mathbb{C})$ mapping the first ordered triple to the second one.
   - We now allow appropriate repetitions. Say our triples are (a_0, a_0, c_0) and (a, b, c). There are certain limitations on b that should be described. For a_0 and a finite, (in this case $b \neq 0$) the requirement for A should be that $A(a_0) = a,\ A'(a_0) = b$, and $A(c_0) = c$, of course. Show that there exists a unique such A and determine its formula. Extend the discussion to allow a_0 to be infinite.
   - Extend the previous analysis to triples (a_0, a_0, a_0) and (a, b, c).

2. We can view $\mathrm{PSL}(2, \mathbb{R})$ as a group of automorphisms of $\widehat{\mathbb{R}}$. Formulate and establish the analogues of the previous part to this setting.
3. More relevant for our work is to study $\mathrm{PSL}(2, \mathbb{R})$ as a group of automorphisms of $\mathbb{H}^2 \cup \widehat{\mathbb{R}}$. Toward this aim, let a_0 and $a \in \mathbb{H}^2$, and b_0 and $b \in \widehat{\mathbb{R}}$.

   - Show that there is a unique $A \in \mathrm{PSL}(2, R)$ mapping the ordered pair (a_0, b_0) to the ordered pair (a, b). What are the appropriate formulae for A?

- Let (a_0, b_0) and (a, b) be pairs of distinct points in $\mathbb{H}^2$. Show that there is a unique $A \in \mathrm{PSL}(2, \mathbb{R})$ mapping the ordered pair (a_0, b_0) to the ordered pair (a, b) iff

$$\rho_{\mathbb{H}^2}(a_0, b_0) = \rho_{\mathbb{H}^2}(a, b).$$

  Determine the formula for A.
- Extend the last claim to allow repetitions.

4. Translate the previous set of exercises on automorphisms of $\mathbb{H}^2$ to automorphisms of $\mathbb{D}$.

8.7. Formulate and prove (as a consequence of Theorem 8.20) the RMT for appropriate simply connected domains $D \subseteq \widehat{\mathbb{C}}$. Include the possibility that $c = \infty$.

8.8. This exercise deals with the construction of specific Riemann maps.

1. Find a conformal map from the disc $\{z \in \mathbb{C}; |z| < 1\}$ onto $\{z \in \mathbb{C}; |z| < 1, \Im z > 0\}$.
2. Let D be the domain in the extended complex plane $\widehat{\mathbb{C}}$ exterior to the circles $|z - 1| = 1$ and $|z + 1| = 1$. Find a Riemann map of D onto the strip $S = \{z \in \mathbb{C};\ 0 < \Im z < 2\}$.
3. Find a conformal map from the domain in $\widehat{\mathbb{C}}$ defined by

$$\{z \in \mathbb{C}; |z - 1| > 1, |z + 1| > 1\} \cup \{\infty\}$$

   onto the upper half plane.
4. For each $n = 1, 2, 3, \ldots$, find a conformal map from the infinite angular sector $0 < \operatorname{Arg} z < \dfrac{\pi}{n}$ onto the unit disc.
5. Find the Riemann map f from the strip $0 < \Im z < 1$ onto the unit disc satisfying $f\left(\frac{\imath}{2}\right) = 0$ and $f'\left(\frac{\imath}{2}\right) > 0$.
6. Find a conformal map from the domain

$$\{z \in \mathbb{C}; |z| < 1 \text{ and } z \neq t \text{ for } 0 \leq t < 1\}$$

   onto $\{w \in \mathbb{C}; |w| < 1\}$.
7. Find a conformal map from the upper half plane onto the unit disc minus the nonnegative real numbers.

8.9. If $f(z) = w$ is a Riemann map from the domain $|\operatorname{Arg} z| < \frac{\pi}{100}$ onto the domain $|w| < 1$ and if $f(1) = 0$ and $f'(1) > 0$, find $f(2)$.

8.10. If $w = g(z)$ maps the quadrant $\{z = x + \imath y \in \mathbb{C}; x > 0, y > 0\}$ conformally onto $|w| < 1$ with $g(1) = 1$, $g(\imath) = -1$ and $g(0) = -\imath$, find $|g'(1 + \imath)|$.

8.11. If f is holomorphic for $|z| < 1$ and satisfies $|f(z)| < 1$ for $|z| < 1$ and $f(0) = f\left(\frac{1}{2}\right) = 0$, show that

$$|f(z)| \le \left| z \cdot \frac{2z-1}{2-z} \right| \quad \text{for all } |z| < 1.$$

8.12. Suppose $\{f_n\}$ is a sequence of holomorphic functions in $|z| < 1$ that satisfy

$$\Re f_n(z) > 0 \text{ and } |f_n(0) - \imath| < \frac{1}{2} \quad \text{for all } n \in \mathbb{Z}_{>0} \text{ and all } |z| < 1.$$

Show that $\{f_n\}$ contains a subsequence that converges uniformly on compact subsets of the unit disc.

8.13. Supply the details of the proof that $\rho_{\mathbb{H}^2}$ defines a metric in $\mathbb{H}^2$ that is invariant under the Möbius group $\mathrm{PSL}(2, \mathbb{R})$.

8.14. Let z and w be points in $\mathbb{H}^2$ with z^* and w^* the corresponding points as in Definition 8.31; set $D(z, w) = |\log(z^*, w^*, w, z)|$. Show that D defines a metric in the upper half plane.

8.15. Let z and w be points in the upper half plane. Prove that $D(z, w) = \log \dfrac{|z - \overline{w}| + |z - w|}{|z - \overline{w}| - |z - w|}$ defines a metric in the upper half plane.

8.16. Prove that

$$\rho_{\mathbb{D}}(0, z) = \log \frac{1 + |z|}{1 - |z|} \text{ for all } z \in \mathbb{D}. \tag{8.17}$$

8.17. Show that for all $z_1 = x_1 + \imath\, y_1$ and $z_2 = x_2 + \imath\, y_2 \in \mathbb{H}^2$,

$$\rho_{\mathbb{H}^2}(z_1, z_2) = \operatorname{arccosh}\left(1 + \frac{(x_2 - x_1)^2 + (y_2 - y_1)^2}{2 y_1 y_2}\right).$$

8.18. This exercise establishes some of the basic facts about geodesics in $\mathbb{H}^2$ with respect to the hyperbolic metric in slightly different form than developed in the text. It should be established using the formula (8.8) for the infinitesimal form for the metric, but ignoring all facts about geodesics that we already established.

1. Show that $\imath \mathbb{R}_{>0}$ is a geodesic in $\mathbb{H}^2$.
2. Let γ be a pdp in $\mathbb{H}^2$. Prove that γ is a geodesic in $\mathbb{H}^2$ iff for all c and $d \in$ range γ, the length of the piece of γ between c and d equals $\rho_{\mathbb{H}^2}(c, d)$.
3. Show that an arbitrary geodesic in $\mathbb{H}^2$ is a semi-circle (this includes infinite line segments) perpendicular to $\mathbb{R}$.
4. Let c and d be two distinct points in $\mathbb{H}^2$. Show that there exists a unique geodesic passing through them.
5. Formulate and prove the corresponding fact about a given single point and a given direction from it.

8.19. This exercise contrasts the transitivity of $\mathrm{PSL}(2,\mathbb{C})$ acting on $\mathbb{C} \cup \{\infty\}$ (established in the text) with that of $\mathrm{PSL}(2,\mathbb{R})$ acting on $\mathbb{H}^2$ (to be established here). It should be compared with Exercise 8.6.

1. Show that $A \in \mathrm{PSL}(2,\mathbb{R})$ fixes $\imath$ iff it is of the form $\pm \begin{bmatrix} a & b \\ -b & a \end{bmatrix}$, with $a, b \in \mathbb{R}$ and $a^2 + b^2 = 1$.
2. Let $\theta \in \mathbb{R}$. Show that there exists a unique $A \in \mathrm{PSL}(2,\mathbb{R})$ with $A(\imath) = \imath$ and $\theta \in \arg A'(\imath)$.
3. Let z and $w \in \mathbb{H}^2$ and $\theta \in \mathbb{R}$. Show that there exists a unique $A \in \mathrm{PSL}(2,\mathbb{R})$ with $A(z) = w$ and $\theta \in \arg A'(z)$.
4. Let $\{a, b\}$ and $\{z, w\}$ be two pairs of distinct points in $\mathbb{H}^2$. Show that there exists a unique $A \in \mathrm{PSL}(2,\mathbb{R})$ with $A(a) = z$ and $A(b) = w$ iff $\rho_{\mathbb{H}^2}(a, b) = \rho_{\mathbb{H}^2}(z, w)$.

8.20. Assume f is a bounded holomorphic function on the unit disc and $f\left(\frac{\imath}{2}\right) = f\left(-\frac{\imath}{2}\right) = 0$. Show that there exists a bounded holomorphic function G on $\mathbb{D}$ such that

$$f(z) = \frac{z - \frac{\imath}{2}}{1 + \frac{\imath}{2}z} \, \frac{z + \frac{\imath}{2}}{1 - \frac{\imath}{2}z} \, G(z), \quad \text{for all } z \in \mathbb{D}.$$

8.21. Let $\mathbb{D} \subset \widehat{\mathbb{C}}$ be a disc. Show that $\mathrm{Aut}(\mathbb{D})$ consists of Möbius transformations.

8.22. Establish the equivalence of Theorem 8.43 and Corollary 8.48.

Chapter 9
Harmonic Functions

This chapter is devoted to the study of harmonic functions. They are closely connected to holomorphic maps, since these as well as their real and imaginary parts are harmonic. The study of harmonic functions is important in physics and engineering, and there are many results in the theory of harmonic functions that are not directly connected with complex analysis. In the early parts of this chapter we consider that part of the theory of harmonic functions that grows out of the Cauchy theory. Mathematically this is quite pleasing. One of the most important aspects of harmonic functions is that they solve a *boundary value problem*, known as the Dirichlet problem. An example is the problem of finding a function that is continuous on a closed disc, that assumes prescribed values on the boundary of the disc, and is harmonic in the interior of the disc. An important tool in the solution is the Poisson formula.

In the first section we define harmonic functions and the Laplacian of a function. In the second we obtain integral representations for harmonic functions that are analogous to the Cauchy integral formula, including the Poisson formula; in the third we use these integral representations to solve the Dirichlet problem. The third section includes three interpretations of the Poisson formula: a geometric interpretation, a Fourier series interpretation, and a classical one that involves differential forms. In the fourth section we characterize harmonic functions by their mean value property (MVP). The fifth section deals with the reflection principle for holomorphic and real-valued harmonic functions, a simple but useful extension tool.

In the second part of this chapter, starting with the sixth section, we study the class of subharmonic functions and use them to give a more complete solution to the Dirichlet problem, introduce the Green's function, and provide an alternate proof of the Riemann mapping theorem (RMT). The first two sections in this part introduce the new class of functions and study their basic properties. Perron families of subharmonic functions are studied in the next section, followed by a section in which the Dirichlet problem is revisited. The final section is devoted to the main aim of the second part of this chapter—existence proofs for Green's function and an alternate proof of the RMT.

R.E. Rodríguez et al., *Complex Analysis: In the Spirit of Lipman Bers*, Graduate Texts in Mathematics 245, DOI 10.1007/978-1-4419-7323-8_9,

9.1 Harmonic Functions and the Laplacian

We begin with

Definition 9.1. Let D be a domain in $\mathbb{C}$ and $g \in \mathbf{C}^2(D)$. We define Δg, the *Laplacian of* g, by

$$\Delta g = \frac{\partial^2 g}{\partial x^2} + \frac{\partial^2 g}{\partial y^2}; \qquad \text{(Laplace)}$$

Δ is also called the *Laplacian* or the *Laplace operator*. If D is a domain in $\mathbb{C}$ and $g \in \mathbf{C}^2(D)$, we say that g is *harmonic* in D if it satisfies *Laplace's equation* $\Delta g = 0$ in D.

Remark 9.2. The analogous of harmonic functions in one real variable are linear and constant functions.

The last two definitions have a number of immediate consequences:

(1) It is obvious from the definition of the Laplacian as a linear operator on $\mathbf{C}^2$ complex-valued functions that it preserves real-valued functions. It is useful to have equivalent formulae for it (Exercise 9.1):

$$\Delta = \frac{\partial^2}{\partial x^2} + \frac{\partial^2}{\partial y^2} = 4\frac{\partial^2}{\partial \bar{z}\partial z} = 4\frac{\partial^2}{\partial z\partial \bar{z}}, \qquad (9.1)$$

and in polar coordinates (r, θ)

$$\Delta = \frac{1}{r^2}\left(r\frac{\partial}{\partial r}\left(r\frac{\partial}{\partial r}\right) + \frac{\partial^2}{\partial \theta^2}\right). \qquad (9.2)$$

(2) Recall that for $f \in \mathbf{C}^1(D)$, f is holomorphic on D if and only if $\frac{\partial f}{\partial \bar{z}} = 0$ in D. Thus, for $g \in \mathbf{C}^2(D)$, g is harmonic if and only if $\frac{\partial g}{\partial z}$ is holomorphic.

In particular, holomorphic functions are harmonic and (2) gives an easy way to construct analytic functions from harmonic ones.

(3) f is harmonic if and only if $\overline{f}$ is.

(4) f is harmonic if and only if $\Re f$ and $\Im f$ are (this follows from the linearity of Δ and (3)).

(5) If f is holomorphic on an open set D, then f, $\Re f$, $\Im f$, and $\overline{f}$ are harmonic on D.

(6) If f is holomorphic or anti-holomorphic on D and g is harmonic on $f(D)$, then $g \circ f$ is harmonic on D.

Proof. Assume that f is holomorphic, let $w = f(z)$, and use the chain rule (see Exercise 2.10) to conclude:

$$(g \circ f)_z = g_w f_z + g_{\overline{w}} \overline{f}_z = g_w f_z$$

and

$$(g \circ f)_{z\overline{z}} = g_{ww} f_{\overline{z}} f_z + g_{w\overline{w}} \overline{f}_{\overline{z}} f_z + g_w f_{z\overline{z}} = 0.$$

The argument in the anti-holomorphic case is similar. □

(7) If $f \in \mathbf{C}^2(D)$ and f is locally on D the real part of an analytic function, then f is harmonic on D.

Example 9.3. $\log |z|$ is harmonic on $D = \mathbb{C}_{\neq 0}$, since it belongs to $\mathbf{C}^2(D)$, and it is locally the real part of $\log z$, a multi-valued but holomorphic function in D.

Proposition 9.4. *If g is real-valued and harmonic in D, then it is locally the real part of an analytic function. The analytic function is unique up to an additive constant.*

Proof. Let $D' \subseteq D$ be a simply connected region. Since g is harmonic in D', it follows from (2) above that $2g_z\, dz$ is closed on D', and hence an exact form on D'. Choose a holomorphic function f on D' with $d f = 2g_z\, dz$. Then

$$d\overline{f} = 2g_{\overline{z}}\, d\overline{z}$$

and hence

$$\frac{1}{2} d(f + \overline{f}) = dg;$$

that is,

$$g = \Re f + \text{constant}$$

on D'. □

Corollary 9.5. *A real-valued harmonic function on a simply connected domain is the real part of a holomorphic function on the same domain.*

Corollary 9.6. *A harmonic function is $\mathbf{C}^\infty$.*

Corollary 9.7. *Harmonic functions have the MVP, and hence they satisfy the maximum modulus principle. Real-valued harmonic functions also satisfy the maximum and minimum principles.*

Proof. See Definition 5.28 and the properties that follow thereof. □

Remark 9.8. The maximum (minimum) principle asserts that if f is real-valued and harmonic on a domain D and if f has a relative maximum (minimum) at a point $c \in D$, then f is constant in a neighborhood of c. Furthermore, if D is bounded and if f is also continuous on the closure of D, with $m \leq f \leq M$ on ∂D for some real constants m and M, then $m \leq f \leq M$ on D.

9.2 Integral Representation of Harmonic Functions

We apply the Cauchy theory towards our present main goal, which is to solve a boundary value problem. Given a harmonic function defined in a disc, we derive an integral formula for it, known as the *Poisson formula*; a major tool in the solution of our boundary value problem.

Proposition 9.9 (The Poisson Formula). *If g is a harmonic function on the domain $|z| < \rho$ for some $\rho > 0$, then, for each $0 < r < \rho$,*

$$g(z) = \frac{1}{2\pi}\int_0^{2\pi} g(re^{\imath\theta}) \cdot \frac{r^2 - |z|^2}{\left|re^{\imath\theta} - z\right|^2}\, d\theta \quad for\ |z| < r. \tag{9.3}$$

Proof. It suffices to assume that g is real-valued. To establish this formula we can thus apply Proposition 9.4 and choose the holomorphic function f on this domain with $\Re f = g$ and $g(0) = f(0)$, noting that there is a unique such f.

The function f has a power series expansion at 0: writing $z = r\,e^{\imath\theta}$,

$$f(r\,e^{\imath\theta}) = f(z) = \sum_{n=0}^{\infty} a_n z^n, \quad \text{with } a_0 = f(0) = g(0) \in \mathbb{R}. \tag{9.4}$$

Then

$$\begin{aligned} g(z) &= \frac{1}{2}\left(f(z) + \overline{f(z)}\right) = \frac{1}{2}\sum_{n=0}^{\infty}\left(a_n z^n + \overline{a_n z^n}\right) \\ &= a_0 + \frac{1}{2}\sum_{n=1}^{\infty} r^n \left(a_n e^{\imath n\theta} + \overline{a_n}\, e^{-\imath n\theta}\right). \end{aligned} \tag{9.5}$$

Integration of (9.5) along the curve $\gamma(\theta) = r\,e^{\imath\theta}$, for $0 \le \theta \le 2\pi$, yields

$$a_0 = \frac{1}{2\pi}\int_0^{2\pi} g(r\,e^{\imath\theta})\, d\theta.$$

Multiplying (9.5) by $e^{-\imath n\theta}$ for $n \in \mathbb{Z}_{>0}$ and integrating along the same curve, we obtain

$$\frac{1}{2} r^n a_n = \frac{1}{2\pi}\int_0^{2\pi} g(re^{\imath\theta}) \cdot e^{-\imath n\theta}\, d\theta;$$

or, equivalently,

$$a_n = \frac{1}{\pi}\int_0^{2\pi} \frac{g(re^{\imath\theta})}{(re^{\imath\theta})^n}\, d\theta\,, \quad \text{for } n \ge 1.$$

Thus, for $|w| < r$, we obtain from (9.4)

$$f(w) = \frac{1}{2\pi}\int_0^{2\pi} g(re^{\imath\,\theta})\cdot\left[1 + 2\sum_{n\geq 1}\left(\frac{w}{re^{\imath\,\theta}}\right)^n\right]\,d\theta\,.$$

But

$$1 + 2\sum_{n\geq 1}\left(\frac{w}{re^{\imath\,\theta}}\right)^n = \frac{re^{\imath\,\theta}+w}{re^{\imath\,\theta}-w},$$

and thus

$$f(w) = \frac{1}{2\pi}\int_0^{2\pi} g(re^{\imath\,\theta})\cdot\frac{re^{\imath\,\theta}+w}{re^{\imath\,\theta}-w}\,d\theta \quad \text{for } |w| < r\,.$$

The last formula gives a representation of a holomorphic function f in terms of its real part g, when the function f is real at 0. Taking the real part of both sides and renaming the variable w to z we obtain equation (9.3), the Poisson formula. □

The function

$$\frac{r^2-|z|^2}{\left|re^{\imath\,\theta}-z\right|^2} = \Re\left(\frac{re^{\imath\,\theta}+z}{re^{\imath\,\theta}-z}\right) \qquad \text{(Poisson kernel)}$$

is known as the *Poisson kernel.* Note that setting $z = 0$ in formula (9.3) we obtain again the *MVP for harmonic functions*.

The original derivation of formula (9.3) assumed that g was harmonic in the closed disc $\{|z| \leq r\}$. However, the result remains true for $|z| < r$ under the weaker assumption that g is harmonic in the open disc $\{|w| < r\}$ and continuous on its closure. In this case, fix t with $0 < t < 1$ and look at the function of z given by $g(tz)$. It is harmonic on the closed disc $\{|w| \leq r\}$ and hence, by the already proven formula (9.3),

$$g(tz) = \frac{1}{2\pi}\int_0^{2\pi}\frac{r^2-|z|^2}{\left|re^{\imath\theta}-z\right|^2}\cdot g(tre^{\imath\theta})\,d\theta.$$

Since the function g is uniformly continuous on the closed disc, we know that $g(tz)$ approaches $g(z)$ uniformly on the circle $\{|w| = r\}$ as t approaches 1. Hence both sides of the last equation converge to the expected quantities.

As a special case we apply the Poisson formula to the function g which is identically 1 and obtain

$$\int_0^{2\pi}\frac{r^2-|z|^2}{\left|re^{\imath\,\theta}-z\right|^2}\,d\theta = 2\pi \quad \text{for all } z\in\mathbb{C} \text{ with } |z| < r. \tag{9.6}$$

Definition 9.10. A *harmonic conjugate* of a real-valued harmonic function u is any real-valued function v such that $u + \imath\, v$ is holomorphic.

Harmonic conjugates always exist locally, and globally on simply connected domains. They are unique up to additive real constants. In fact, it is easy to see that they are given locally as follows.

Proposition 9.11. *If g is harmonic and real-valued in $|z| < \rho$ for some $\rho > 0$, then the harmonic conjugate of g vanishing at the origin is given by*

$$\frac{1}{2\pi\imath}\int_0^{2\pi} g(re^{\imath\theta})\cdot\frac{re^{-\imath\theta}z - re^{\imath\theta}\bar{z}}{\left|re^{\imath\theta} - z\right|^2}\,d\theta, \quad \text{for } |z| < r < \rho.$$

The following result is interesting and useful.

Theorem 9.12 (Harnack's Inequalities). *If g is a positive harmonic function on $|z| < r$ that is continuous on $|z| \le r$, then*

$$\frac{r-|z|}{r+|z|}\cdot g(0) \le g(z) \le \frac{r+|z|}{r-|z|}\cdot g(0)\,, \quad \text{for all } |z| < r.$$

Proof. Our starting point is (9.3). We use elementary estimates for the Poisson kernel:

$$\frac{r-|z|}{r+|z|} = \frac{r^2-|z|^2}{(r+|z|)^2} \le \frac{r^2-|z|^2}{\left|re^{\imath\theta}-z\right|^2} \le \frac{r^2-|z|^2}{(r-|z|)^2} = \frac{r+|z|}{r-|z|}.$$

Multiplying these inequalities by the positive number $g(w) = g(r\,e^{\imath\theta})$ and then averaging the resulting function over the circle $|w| = r$, we obtain

$$\frac{r-|z|}{r+|z|}\cdot\frac{1}{2\pi}\int_0^{2\pi} g(re^{\imath\theta})\,d\theta \le \frac{1}{2\pi}\int_0^{2\pi} g(re^{\imath\theta})\cdot\frac{r^2-|z|^2}{\left|re^{\imath\theta}-z\right|^2}\,d\theta$$
$$\le \frac{r+|z|}{r-|z|}\cdot\frac{1}{2\pi}\int_0^{2\pi} g(re^{\imath\theta})\,d\theta.$$

The middle term in the above inequalities is $g(z)$ as a consequence of (9.3), while the extreme averages are equal to $g(0)$ by the MVP. □

Remark 9.13. Exercise 9.6 gives a remarkable consequence of Harnack's inequalities that we use in establishing our next result.

Theorem 9.14 (Harnack's Convergence Theorem). *Let D be a domain and let $\{u_j\}$ be a nondecreasing sequence of real-valued harmonic functions on D. Then*

(a) *Either* $\lim_{j\to\infty} u_j(z) = +\infty$ *for all* $z \in D$

(b) *The function on D defined by* $U(z) = \lim_{j\to\infty} u_j(z)$ *is harmonic in D.*

Proof. Since a nondecreasing sequence of real numbers converges if and only if it is bounded, the assumption that $\lim_{j\to\infty} u_j(z)$ is not $+\infty$ for all $z \in D$ allows us to conclude that there exist z_0 in D and a real number M such that $u_j(z_0) < M$ for all j. Then $\lim_{j\to\infty} u_j(z_0)$ exists, and it equals the value of the series

$$u_1(z_0) + \sum_{n=1}^{\infty} [u_{n+1}(z_0) - u_n(z_0)],$$

which is therefore convergent.

Let K denote a compact subset of D. By enlarging K if necessary, we may assume that $z_0 \in K$. It follows from Harnack's inequalities (see Exercise 9.6) that there exists a real constant c such that

$$0 \le u_{n+1}(z) - u_n(z) \le c\,[u_{n+1}(z_0) - u_n(z_0)]$$

for all z in K and all n in $\mathbb{N}$.

It follows immediately that the series $u_1(z) + \sum_{n=1}^{\infty} [u_{n+1}(z) - u_n(z)]$ converges uniformly on K; that is, u_j converges uniformly to a function U on compact subsets of D. It is now easy to show that U is harmonic in D. □

9.3 The Dirichlet Problem

Let D be a bounded region in $\mathbb{C}$ and let $f \in \mathbf{C}(\partial D)$. The *Dirichlet problem* is to find a continuous function u defined on the closure of D that agrees with f on the boundary of D and whose restriction to D is harmonic.

We will consider, for the moment, only the special case where D is a disc; without loss of generality we may assume that the disc has radius one and center at zero.

For a piecewise continuous function u on S^1 and $z \in \mathbb{C}$ with $|z| < 1$, we define (compare with (9.3))

$$P[u](z) = \frac{1}{2\pi}\int_0^{2\pi} u(e^{\imath\theta}) \cdot \Re\left(\frac{e^{\imath\theta} + z}{e^{\imath\theta} - z}\right) d\theta \tag{9.7}$$

or, equivalently,

$$P[u](z) = \frac{1}{2\pi}\int_0^{2\pi} u(e^{\imath\theta}) \cdot \frac{1 - |z|^2}{|e^{\imath\theta} - z|^2}\, d\theta. \tag{9.8}$$

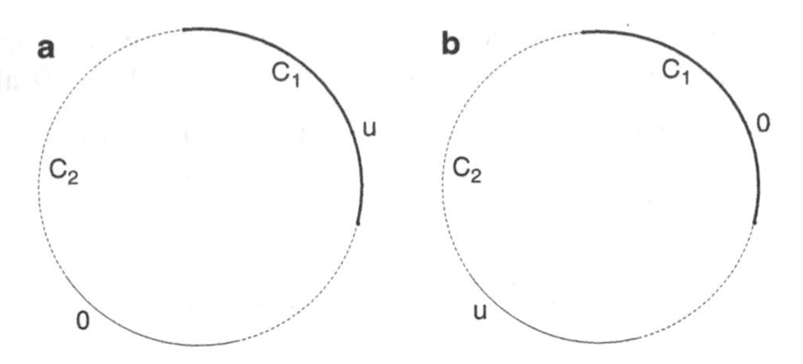

Fig. 9.1 u_1 and u_2. (**a**) The function u_1. (**b**) The function u_2.

The following properties of the operator P are easily established:

1. $P[u]$ is a well-defined function on the open unit disc. Hence we view P as an operator that assigns the function $P[u]$, on the open unit disc to each piecewise continuous function u on the unit circle.
2. $P[u+v] = P[u] + P[v]$ and $P[cu] = c \cdot Pu$ for all piecewise continuous functions u and v on S^1 and every constant c (thus P is a linear operator).
3. If u is a real *nonnegative* piecewise continuous function on S^1, then $P[u]$ is a real-valued nonnegative function on the open unit disc.
4. $P[u]$ is harmonic in the open unit disc. To establish this claim we may assume (by linearity of the operator P) that u is real-valued. In this case, $P[u]$ is obviously the real part of an analytic function on the disc.
5. For all constants c,
$$P[c] = c,$$
as follows from (9.6) (or directly because constant functions are harmonic).
6. Properties 5 and 3 imply that any bound on u yields the same bound on Pu. For example, for real-valued function u satisfying $m \le u \le M$ for some real constants m and M, we have $m \le P[u] \le M$.

We now establish the solvability of the Dirichlet problem for discs.

Theorem 9.15 (H. A. Schwarz). *If u is a piecewise continuous function on the unit circle S^1, then the function $P[u]$ is harmonic on $\{|z| < 1\}$; furthermore, for $\theta_0 \in \mathbb{R}$, its limit as z approaches $e^{\imath\theta_0}$ is $u(e^{\imath\theta_0})$ provided u is continuous at $e^{\imath\theta_0}$. In particular, the Dirichlet problem is solvable for discs.*

Proof. We only have to study the boundary values for $P[u]$.

Let C_1 and C_2 be complementary arcs on the unit circle. Let u_1 be the function which coincides with u on C_1 and vanishes on C_2; let u_2 be the corresponding function for C_2 (see Fig. 9.1). Clearly $P[u] = P[u_1] + P[u_2]$.

The function $P[u_1]$ can be regarded as an integral over the arc C_1; hence it is harmonic on $\mathbb{C} - C_1$. The expression

$$\Re\left(\frac{e^{\imath\theta}+z}{e^{\imath\theta}-z}\right)=\frac{1-|z|^2}{\left|e^{\imath\theta}-z\right|^2}$$

vanishes on $|z|=1$ for $z\neq e^{\imath\theta}$. It follows that $P[u_1]$ is zero on the one-dimensional interior of the arc C_2. By continuity $P[u_1](z)$ approaches zero as z approaches a point in the interior of C_2.

In proving that $P[u]$ has limit $u(e^{\imath\theta_0})$ at $e^{\imath\theta_0}$, we may assume that $u(e^{\imath\theta_0})=0$ (if not replace u by $u-u(e^{\imath\theta_0})$). Under this assumption, given an $\epsilon>0$, we can find complementary arcs C_1 and C_2 on the unit circle such that $e^{\imath\theta_0}$ is an interior point of C_2 and $\left|u(e^{\imath\theta})\right|<\frac{\epsilon}{2}$ for $e^{\imath\theta}\in C_2$. This last condition implies that $\left|u_2(e^{\imath\theta})\right|<\frac{\epsilon}{2}$ for all $e^{\imath\theta}$, and hence $|Pu_2(z)|<\frac{\epsilon}{2}$ for all $|z|<1$.

But we also have that u_1 is continuous and vanishes at $e^{\imath\theta_0}$. Since $P[u_1]$ is continuous at $e^{\imath\theta_0}$ and agrees with u_1 there, there exists a $\delta>0$ such that $|P[u_1](z)|<\frac{\epsilon}{2}$ for $\left|z-e^{\imath\theta_0}\right|<\delta$. It follows that

$$|P[u](z)|\leq|P[u_1](z)|+|P[u_2](z)|<\epsilon$$

as long as $|z|<1$ and $\left|z-e^{\imath\theta_0}\right|<\delta$. This is the required continuity statement. □

9.3.1 Geometric Interpretation of the Poisson Formula

This interpretation is due to Schwarz; the presentation follows Ahlfors.[1]

Recipe. *To find $P[u](z)$ we replace $u(e^{\imath\theta})$—the value of the function u at the point $e^{\imath\theta}$—by its value $u(e^{\imath\theta^*})$ at the point $e^{\imath\theta^*}$ on the unit circle opposite to the point to $e^{\imath\theta}$ with respect to z (i.e., on the intersection of the line through $e^{\imath\theta}$ and z with the unit circle), and average these values over the unit circle (see Fig. 9.2).*

This conclusion comes from reinterpreting the second formula (9.8) defining $P[u](z)$ as follows: fix a point z inside the unit circle and a point $e^{\imath\theta}$ on the unit circle. Let $e^{\imath\theta^*}$ be the unique point on the unit circle which also lies on the straight line through z and $e^{\imath\theta}$.

High school geometry (similar triangles, see Fig. 9.2) or a calculation (i.e., using the law of cosines) yields

$$1-|z|^2=\left|e^{\imath\theta}-z\right|\cdot\left|e^{\imath\theta^*}-z\right|.$$

[1] See [1].

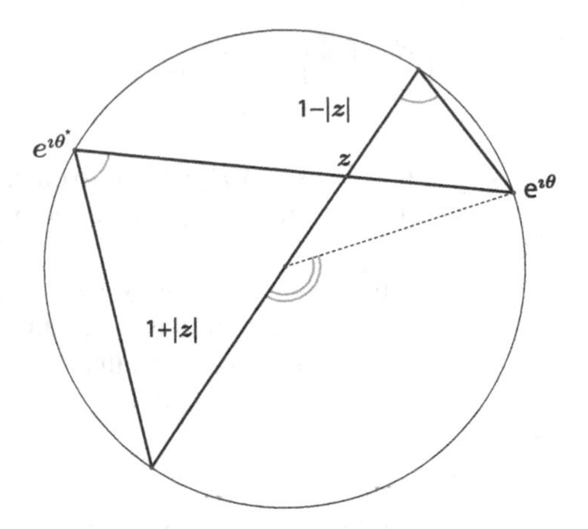

Fig. 9.2 The similar triangles

Since the three points z, $e^{\imath\theta}$, and $e^{\imath\theta^*}$ lie on a straight line and z is between the other two points, the ratio $\dfrac{e^{\imath\theta}-z}{e^{\imath\theta^*}-z}$ is a negative real number; it follows from this observation that

$$1-|z|^2 = -(e^{\imath\theta}-z)\cdot(e^{-\imath\theta^*}-\overline{z}). \tag{9.9}$$

To verify this last equality note that the real number

$$\frac{e^{\imath\theta}-z}{e^{\imath\theta^*}-z}\cdot\left|e^{\imath\theta^*}-z\right|^2 = \frac{e^{\imath\theta}-z}{e^{\imath\theta^*}-z}\cdot(e^{\imath\theta^*}-z)\cdot(e^{-\imath\theta^*}-\overline{z})$$

is negative and has the same absolute value as $(e^{\imath\theta}-z)\cdot(e^{\imath\theta^*}-z)$.

An alternate argument to obtain (9.9) follows. By definition there is a real number t such that

$$e^{\imath\theta^*} = t\,e^{\imath\theta} + (1-t)\,z;$$

it is easy to compute that $t = \dfrac{|z|^2-1}{\left|e^{\imath\theta}-z\right|^2}$.

It now follows from

$$e^{-\imath\theta^*}-\overline{z} = t\,(e^{-\imath\theta}-\overline{z})$$

that

$$(e^{\imath\theta}-z)\cdot(e^{-\imath\theta^*}-\overline{z}) = (e^{\imath\theta}-z)\cdot t\,(e^{-\imath\theta}-\overline{z}) = t\,\left|e^{\imath\theta}-z\right|^2 = |z|^2-1.$$

We now regard θ^* as a function of θ, with z fixed, and differentiate equality (9.9) logarithmically to obtain

$$\frac{e^{\imath\theta}}{e^{\imath\theta}-z}\,d\theta = \frac{e^{-\imath\theta^*}}{e^{-\imath\theta^*}-\overline{z}}\,d\theta^*.$$

Hence we see that (because θ^* is an increasing function of θ) that

$$\frac{d\theta^*}{d\theta} = \left|\frac{e^{\imath\theta^*}-z}{e^{\imath\theta}-z}\right| = \frac{1-|z|^2}{\left|e^{\imath\theta}-z\right|^2}.$$

We have thus shown that

$$P[u](z) = \frac{1}{2\pi}\int_0^{2\pi} u(e^{\imath\theta})\,d\theta^* = \frac{1}{2\pi}\int_0^{2\pi} u(e^{\imath\theta^*})\,d\theta.$$

The recipe follows.

9.3.2 Fourier Series Interpretation of the Poisson Formula

We again consider the case of the unit disc and proceed to compute the power series expansion of $P[u](z)$ at the origin.

Noting that

$$P[u](z) = \frac{1}{2\pi}\int_0^{2\pi} \frac{1-z\overline{z}}{(e^{\imath\theta}-z)\cdot(e^{-\imath\theta}-\overline{z})}\cdot u(e^{\imath\theta})\,d\theta \text{ for } |z|<1,$$

we start with an expansion of the Poisson kernel

$$\begin{aligned}\frac{1-z\overline{z}}{(e^{\imath\theta}-z)\cdot(e^{-\imath\theta}-\overline{z})} &= \frac{1-z\overline{z}}{(1-e^{-\imath\theta}z)\cdot(1-e^{\imath\theta}\overline{z})}\\ &= (1-z\overline{z})\sum_{n,m\ge 0} e^{-\imath n\theta}\cdot z^n\cdot e^{\imath m\theta}\overline{z}^m\\ &= (1-z\overline{z})\sum_{n,m\ge 0} e^{\imath(m-n)\theta}z^n\overline{z}^m\\ &= 1+\sum_{n=1}^{\infty} e^{-\imath n\theta}z^n + \sum_{m=1}^{\infty} e^{\imath m\theta}\overline{z}^m\end{aligned}$$

and since the last two series converge uniformly and absolutely on all compact subsets of the unit disc, we conclude

$$P[u](z) = a_0 + \sum_{n=1}^{\infty} a_n z^n + \sum_{m=1}^{\infty} b_m \overline{z}^m,$$

where, for $n \in \mathbb{Z}_{\geq 0}$ and $m \in \mathbb{Z}_{>0}$,

$$a_n = \frac{1}{2\pi} \int_0^{2\pi} e^{-\imath n \theta} \cdot u(e^{\imath \theta})\, d\theta,\ b_m = \frac{1}{2\pi} \int_0^{2\pi} e^{\imath m \theta} \cdot u(e^{\imath \theta})\, d\theta. \tag{9.10}$$

We thus have the following procedure for extending a given continuous function u on the unit circle to a continuous function on the closed unit disc that is harmonic on the interior of the disc. First compute the Fourier series of u:

$$u(e^{\imath\theta}) = \sum_{n=0}^{\infty} a_n e^{\imath n\theta} + \sum_{m=1}^{\infty} b_m e^{-\imath m\theta},$$

where the Fourier coefficients a_n and b_m are given by (9.10). In this series replace $e^{\imath n\theta}$ by z^n, for each $n \in \mathbb{Z}_{\geq 0}$, and $e^{-\imath m\theta}$ by $\bar{z}^m$, for each $m \in \mathbb{Z}_{>0}$.

9.3.3 Classical Reformulation of the Poisson Formula

For the next reformulation, we start with

Definition 9.16. If $\omega = P\, dx + Q\, dy$ is a differential form, we define $^*\omega$, the *conjugate differential of* ω, by

$$^*\omega = -Q\, dx + P\, dy.$$

If D is a simply connected domain in $\mathbb{C}$ and $u \in \mathbf{C}^2(D)$ is real-valued, we know that u is harmonic on D if and only if u is the real part of an analytic function f on D. In this case,

$$\begin{aligned} df &= f'(z)\, dz = (u_x - \imath u_y)(dx + \imath\, dy) \\ &= (u_x\, dx + u_y\, dy) + \imath\, (-u_y\, dx + u_x\, dy) \\ &= du + \imath\ ^*du \end{aligned}$$

is an exact differential on D and $^*du = dv$, where v is a harmonic conjugate of u on D. Thus du and $^*du$ are exact differential forms on D whenever u is a real-valued harmonic function on a simply connected domain D.

In what follows we work with cycles rather than curves. The definitions of cycles and cycles homologous to zero can be found in Sect. 5.2. In the general case, for a harmonic (including complex-valued) function u on an arbitrary (not necessarily simply connected) domain D, the form $du = u_x dx + u_y dy$ is always exact on D, and its conjugate differential $^*du = -u_y dx + u_x dy$ is closed since $(-u)_{yy} = u_{xx}$. We conclude that

$$\int_\gamma {}^*\mathrm{d}u = 0 \tag{9.11}$$

for all harmonic functions u on D and all cycles γ in D that are homologous to zero on that domain.

We can now turn to the classical reformulation of the Poisson formula.

Assume that γ is a *regular* curve with equation $z = z(t)$ (regular means that $z'(t) \neq 0$ for all t). The direction of the *tangent* line to the curve at $z(t)$ is determined by the angle $\alpha = \operatorname{Arg} z'(t)$ and

$$\mathrm{d}x = |\mathrm{d}z| \cos\alpha, \ \mathrm{d}y = |\mathrm{d}z| \sin\alpha.$$

The *normal line* at $z(t)$, which points to the right of the tangent line, has direction $\beta = \alpha - \frac{\pi}{2}$. The *normal derivative* of u is the directional derivative of u in the direction β:

$$\frac{\partial u}{\partial n} = u_x \cos\beta + u_y \sin\beta = u_x \sin\alpha - u_y \cos\alpha.$$

Thus we see that ${}^*\mathrm{d}u = \dfrac{\partial u}{\partial n}\ |\mathrm{d}z|$, and (9.11) can be rewritten as

$$\int_\gamma \frac{\partial u}{\partial n}\ |\mathrm{d}z| = 0.$$

It is important to realize that if γ is the circle $\{|z| = r\}$, then $\dfrac{\partial u}{\partial n} = \dfrac{\partial u}{\partial r}$. We prove an important generalization of (9.11).

Theorem 9.17. *If u_1 and u_2 are harmonic functions on D, then*

$$u_1\, {}^*\mathrm{d}u_2 - u_2\, {}^*\mathrm{d}u_1$$

is a closed form on D.

Proof. To establish this assertion, it involves no loss of generality to assume that the functions are real-valued (see Exercise 9.4), and hence we may also assume (because the issue is local) that each function u_j has a single-valued harmonic conjugate v_j; thus

$$u_1\, {}^*\mathrm{d}u_2 - u_2\, {}^*\mathrm{d}u_1 = u_1\, \mathrm{d}v_2 - u_2\, \mathrm{d}v_1 = u_1\, \mathrm{d}v_2 + v_1\, \mathrm{d}u_2 - \mathrm{d}(u_2 v_1).$$

The last expression $\mathrm{d}(u_2 v_1)$ is, of course, exact, and

$$u_1\, \mathrm{d}v_2 + v_1\, \mathrm{d}u_2 = \Im\left((u_1 + \imath\, v_1)(\mathrm{d}u_2 + \imath\, \mathrm{d}v_2)\right).$$

Now $u_1 + \imath v_1$ is an analytic function and $\mathrm{d}u_2 + \imath\, \mathrm{d}v_2$ is the total differential of an analytic function. By Cauchy's theorem, their product is a closed form, and hence we have shown that

$$\int_\gamma u_1 {}^*\mathrm{d}u_2 - u_2 {}^*\mathrm{d}u_1 = 0$$

for all cycles γ which are homologous to zero in D. In classical language the above formula reads

$$\int_\gamma \left(u_1 \frac{\partial u_2}{\partial n} - u_2 \frac{\partial u_1}{\partial n} \right) |\mathrm{d}z| = 0.$$

□

Let us take for D the annulus $\{z \in \mathbb{C}; R_1 < |z| < R_2\}$ and apply the above formula to the functions $z \mapsto u_1(z) = \log r$ (in polar coordinates) and an arbitrary harmonic function $u_2 = u$ on D. We take for γ the cycle $C_1 - C_2$ where C_j is the circle $\{|z| = r_j\}$ oriented counter clockwise; here $R_1 < r_1 < r_2 < R_2$. On any circle $\{|z| = r\}$, with $R_1 < r < R_2$, we have ${}^*\mathrm{d}u = r \cdot \frac{\partial u}{\partial r}\, \mathrm{d}\theta$. Hence we also see that

$$\log r_1 \int_{C_1} r_1 \cdot \frac{\partial u}{\partial r}\, \mathrm{d}\theta - \int_{C_1} u\, \mathrm{d}\theta = \log r_2 \int_{C_2} r_2 \cdot \frac{\partial u}{\partial r}\, \mathrm{d}\theta - \int_{C_2} u\, \mathrm{d}\theta$$

or

$$\log r \int_{|z|=r} r \cdot \frac{\partial u}{\partial r}\, \mathrm{d}\theta - \int_{|z|=r} u\, \mathrm{d}\theta = -B$$

is a constant (independent of r).

Applying the same argument to the functions $u_1 = 1$ (constant function) and $u_2 = u$, we obtain that

$$\int_{|z|=r} r \cdot \frac{\partial u}{\partial r}\, \mathrm{d}\theta = A$$

is constant over the annulus D, and hence is equal to zero (let $r \to 0$) if u is harmonic in the disc $\{z \in \mathbb{C}; |z| < R_2\}$.

Thus, for a function u harmonic in an annulus, the arithmetic mean over concentric circles $|z| = r$ is a linear function of $\log r$

$$\frac{1}{2\pi} \int_{|z|=r} u\, \mathrm{d}\theta = A \log r + B\,;$$

if u is harmonic in the disc $|z| < R_2$ or bounded in the punctured disc $0 < |z| < R_2$, then $A = 0$ and the arithmetic mean is constant. In the latter case, if u is harmonic in the disc, then $B = u(0)$ by continuity (the reader should know other proofs of this fact.)

The fact that A may be 0 in the above discussion has consequences:

Theorem 9.18. *If u is a bounded harmonic function on the punctured disc $0 < |z-a| < R$, then*

(a) *u extends to be harmonic on the disc $|z-a| < R$*
(b) *For real-valued u, the extension has a harmonic conjugate.*

We leave the proof to the reader, see Exercise 9.17.

Changing the origin, we see that if u is harmonic in $U(z_0, R)$, then for $0 < r < R$,

$$u(z_0) = \frac{1}{2\pi} \int_0^{2\pi} u(z_0 + r\, e^{\imath\theta})\, d\theta; \tag{9.12}$$

this is the MVP for harmonic functions, that was already established in Corollary 9.7 as a consequence of the fact that real-valued harmonic functions are locally real parts of analytic functions. From it one also obtains the *area MVP*

$$u(z_0) = \frac{1}{2\pi\imath r^2} \iint_{|z-z_0|\le r} u(z)\, dz\, d\bar{z}. \tag{9.13}$$

Remark 9.19. If $u : S^1 \to S^1$ is a homeomorphism, then $P[u]$ is also a homeomorphism, from $\{z; |z| < 1\}$ onto itself. This useful observation is not at all obvious, and not established here.

9.4 The Mean Value Property: A Characterization of Harmonicity

Harmonic functions satisfy the MVP, as we have seen in Corollary 9.7. As a matter of fact this property characterizes harmonic functions. The proof below is based on the solution to the Dirichlet problem.

Theorem 9.20. *A continuous complex-valued function that satisfies the* MVP *is harmonic.*

Proof. Let f be a continuous function on a domain D, let $c \in D$ and let $r_0 > 0$ be sufficiently small so that $\text{cl}\, U(c, r_0) \subset D$ and f satisfies (5.7) for all $r \le r_0$.

It suffices to assume that f is real-valued. Let v be the continuous function on $\{|z-c| \le r_0\}$ that is harmonic on $\{|z-c| < r_0\}$ and agrees with f on $\{|z-c| = r_0\}$. Then $f - v$ has the MVP in $\{|z-c| < r_0\}$, and thus attains its maximum and minimum on $\{|z-c| = r_0\}$. Since $f = v$ on $\{|z-c| = r_0\}$, we conclude that $f = v$ on $\{|z-c| \le r_0\}$ and thus that f is harmonic there. □

9.5 The Reflection Principle

We start with the simplest form of the general principle we are to establish. Let Ω be a nonempty region in the complex plane which is *symmetric* about the real axis; that is, $\bar{z} \in \Omega$ if and only if $z \in \Omega$ (see Fig. 9.3). Such a region must intersect the real axis nontrivially, and it is a disjoint union of three sets:

$$\Omega = \Omega^+ \cup \sigma \cup \Omega^-,$$

where

$$\Omega^+ = \{z \in \Omega;\ \Im z > 0\},\ \ \sigma = \Omega \cap \mathbb{R},\ \text{ and }\ \Omega^- = \{z \in \Omega;\ \Im z < 0\}.$$

Remark 9.21. A function $z \mapsto f(z)$ on a symmetric region Ω is harmonic (analytic) if and only if the function $z \mapsto \overline{f(\bar{z})}$ is (see Exercise 9.2).

We concentrate on the holomorphic case. Assume that $f \in \mathbf{H}(\Omega)$ and f is real on at least one segment of σ; then $f(z) = \overline{f(\bar{z})}$ for all $z \in \Omega$.

Proof. The function $z \mapsto g(z) = f(z) - \overline{f(\bar{z})}$ is analytic on Ω and vanishes on a subset of Ω with a limit point in Ω (namely, on σ); g is thus identically zero on Ω. □

The same conclusion holds if we merely assume that $f \in \mathbf{C}(\Omega^+ \cup \sigma)$, is analytic on Ω^+ and real on σ, since in this case the extension of f to Ω defined by $f(z) = \overline{f(\bar{z})}$ for $z \in \Omega^-$ satisfies the previous hypothesis. We now strengthen this statement considerably.

Theorem 9.22. *Let Ω be a nonempty region in the complex plane that is symmetric about the real axis.*

If v is a real-valued and continuous function on $\Omega^+ \cup \sigma$, and it is harmonic on Ω^+ and zero on σ, then v has a harmonic extension to Ω that satisfies the symmetry condition $v(z) = -v(\bar{z})$.

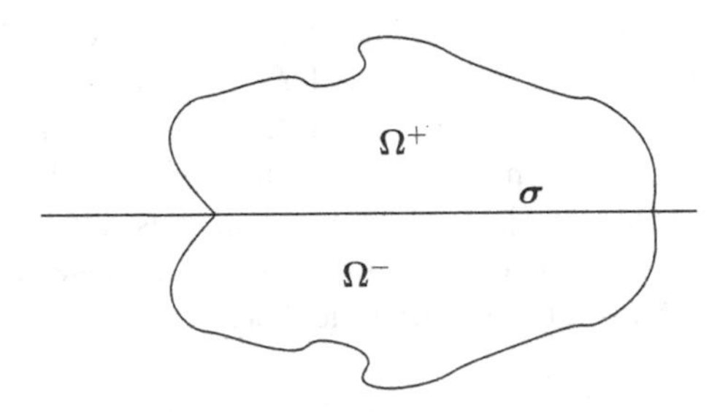

Fig. 9.3 A symmetric region

Moreover, if v is the imaginary part of an analytic function f in $\mathbf{H}(\Omega^+)$, then f has an analytic extension to Ω that satisfies the symmetry condition $f(z) = \overline{f(\bar{z})}$.

Proof. We use the symmetry to extend v to all of Ω. We show that the resulting extension (also called v) is continuous on Ω, harmonic on $\Omega^+ \cup \Omega^-$, and vanishes on σ.

Harmonicity of v in Ω is a local property; therefore we only need to show that v is harmonic in a neighborhood of each point $x \in \sigma$. For this, consider an open disc D with center at x whose closure is contained in Ω. Let V be the unique function that is continuous on cl D, harmonic on its interior, and agrees with v on the boundary of D. Since v restricted to ∂D satisfies the symmetry condition $v(z) = -v(\bar{z})$, so does the function V (on cl D). Hence V vanishes on cl $D^+ \cap \mathbb{R}$. The function $V - v$ is continuous on cl D^+, harmonic on D^+, and vanishes on its boundary; hence it is identically zero on cl D^+. Similarly, $V - v = 0$ on cl D^-. We conclude that $V = v$ on D, and we have shown that v is harmonic on Ω.

The function f has a symmetric extension to $\Omega^+ \cup \Omega^-$ (that satisfies $f(z) = \overline{f(\bar{z})}$). We only know that its imaginary part can be extended to all of Ω (and that the extension vanishes on σ). We must use information on the imaginary part of f to draw conclusions about its real part. Again, the problem is local, and we work with the disc D defined above. The real-valued harmonic function v on D has a harmonic conjugate $-u$ on this disc. The fact that harmonic conjugates are unique up to addition of real constants allows us to normalize so that $u = \Re f$ in D^+. We study the function

$$U(z) = u(z) - u(\bar{z}),\ z \in D.$$

This function vanishes on $D \cap \sigma$; hence

$$\frac{\partial U}{\partial x} = 0 \text{ for } (x, 0) \in D \cap \sigma.$$

Also, from the definition of U and the CR equations for the analytic function $u + \imath v$ on D,

$$\frac{\partial U}{\partial y} = 2\frac{\partial u}{\partial y} = -2\frac{\partial v}{\partial x} = 0 \text{ for } (x, 0) \in D.$$

Thus the analytic function $2U_z = U_x - \imath U_y$ vanishes on $D \cap \sigma$ and is hence identically zero on D; therefore U is constant on D. Since it vanishes on $D \cap \sigma$, this constant must be equal to zero. We have shown that on $D^+ \cup D^-$ the functions f and $u + \imath v$ agree. Since $u + \imath v$ is analytic on all of D, so is f. □

9.6 Subharmonic Functions

Definition 9.23. Consider a domain D in $\mathbb{C}$. A continuous real-valued function u on D is *subharmonic* if whenever G is a bounded subdomain of D such that $\partial G \subset D$ and φ is a continuous real-valued function in $G \cup \partial G$ such that

(i) φ is harmonic in G
(ii) $u(z) \le \varphi(z)$ for all z in ∂G

then

$$u(z) \le \varphi(z) \quad \text{for all } z \text{ in } G. \tag{9.14}$$

A function u is *superharmonic* if the function $-u$ is subharmonic.

Remark 9.24. The analogous of subharmonic (superharmonic) functions in one real variable are convex (concave) functions. We leave it to the reader to translate results that hold for subharmonic functions to the analogous results for superharmonic functions.

We discuss a number of key properties of the class of subharmonic functions:

1. A continuous real-valued function is harmonic if and only if it is both subharmonic and superharmonic. This local property is an immediate consequence of the definitions.
2. Harmonic functions are characterized by the mean value property (Theorem 9.20); that is, if u is continuous in D, then u is harmonic in D if and only if for all $z_0 \in D$ and all $r_0 > 0$ such that cl $U(z_0, r_0) \subset D$, u satisfies equality (9.12).

   Subharmonic functions are similarly characterized: If u is real-valued and continuous in D, then u is subharmonic in D if and only if for all $z_0 \in D$ and all $r_0 > 0$ such that cl $U(z_0, r_0) \subset D$, u satisfies the *mean value inequality*

$$u(z_0) \le \frac{1}{2\pi} \int_0^{2\pi} u(z_0 + r\,e^{\imath\theta})\,d\theta \quad \text{for all } 0 \le r < r_0. \tag{9.15}$$

Proof. First assume u is subharmonic in D, and consider z_0 and r_0 as above. Let φ denote the function (solution to the Dirichlet problem) that is continuous in cl $U(z_0, r_0)$, harmonic in $U(z_0, r_0)$, and coincides with u on $\partial U(z_0, r_0)$.

Then $u(z) \le \varphi(z)$ for all z with $|z - z_0| < r_0$. In particular, $u(z_0) \le \varphi(z_0)$, and $\varphi(z_0)$ is precisely the RHS of (9.15) by the MVP for harmonic functions.

We have established the "only if" part of the claim. To establish the "if" part, assume u is real-valued and continuous in D, and for all z_0 and r_0 as above, u satisfies (9.15). Let G be a bounded subdomain of D such that $\partial G \subset D$, and let φ denote a continuous real-valued function on $G \cup \partial G$ such that φ is harmonic in G, and $u(z) \le \varphi(z)$ for all z in ∂G. We need to show that $u(z) \le \varphi(z)$ for all z in G. If this were not so, then the function $v = u - \varphi$, being continuous in cl G, would attain a positive maximum M at some point in G, and the set

$$S = \{z \in \text{cl } G : v(z) = M\}$$

would be a nonempty closed set contained in G. Since both S and ∂G are compact, there exists a point c in S minimizing the distance from S to ∂G. Furthermore, on every circumference $\{|z - c| = r\}$, for small positive values of r, there would exist an arc α_r such that $v < M$ on α_r. Therefore

$$\frac{1}{2\pi}\int_0^{2\pi} u(c + r\,e^{\imath\theta})\,d\theta - \varphi(c) = \frac{1}{2\pi}\int_0^{2\pi} v(c + r\,e^{\imath\theta})\,d\theta$$
$$< M = u(c) - \varphi(c);$$

from this it follows that

$$\frac{1}{2\pi}\int_0^{2\pi} u(c + r\,e^{\imath\theta})\,d\theta < u(c)$$

for all small positive values of r, a contradiction. □

3. It follows from the mean value inequality (9.15) that a subharmonic function satisfies the maximum principle. It also follows from 2 that subharmonicity is a local property.
4. Let u be a subharmonic function in D with G and φ as in Definition 9.23. Condition (ii) of the definition can now be strengthened to:

   (ii)' If $u(z) \le \varphi(z)$ for all z in ∂G, then either

$$u(z) = \varphi(z) \text{ or } u(z) < \varphi(z) \text{ for all } z \text{ in } G.$$

   For suppose there exists c in G such that $u(c) = \varphi(c)$. Then

$$\varphi(c) = u(c) \le \frac{1}{2\pi}\int_0^{2\pi} u(c + r\,e^{\imath\theta})\,d\theta$$
$$\le \frac{1}{2\pi}\int_0^{2\pi} \varphi(c + r\,e^{\imath\theta})\,d\theta = \varphi(c)$$

   for all small positive values of r, and hence $u = \varphi$ near c. Thus $\{z \in G : u(z) = \varphi(z)\}$ is an open set, that we are assuming to be nonempty. On the other hand, u and φ are continuous functions on cl G, and hence $\{z \in G : u(z) < \varphi(z)\}$ is an open set, which must then be empty.
5. Let D be a domain in $\mathbb{C}$. If $\{u_n\}_{n\in\mathbb{N}}$ is a sequence of subharmonic functions on D that converges to a function u on D uniformly on all compact subsets of D, then u is subharmonic in D.

   We leave the proof as an exercise for the reader.
6. If u_1 and u_2 are subharmonic functions in D, then the function $u = \max\{u_1, u_2\}$ is also subharmonic in D.

   Proof. It is clear that u is continuous in D. Let $c \in D$. Without loss of generality we assume $u(c) = u_1(c)$. Then

$$u(c) = u_1(c) \le \frac{1}{2\pi}\int_0^{2\pi} u_1(c + r\,e^{\imath\theta})\,d\theta$$
$$\le \frac{1}{2\pi}\int_0^{2\pi} u(c + r\,e^{\imath\theta})\,d\theta$$

for all small positive values of r, and the result follows. □

This result provides many examples of subharmonic functions (that are not harmonic): If h is a real-valued harmonic function, then so is $-h$, and therefore $h^+ = \max\{h, 0\}$ and $|h| = \max\{h, -h\}$ are subharmonic. In particular, $u(z) = |z|^2$ is an example of a subharmonic function in $\mathbb{C}$ (that is not harmonic).

7. Harmonic functions are of class[2] $\mathbf{C}^2$, and they are characterized by the vanishing of their Laplacian. We have seen above that subharmonic functions need not be differentiable, but we now prove that a real-valued function u of class $\mathbf{C}^2$ in a domain D is subharmonic if and only if $\Delta u \ge 0$ in D.

Proof. First assume that $u \in \mathbf{C}^2_{\mathbb{R}}(D)$ and $\Delta u(z) > 0$ for all z in D. Let G be a bounded subdomain of D such that cl $G \subset D$, and let φ be a continuous real-valued function in cl G such that φ is harmonic in G, and $u(z) \le \varphi(z)$ for all z in ∂G. Assume $u(z) \not\le \varphi(z)$ for all z in G. Then $v = u - \varphi$ is a continuous function in cl G, and it attains its maximum at a point c in G. But v is of class $\mathbf{C}^2$ in G, and hence $\Delta v(c) \le 0$; but $\Delta v = \Delta u - \Delta\varphi = \Delta u$ and we have obtained a contradiction. Therefore u is subharmonic.

In the general case $u \in \mathbf{C}^2_{\mathbb{R}}(D)$ and $\Delta u(z) \ge 0$ for all z in D, set $u_\epsilon(z) = u(z) + \epsilon\,|z|^2$ for each positive ϵ and all z in D. Then u_ϵ is of class $\mathbf{C}^2$ in D, with $\Delta u_\epsilon > 0$ in D, and hence subharmonic in D for each positive ϵ. It is also clear that u_ϵ converges uniformly to u on compact subsets of D as ϵ goes to zero, and therefore u is subharmonic in D.

For the converse assume $u \in \mathbf{C}^2_{\mathbb{R}}(D)$ and u is subharmonic in D. Suppose that there exists c in D such that $\Delta u(c) < 0$. Then u is superharmonic in a neighborhood of c, and hence harmonic in that neighborhood, from where it follows that $\Delta u(c) = 0$, a contradiction. □

Example 9.25. The complex plane does not carry any negative nonconstant subharmonic function (Exercise 9.21). Whether a domain does or does not carry such a function is an important distinguishing characteristic. It classifies domains in $\widehat{\mathbb{C}}$ into three *types*, and serves as the basis for the next definition.

Definition 9.26. The Riemann sphere is called *elliptic*. A proper domain $D \subset \mathbb{C} \cup \{\infty\}$ is called *parabolic* if it does not carry a nonconstant bounded from above subharmonic function; otherwise it is called *hyperbolic*.

Remark 9.27. It is unfortunate that at times terminology, including our last definitions, can be confusing.

---

[2]They are, of course, much smoother than that, but this suffices for the discussion that follows.

- It is easy to see that the type of a domain is a conformal invariant in the sense that for all domains D and all conformal maps f, D and $f(D)$ have the same type. In Sect. 8.4 we gave a different definition of hyperbolic simply connected domains. We will see in Theorem 9.47 that both notions coincide for simply connected plane domains.
- However, according to the above definitions, the extended complex plane $\mathbb{C} \cup \{\infty\}$ punctured at a positive number n of points is parabolic. Whereas in the classification of domains in relation to what their simply connected holomorphic cover is, such domains are regarded to be hyperbolic if $n \geq 3$.

Example 9.28. Every bounded domain $D \subset \mathbb{C}$ is hyperbolic. Choose any $c \notin D$ and $R > 0$ so that D is contained in $U(c, R)$, and observe that the function $z \mapsto \log \left| \frac{z-c}{R} \right|$ is nonconstant, negative, and harmonic (hence subharmonic) on D.

9.7 Perron Families

To prove that the Dirichlet problem has a solution (in fact, a unique one) for certain domains (in addition to closed discs), we need to consider families of subharmonic functions that are closed under certain operations. We start by establishing some terminology.

Definition 9.29. Let D be a domain in $\mathbb{C}$, $u : D \to \mathbb{R}$ be a continuous function, and U be an open disc such that $\operatorname{cl} U \subset D$. The continuous function u_U defined in D by being harmonic in U and coinciding with u in $D - U$ is called the *harmonization* of u in U.

Lemma 9.30. *Let D be a domain in $\mathbb{C}$ and U be an open disc such that* $\operatorname{cl} U \subset D$. *If u is subharmonic in D, then so is u_U.*

Proof. It suffices to show that u_U satisfies the mean value inequality (9.15) at every point c in ∂U. Since u is subharmonic in D, $u(z) \leq u_U(z)$ for all z in D. But then

$$u_U(c) = u(c) \leq \frac{1}{2\pi} \int_0^{2\pi} u(c + r\,\mathrm{e}^{\imath\theta})\,\mathrm{d}\theta \leq \frac{1}{2\pi} \int_0^{2\pi} u_U(c + r\,\mathrm{e}^{\imath\theta})\,\mathrm{d}\theta$$

for all small positive values of r, and the result follows. □

Remark 9.31. We have shown that the family of subharmonic function on a domain D is a cone (i.e., the family is closed under addition and multiplication by positive constants) in the vector space of continuous functions on D. It is also closed under maximization (by property 6 above) and harmonization (by Lemma 9.30). These last two properties are the key to progress.

Definition 9.32. Let D be a domain in $\mathbb{C}$. A *Perron family* $\mathcal{F}$ in D is a nonempty collection of subharmonic functions in D such that

(a) If u, v are in $\mathcal{F}$, then so is $\max\{u, v\}$.
(b) If u is in $\mathcal{F}$, then so is u_U for every disc U with $\operatorname{cl} U \subset D$.

The following result, due to Perron, is useful for constructing harmonic functions.

Theorem 9.33 (Perron's Principle). *If $\mathcal{F}$ is a uniformly bounded from above Perron family in D, then the function defined for $z \in D$ by*

$$V(z) = \sup\{u(z) : u \in \mathcal{F}\} \tag{9.16}$$

is harmonic in D.

Proof. First note that by definition a Perron family is never empty. Since we are assuming that there exists a constant M such that $u(z) < M$ for all z in D and all u in $\mathcal{F}$, the function V is clearly well defined and real-valued.

Let U be any disc such that $\operatorname{cl} U \subset D$. It is enough to show that V is harmonic in U. For any point z_0 in U, there exists a sequence $\{u_j : j \in \mathbb{N}\}$ of functions in $\mathcal{F}$ such that

$$\lim_{j \to \infty} u_j(z_0) = V(z_0). \tag{9.17}$$

Without loss of generality, we may assume $u_{j+1} \geq u_j$ for all j in $\mathbb{N}$, since if $\{u_j\}$ is any sequence in $\mathcal{F}$ satisfying (9.17), then the new sequence given by $v_1 = u_1$ and $v_{j+1} = \max\{u_{j+1}, v_j\}$ for $j \geq 1$ is also contained in $\mathcal{F}$, satisfies (9.17) (with u_j replaced by v_j, of course), and is nondecreasing, as needed.

The sequence $\{w_j = (u_j)_U\}$ of harmonizations of the u_j in U consists of subharmonic functions with the following properties:

1. $w_j \geq u_j$ for all j
2. $w_j \leq w_{j+1} < M$ for all j,
   since the two inequalities clearly hold outside U and on the boundary of U, from which it follows that they also hold in U.

Thus the sequence $\{w_j\}$ lies in $\mathcal{F}$, is nondecreasing, and satisfies $\lim_{j \to \infty} w_j(z_0) = V(z_0)$.

It follows from the Harnack's convergence Theorem 9.14 that the function defined by

$$\Phi(z) = \lim_{j \to \infty} w_j(z) = \sup\{w_j(z) : j \in \mathbb{N}\}$$

is harmonic in U.

We will now show that $\Phi = V$ in U.

Let c denote any point in U. As before, we can find a nondecreasing sequence $\{s_j\}$ in $\mathcal{F}$ such that $V(c) = \lim_{j \to \infty} s_j(c)$.

By setting $t_1 = \max\{s_1, w_1\}$ and $t_{j+1} = \max\{s_{j+1}, w_{j+1}, t_j\}$ for all $j \geq 1$, we obtain a nondecreasing sequence $\{t_j\}$ in $\mathcal{F}$ such that $t_j \geq w_j$ for all j, and such that $\lim_{j \to \infty} t_j(z) = V(z)$ for $z = c$ and $z = z_0$.

The harmonizations of the t_j in U give a nondecreasing sequence $\{r_j = (t_j)_U\}$ in $\mathcal{F}$ satisfying $M > r_j \geq t_j \geq w_j$ for all j. As before, the function defined by

$$\Psi(z) = \sup\{r_j(z) : j \in \mathbb{N}\}$$

is harmonic in U, and coincides with V at c and z_0.

But $\Psi \geq \Phi$, since $r_j \geq w_j$ for all j, and hence $\Psi - \Phi$ is a nonnegative harmonic function in U. Since it is equal to zero at z_0, by the minimum principle for harmonic functions, it is identically zero in U, and the result follows. □

9.8 The Dirichlet Problem (Revisited)

This section has two parts. The first describes a method for obtaining the solution to the Dirichlet problem, provided it is solvable. In the second part, we offer a solution. Recall that the *Dirichlet problem* for a bounded region D in $\mathbb{C}$ and a function $f \in \mathbf{C}(\partial D)$ is to find a continuous function U on the closure of D whose restriction to D is harmonic and which agrees with f on the boundary of D. Under these conditions, let $\mathcal{F}$ denote the family of all continuous functions u on cl D such that u is subharmonic in D and $u \leq f$ on ∂D. Then $\mathcal{F}$ is a Perron family of functions uniformly bounded from above. Note that the constant function $u = \min\{f(z) : z \in \partial D\}$ belongs to $\mathcal{F}$, hence $\mathcal{F}$ is nonempty. The other conditions for $\mathcal{F}$ to be a Perron family are also easily verified. Therefore, by Theorem 9.33, the function V defined by (9.16) is harmonic in D.

Now, if we *assume* that there is a solution U to the Dirichlet problem for D and f, then we can show that $U = V$. Indeed, for each u in $\mathcal{F}$ the function $u - U$ is subharmonic in D, and satisfies $u - U = u - f \leq 0$ on ∂D, from where it follows that $u - U \leq 0$ in D, and hence $V \leq U$ in D. But U belongs to $\mathcal{F}$, and it follows that $U \leq V$, and therefore $U = V$.

The Dirichlet problem does not always have a solution. A very simple example is given by considering the domain $D = \{0 < |z| < 1\}$ and the function

$$f(z) = \begin{cases} 0, & \text{if } |z| = 1, \\ 1, & \text{if } z = 0. \end{cases}$$

The corresponding function V given by Theorem 9.33 is harmonic in the punctured disc D. If the Dirichlet problem were solvable in our case, then V would extend to a continuous function on $|z| \leq 1$ that is harmonic in $|z| < 1$ (see Exercise 9.17). But then the maximum principle would imply that V is identically zero, a contradiction.

To solve the Dirichlet problem, we start with a bounded domain $D \subset \mathbb{C}$, with boundary ∂D, and the following definition.

Definition 9.34. A function β is a *barrier* at $z_0 \in \partial D$, and z_0 is a *regular* point for the Dirichlet problem provided there exists an open neighborhood N of z_0 in $\mathbb{C}$ such that

(1) $\beta \in \mathbf{C}(\text{cl } D \cap N)$.
(2) $-\beta$ is subharmonic in $D \cap N$.
(3) $\beta(z) > 0$ for $z \neq z_0$, $\beta(z_0) = 0$.
(4) $\beta(z) = 1$ for $z \notin N$.

Remark 9.35. A few observations are in order.

1. Condition (4) is easily satisfied by adjusting a function β that satisfies the other three conditions for being a barrier. To see this we may assume that N is relatively compact in $\mathbb{C}$, and choose a smaller neighborhood N_0 of z_0 with cl $N_0 \subset N$. Then let

$$m = \min\{\beta(z); z \in \text{cl}(N - N_0) \cap \text{cl } D\},$$

note that $m > 0$, and define

$$\beta_1(z) = \begin{cases} \min\{m, \beta(z)\} & \text{for } z \in N \cap D, \\ m & \text{for } z \in \text{cl}(D - N). \end{cases}$$

Finally set $\beta_2 = \dfrac{\beta_1}{m}$, and observe that β_2 satisfies all the conditions for being a barrier at z_0.

Thus, to prove the existence of a barrier, it suffices to produce a function that satisfies the first three conditions.

2. The existence of barriers is a local property. If a point $z_0 \in \partial D$ can be reached by an *analytic* arc (a curve that is the image of [0, 1] under an injective analytic map defined in a neighborhood of [0, 1]) with no points in common with cl $D - \{z_0\}$, then a barrier exists at this point. To establish this we may, without loss of generality, assume that $z_0 = 0$, that the closure of D lies in the right half plane, and that the analytic arc consists of the negative real axis including the origin. Using polar coordinates $z = re^{\imath\theta}$, we see that $\beta(z) = r^{\frac{1}{2}} \cos \frac{\theta}{2}$, $-\pi < \theta < \pi$, satisfies the first three conditions for a barrier function.

Definition 9.36. Let D be a nonempty domain in $\mathbb{C}$. A solution u to the Dirichlet problem for $f \in \mathbf{C}_{\mathbb{R}}(\partial D)$ is *proper* provided

$$\inf\{f(w); w \in \partial D\} \leq u(z) \leq \sup\{f(w); w \in \partial D\}$$

for all z in D.

A far-reaching generalization of Schwartz's Theorem 9.15 is provided by our next result.

Theorem 9.37. *Let D be a nonempty domain in $\mathbb{C}$. There exists a proper solution to the Dirichlet problem for D for every bounded continuous real-valued function on ∂D if and only if every point on ∂D is a regular point for the Dirichlet problem.*

Proof. We leave it to the reader to discuss the cases where the boundary of D is empty or consists of a single point. So assume from now on that ∂D has two or more points.

First we assume that there exists a proper solution to the Dirichlet problem for D for every bounded continuous real-valued function on ∂D; we prove that every $z_0 \in \partial D$ is a regular point for the Dirichlet problem.

Define a continuous function f on $\mathbb{C}$ as follows:

$$f(z) = \begin{cases} |z - z_0| & \text{if } |z - z_0| \leq 1 \\ 1 & \text{if } |z - z_0| \geq 1 \end{cases}$$

and restrict it to ∂D. Let u be a proper solution to the Dirichlet problem for $f \in \mathbf{C}_{\mathbb{R}}(\partial D)$. Thus $0 \leq u \leq 1$. Since f is not identically zero, $u > 0$ by the minimum principle and is thus a barrier at z_0.

To prove the converse, let f be a bounded continuous real-valued function on ∂D, and let $\mathcal{F}$ consist of all functions v that satisfy the following conditions:

- v is continuous on cl D
- v is subharmonic on D
- v satisfies $m = \inf f \leq v \leq \sup f = M$ in D
- $v(z) \leq f(z)$ for all $z \in \partial D$.

Now observe that the constant function m belongs to $\mathcal{F}$; this family is obviously closed under maximization and harmonization, and is thus a Perron family, uniformly bounded from above by M.

It follows from Theorem 9.33 that the function V defined by (9.16) is harmonic in D and satisfies $m \leq V \leq M$ there.

Let $z \in \partial D$. To establish that $\lim_{w \to z} V(w) = f(z)$ we verify that

(a) $\liminf_{w \to z} V(w) \geq f(z)$
(b) $\limsup_{w \to z} V(w) \leq f(z)$

The arguments that follow are generic for this type of problem: for part (a) we construct a particular function $w \in \mathcal{F}$ that helps, while the proof of (b) involves arguing about all $v \in \mathcal{F}$.

Proof of (a): If $f(z) = m$, there is nothing to prove. Assume hence that $f(z) > m$. Choose a positive real ϵ such that $f(z) - \epsilon > m$, and note that ϵ can be chosen arbitrarily small. There exists a neighborhood N of z such that $f(\zeta) \geq f(z) - \epsilon$ for all $\zeta \in N \cap \partial D$. Let β be a barrier at z that is identically equal to 1 outside of N, and define

$$w(\zeta) = -(f(z) - m - \epsilon)\beta(\zeta) + f(z) - \epsilon \quad \text{for } \zeta \in \text{cl } D.$$

We show that the function $w \in \mathcal{F}$. Observe that $w \in \mathbf{C}_{\mathbb{R}}(\text{cl } D)$ is subharmonic on D.

For $\zeta \in \text{cl } D$,

$$w(\zeta) \le f(z) - \epsilon < M,$$

and

$$w(\zeta) = m\beta(\zeta) + (f(z) - \epsilon)(1 - \beta(\zeta)) > m\beta(\zeta) + m(1 - \beta(\zeta)) = m.$$

For $\zeta \in \partial D$,

$$w(\zeta) = m \le f(\zeta) \quad \text{if } \zeta \notin N$$

and

$$w(\zeta) \le f(z) - \epsilon \le f(\zeta) \quad \text{if } \zeta \in N.$$

We have completed the proof that $w \in \mathcal{F}$. Hence $w(\zeta) \le V(\zeta)$ for all $\zeta \in D$, and therefore

$$\liminf_{\zeta \to z} V(\zeta) \ge w(z) = f(z) - \epsilon.$$

Since ϵ is arbitrary, (a) follows.

Proof of (b): As before, there is nothing to verify under certain conditions: in this case, if $f(z) = M$. So we assume that $f(z) < M$ and choose $\epsilon > 0$ such that $f(z) + \epsilon < M$, and a relatively compact neighborhood N of z such that $f(\zeta) \le f(z) + \epsilon$ for $\zeta \in N \cap \partial D$. Let, as above, β be a barrier at z.

Fix an arbitrary $v \in \mathcal{F}$. We claim that

$$v(\zeta) - (M - f(z) - \epsilon)\beta(\zeta) \le f(z) + \epsilon \quad \text{for all } \zeta \in N \cap D. \tag{9.18}$$

Observe that the function on the LHS of (9.18) is subharmonic; hence it suffices to establish the inequality for $\zeta \in \partial(N \cap D)$; we consider two cases:

(i) $\zeta \in \partial N \cap \text{cl } D$, and
(ii) $\zeta \in \text{cl } N \cap \partial D$.

In case (i), the LHS of (9.18) satisfies

$$v(\zeta) - (M - f(z) - \epsilon)\beta(\zeta) = v(\zeta) - M + f(z) + \epsilon \le f(z) + \epsilon;$$

while in case (ii) we estimate it by

$$v(\zeta) \le f(\zeta) \le f(z) + \epsilon.$$

Hence, we obtain, for all $\zeta \in N \cap D$, the estimate on v given by

$$v(\zeta) \le f(z) + \epsilon + (M - f(z) - \epsilon)\beta(\zeta);$$

since $v \in \mathcal{F}$ is arbitrary, the same estimate holds for the function V. Thus

$$\limsup_{\zeta \to z} V(\zeta) \leq f(z) + \epsilon.$$

Since $\epsilon > 0$ can be chosen to be arbitrarily small, (b) follows. □

As a second application of Perron's method, we establish the following result.

Theorem 9.38. *Let D be a hyperbolic domain in $\mathbb{C}$ and K a compact subset of D with $D - K$ connected and ∂K regular (for the Dirichlet problem). There exists a unique function $\omega \in \mathbf{C}_{\mathbb{R}}((D - K) \cup \partial K)$ such that*

- *ω is harmonic on $D - K$.*
- *$\omega = 1$ on ∂K.*
- *$0 < \omega < 1$ on $D - K$.*
- *If $\omega_1 \in \mathbf{C}_{\mathbb{R}}((D - K) \cup \partial K)$ is a competing function that satisfies the above three properties, then $\omega_1 \geq \omega$.*

Proof. Let ψ_0 be any nonconstant positive superharmonic function on D, whose existence is guaranteed by the hyperbolicity of D. We now adjust this function to obtain a more suitable one.

Let m_0 be the minimum of ψ_0 on K, and set $\psi_1 = \dfrac{\psi_0}{m_0}$. Then ψ_1 is a positive, nonconstant, superharmonic function on D, with $\psi_1|_K \geq 1$.

There clearly exists a point $z \in K$ such that $\psi_1(z) = 1$. If z is in the interior of K, then $\psi_1 = 1$ on the connected component of K containing the point z, thus also on the boundary of that component. In particular, there exists a $z \in \partial K$ with $\psi_(z) = 1$. There also exists a point $w \in D - K$ with $\psi_1(w) < 1$, since otherwise $\psi_1 \geq 1$, and the fact that $\psi_1(z) = 1$ would imply that ψ_1 is constant.

Setting $\psi = \min\{1, \psi_1\}$ we obtain a superharmonic function on D that satisfies $0 < \psi \leq 1$, $\psi(w) < 1$, and $\psi|_K = 1$.

Let $\mathcal{F}$ consist of all functions $v \in \mathbf{C}_{\mathbb{R}}((D - K) \cup \partial K)$ satisfying the following conditions:

1. v is subharmonic on $D - K$.
2. $v \leq \psi|(D - K)$.
3. v has compact support.

The family we have defined is closed under maximization and harmonization, and thus a Perron family if not empty; we proceed to establish this next.

Choose a small disc in D so that its complement Δ in D contains K. Let v_0 be the solution to the Dirichlet problem on $\Delta - K$ with boundary values 1 on ∂K and 0 on $\partial \Delta$; then $0 \leq v_0 \leq 1$ on $\mathrm{cl}(\Delta - K)$. Extending v_0 to be 0 on $D - \Delta$ yields a subharmonic function on $D - K$. By the maximum principle, $v_0 - \psi \leq 0$ on $D - K$. (This inequality holds on $D - \Delta$ since $v_0|(D - \Delta) = 0$ and $\psi > 0$, hence also on $\partial \Delta$. On ∂K both functions v_0 and ψ have the constant value 1; thus $(v_0 - \psi)|(\Delta - K) \leq 0$.)

Defining

$$\omega(z) = \sup_{v \in \mathcal{F}}\{v(z)\} \quad \text{for all } z \in (D - K) \cup \partial K,$$

we see that

- $v_0 \le \omega \le \psi$
- ω is harmonic on $D - K$
- $\omega = 1$ on ∂K
- $\omega \in \mathbf{C}_{\mathbb{R}}((D - K) \cup \partial K)$
- ω is not constant, since $\omega(w) \le \psi(w) < 1$
- $0 \le \omega \le 1$ on $(D - K) \cup \partial K$
- $0 < \omega < 1$ on $D - K$

Finally let ω' be a competing function satisfying the above conditions. Let $v \in \mathcal{F}$ and consider the function $\omega' - v$. Assume that the compact set $K' \subset D$ contains the support of v. Since $\omega' \ge 0$ and $v = 0$ on $\partial K'$, $\omega' - v \ge 0$ on $\partial(K' - K)$, hence also on $(K' - K) \cup (D - K') = D - K$ and on ∂K. Since $v \in \mathcal{F}$ is arbitrary, $\omega' \ge \omega$. □

Definition 9.39. For D and K as above, the function ω described by the theorem is called the *harmonic measure of* K.

9.9 Green's Function and RMT Revisited

In this section we introduce a new function and use it to give a second proof of the RMT.

Definition 9.40. Let $D \subset \mathbb{C}$ be a nonempty domain and let $z \in D$. A real-valued function g on $D - \{z\}$ is the *Green's function for D with singularity at z* if

(1) g is harmonic and positive on $D - \{z\}$.
(2) The function $w \mapsto g(w) + \log|w - z|$ is (extends to be) harmonic on D
(3) $\tilde{g} \in \mathbf{C}_{\mathbb{R}}(D - \{z\})$ is another function satisfying the above two conditions, then $\tilde{g} \ge g$.

Remark 9.41. (a) The last condition in the definition guarantees that the Green's function is unique if it exists.
(b) The entire plane $\mathbb{C}$ does not have a Green's function with singularity at any point. Assume to the contrary that g is a Green's function for $\mathbb{C}$ with singularity at zero (this latter assumption involves no loss of generality). It follows that $z \mapsto g(z) + \log|z|$ extends to be a real-valued harmonic function u on $\mathbb{C}$.
Then u has a (single-valued) harmonic conjugate v, and thus $e^{-(u+\imath v)}$ is an entire function. Let $R > 0$; we estimate for $|z| = R$

$$\left|e^{-u(z)-\imath v(z)}\right| = \left|e^{-g(z)-\log|z|}\right| \le \frac{1}{|z|} = \frac{1}{R}.$$

and conclude that the inequality holds for all $|z| \leq R$. Since R may be arbitrarily large, we have arrived at a contradiction, the contradiction that our entire function (an exponential!) vanishes identically.

(c) The Green's function for the unit disc $\mathbb{D}$ with singularity at 0 is given by $z \mapsto -\log|z|$. For proper simply connected domains in $\mathbb{C}$, we may use the RMT to construct their Green's functions (see Exercise 9.24). We handle the reverse direction by giving a second proof of the RMT based on the existence of Green's functions—which we now proceed to prove.

The value of the Green's function for a (hyperbolic) domain D with singularity at $c \in D$ at any point $d \neq c$ in D will be denoted by $g_D(d, c)$, and by $g(d, c)$ when the domain is clear from the context.

Theorem 9.42. *Let D be a nonempty domain in $\mathbb{C}$. There exists a Green's function on D with singularity at some point $c \in D$ if and only if D is hyperbolic. In the latter case D has a Green's function with a singularity at any arbitrary point of D.*

Proof. Assume that there exists a Green's function on D with singularity at some point $c \in D$. It involves no loss of generality (by Exercise 9.25) to assume that D contains the unit disc and that $c = 0$. Let g be the Green's function for D with singularity at 0. Then $\lim_{z \to 0} g(z) = +\infty$.

Let $m > 0$. Then $f = \min\{g, m\}$ is positive and superharmonic on $D - \{0\}$, and f is constant (equal to m) near 0; therefore f is positive and superharmonic on D. By choosing m sufficiently large, we can make certain that f is not constant. Thus D is hyperbolic, since it carries the nonconstant negative subharmonic function $-f$.

To establish the converse, we define $\mathcal{F}$ to consist of all real-valued functions v satisfying the following conditions:

1. $v \geq 0$ is subharmonic on $D - \{0\}$.
2. $K_v = \{w \in D - \{0\}; v(w) \neq 0\} \cup \{0\}$ is compact ("roughly expressed,"v has compact support)
3. $w \mapsto v(w) + \log|w|$ is (extends to be) a subharmonic function on D.

To see that $\mathcal{F}$ is nonempty we define the function

$$v(w) = \max\{-\log|w|, 0\}$$

for $w \neq 0$. It is subharmonic on the plane punctured at the origin, hence certainly on $D - \{0\}$, and it clearly belongs to the family $\mathcal{F}$. It is quite obvious that this family is closed under maximization and harmonization; hence it is a Perron family.

We show next that the family $\mathcal{F}$ is uniformly bounded outside every neighborhood of 0. We have not yet used the hypothesis that D is hyperbolic. It is this last assertion that requires hyperbolicity (in the form of existence of harmonic measures, see Theorem 9.38).

Let $0 < r < 1$ and let ω_r be the harmonic measure of $\{|z| \leq r\}$. Thus ω_r is harmonic on $D - \{|z| \leq r\}$, $0 < \omega_r < 1$ on this set, and $\omega_r(z) = 1$ for $|z| = r$. Let $\lambda_r = \max\{\omega_r(z); |z| = 1\}$; hence $0 < \lambda_r < 1$.

For $u \in \mathcal{F}$, let $u_r = \max\{u(z); |z| = r\}$. We claim that

$$u_r\omega_r(z) - u(z) \geq 0 \text{ for all } z \in D \text{ such that } |z| \geq r. \tag{9.19}$$

Hence, in particular, we record for later use that

$$u_r\lambda_r - u(z) \geq 0 \text{ for } |z| = 1. \tag{9.20}$$

To verify (9.19), note that the function $u_r\omega_r - u$ is superharmonic on $D \cap \{|z| > r\}$ and is nonnegative on $\{|z| = r\}$. Let $K = K_u$ be the compact support of u. Since $u = 0$ on ∂K, $u_r\omega_r - u \geq 0$ there. Thus (9.19) holds for $K - \{|z| \leq r\}$; this inequality certainly holds on $D - K$, and the claim is verified.

Next we use the fact that $z \mapsto u(z) + \log|z|$ is continuous on $\{|z| \leq 1\}$ and subharmonic on $\{|z| < 1\}$. Thus

$$\begin{aligned} u_r + \log r &= \max\{u(z) : |z| = r\} + \log r \\ &\leq \max\{u(z) : |z| = 1\} + \log 1 \\ &\leq \lambda_r u_r \end{aligned}$$

where the last inequality follows from (9.20).

We conclude that

$$u_r \leq \frac{-\log r}{1 - \lambda_r},$$

and since $u = 0$ off the compact set K, we see that

$$\max\{u(z); z \in D \text{ and } |z| \geq r\} \leq \frac{\log r}{\lambda_r - 1}.$$

By Perron's principle (Theorem 9.33)

$$g(z) = \sup_{u \in \mathcal{F}} u(z), \ z \in D - \{0\}$$

defines a nonnegative harmonic function. This function is actually positive since it would otherwise be constant. Finally we must show that $z \mapsto g(z) + \log|z|$ is harmonic on $\{|z| < r\}$. For any such z and for every $u \in \mathcal{F}$,

$$u(z) + \log|z| \leq u_r + \log r \leq \frac{\lambda_r \log r}{\lambda_r - 1}.$$

Thus the same inequality holds when we replace u by g.

To complete the proof that g is the Green's function of D with singularity at 0, we must show that if $\tilde{g}$ is a competing candidate, then $\tilde{g} \geq g$ (from this it will follow that either $\tilde{g} > g$ or $\tilde{g} = g$). Now if $u \in \mathcal{F}$, then $U = \tilde{g} - u$ is superharmonic on

D. If K is the support of u, then $U \geq 0$ on $D - K$, and by the minimum principle for superharmonic functions also on K. Since $u \in \mathcal{F}$ is arbitrary, $\tilde{g} \geq g$. □

Remark 9.43. The above existence proof for Green's function generalizes to Riemann surfaces. We present next a second, simpler, argument that is valid only for plane domains and also yields additional information.

Theorem 9.44. *Let $D \subset \mathbb{C}$ be a domain with nonempty ∂D that is regular for the Dirichlet problem. Then D is hyperbolic.*

In particular, for every $c \in D$, let $u \in \mathbf{C}(D \cup \partial D)$ be a harmonic function in D with $u(z) = \log|z - c|$ for $z \in \partial D$. Then

$$g_D(z, c) = u(z) - \log|z - c| \quad \textit{for all } z \in D - \{c\}.$$

Proof. It suffices, of course, to prove the particular claim. Without loss of generality $\mathbb{D} = \{z \in \mathbb{C}; |z| < 1\} \subseteq D$ and $c = 0$.

Let $G(z) = u(z) - \log|z|$ for $z \in D \cup \partial D - \{0\}$. Observe that the hypothesis on D guarantees that u exists and that $u(z) - \log|z| = 0$ for $z \in \partial D$. For $|z| = 1$, $u(z) > 0$ and $\log|z| = 0$. It follows that $G(z) \geq 0$ for $z \in D - \mathbb{D}$. For $0 < |z| < 1$, $u(z) > 0$ and $\log|z| < 0$, hence $G(z) > 0$. Hence G is a positive harmonic function on $D - \{0\}$. Obviously $z \mapsto G(z) + \log|z|$ defines a harmonic function on D. So G is a candidate for the Green's function of D with singularity at 0.

Let $\tilde{g}$ be a competitor, and let $d \in D - \{0\}$ and $\epsilon > 0$. Choose a neighborhood N of ∂D in cl D such that $d \notin N$ and $G(z) < \epsilon$ for all $z \in N$. Then $\tilde{g} - G \geq -\epsilon$ on $\partial(D - N) = \partial N$, and thus also on $D - N$; in particular, $\tilde{g}(d) - G(d) \geq -\epsilon$. Since $d \in D$ is arbitrary and so is $\epsilon > 0$, $\tilde{g} \geq G$ and $G = g_D(\cdot, 0)$. □

The proofs of the next two results are left as exercises for the reader.

Theorem 9.45. *The complex plane punctured at $n \geq 0$ points is parabolic.*

Theorem 9.46. *Let D be a hyperbolic domain in $\mathbb{C}$. Then*

$$g(z, w) = g(w, z) \quad \textit{for all } z \neq w \in D.$$

The final application of our work on subharmonic functions is a second proof of

Theorem 9.47 (Riemann Mapping Theorem, Version 2). *Every hyperbolic simply connected domain in $\mathbb{C}$ is conformally equivalent to the unit disc $\mathbb{D}$.*

Proof. Let $c \in D$. We want to use the Green's function $g(\cdot, c)$ to produce a Riemann map $f(\cdot, c)$. If g had a harmonic conjugate $\tilde{g}$, we could define $f = e^{-(g + \imath\tilde{g})}$. The problem is that g has a singularity at c.

To get around this problem, we note that the function $u(\cdot, c)$ defined on D by $u(z, c) = g(z, c) + \log|z - c|$ is harmonic and real-valued. Since D is simply connected, $u(\cdot, c)$ has a harmonic conjugate $v(\cdot, c)$ in D. It follows that $F(z, c) = e^{-(u(z,c) + \imath v(z,c))}$ defines (for fixed c) a holomorphic function of z on D. Now

$$-u(z,c) = -g(z,c) - \log|z-c| = \Re \log(F(z,c)) = \log|F(z,c)|$$

or, equivalently,

$$-g(z,c) = \log|(z-c)F(z,c)|\,.$$

We define

$$f(z,c) = (z-c)F(z,c).$$

Obviously $f(\cdot,c)$ is a holomorphic function on D that vanishes (has a simple zero) only at c. Since $\log|f(z,c)| = -g(z,c) < 0$, $f(z,c) \in \mathbb{D}$ for all $z \in D$.

For any three points $c,\ d,\ z \in D$, set

$$\varphi(z) = \varphi(c,d,z) = \frac{f(d,c) - f(z,c)}{1 - \overline{f(d,c)}f(z,c)}. \tag{9.21}$$

Now $\varphi(c,d,\cdot)$ is $f(\cdot,c)$ followed by a Möbius transformation that leaves $\mathbb{D}$ invariant, and hence a holomorphic map of D into $\mathbb{D}$, with $\varphi(d) = \varphi(c,d,d) = 0$. In the above argument, we fixed c and d and regarded φ as a function of z. We need to relate φ to some f; for these purposes we may regard φ as a function of one of the three variables c, d, and z for fixed values of the other two. However, only for fixed c and d is φ a holomorphic function of z. Thus the emphasis on the variable z in the left-hand side of (9.21).

We study the function $\varphi(\cdot)$ in a neighborhood of $z = d$. Let $n \geq 1$ be the order of vanishing of φ at d; then we have the power series expansion

$$\varphi(z) = \alpha(z-d)^n\left(1 + a_1(z-d) + a_2(z-d)^2 + \cdots\right),$$

with $\alpha \neq 0$, for all z with $|z-d|$ sufficiently small, and if we set

$$v(z) = -\frac{1}{n}\log|\varphi(z)|,$$

we obtain a positive harmonic function on D punctured at the (isolated) zeros of φ. We may regard v to be defined on all of D with value $+\infty$ at the zeros of φ.

We proceed to compare v to $g(\cdot,d) = -\log|f(\cdot,d)|$. Let $\mathcal{F}$ be the family of functions used to define $g(\cdot,d)$. Let $w \in \mathcal{F}$ and let K be the compact support of w. By enlarging K if necessary, we may assume that none of the zeros of φ lie on ∂K. Let K_0 be the set K from which we delete small discs about each of the finitely many zeros excluding d of φ in the compact set K.

We claim that $v(z) - w(z) \geq 0$ for all $z \in K$. Let us first establish this claim for $z \in \partial K_0$. The boundary of K_0 consists of ∂K and finitely many circles about zeros of φ. The circles do not cause any problems since w is bounded on each of them (by a bound independent of the radius of the circle if we a priori specify that the radii of these circles be less than some fixed number) and $v \to +\infty$ as we shrink the circles further (which is clearly permissible). Now on the boundary of ∂K, $v \geq 0$ and $w = 0$. Since the superharmonic function $v - w$ on K is ≥ 0 on ∂K_0, by the minimum principle for superharmonic functions, it is also nonnegative on K_0. It follows that $v \geq w$ on $D - \{c\}$. Hence for $z \in D - \{c\}$ we have

$$-\frac{1}{n}\log|\varphi(z)| = v(z) \geq \sup_{h \in \mathcal{F}} h(z) = g(z,d) = -\log|f(z,d)|$$

or

$$|\varphi(z)| \leq |\varphi(z)|^{\frac{1}{n}} \leq |f(z,d)|\,. \tag{9.22}$$

Setting $z = c$ (since by continuity the last set of inequalities are also valid at c), we conclude that $|f(d,c)| \leq |f(c,d)|$, and since c and d are arbitrary, we may interchange them to obtain

$$|f(c,d)| = |f(d,c)|\,,$$

an equality that also follows from Theorem 9.46 (which we did not prove and hence are not using). We consider another holomorphic function from D into the closed unit disc. We define

$$h(z) = \frac{\varphi(z)}{f(z,d)},\ z \in D.$$

From (9.22), we conclude that $|h(z)| \leq 1$; since

$$|h(c)| = \left|\frac{\varphi(c)}{f(c,d)}\right| = \left|\frac{f(d,c) - f(c,c)}{1 - \overline{f(d,c)}f(c,c)}\frac{1}{f(c,d)}\right| = \left|\frac{f(d,c)}{f(c,d)}\right| = 1,$$

h is constant, and there exists a $\lambda \in \mathbb{C}$ with $|\lambda| = 1$ and such that $\varphi = \lambda f(\cdot,d)$, or, equivalently,

$$\lambda f(z,d) = \frac{f(d,c) - f(z,c)}{1 - \overline{f(d,c)}f(z,d)} \quad \text{for all } z \in D.$$

We conclude that

$$f(d,c) = f(z,c) \text{ iff } f(z,d) = 0 \text{ iff } z = d;$$

that is, $f(\cdot,c)$ is an injective holomorphic map of D into $\mathbb{D}$ that takes c to 0.

The proof concludes by showing that $f(\cdot,c)$ is surjective, in a manner quite similar to that used in the earlier proof of the RMT. Assume that we do not have surjectivity, and let $\Delta = f(\cdot,c)(D) \subset \mathbb{D}$. The domain Δ is simply connected and contains 0. Let t^2 be any point in $\mathbb{D} - \Delta$. We construct a function $h : \Delta \to \mathbb{D}$ by the formula

$$h(z) = \frac{\sqrt{\frac{z-t^2}{1-\bar{t}^2 z}} - \imath t}{1 + \imath\bar{t}\sqrt{\frac{z-t^2}{1-\bar{t}^2 z}}},\ z \in \Delta.$$

The square root is defined (and holomorphic) since Δ is simply connected and $\frac{z-t^2}{1-\bar{t}^2 z} \neq 0$ for $z \in \Delta$; we choose the square root so that $\sqrt{-t^2} = \imath t$. Straightforward calculations now show that

$$h(0) = 0, |h'(0)| = \frac{1 + |t|^2}{2|t|} > 1, |h(z)| < 1, \quad \text{for all } z \in \Delta.$$

We now define $W = h \circ f(\cdot, c)$ and observe that W is a holomorphic function from D into $\mathbb{D}$, with a simple zero at c and no other zeros. Hence $-\log|W|$ is a competing function for the Green's function $g(\cdot, c)$. Thus

$$-\log|W| \geq g(\cdot, c) = -\log|f(\cdot, c)| \text{ or } |W| \leq |f(\cdot, c)|.$$

We thus see that

$$\left|\frac{h(z)}{z}\right| \leq 1 \quad \text{for small positive } |z| \text{ and } |h'(0)| \leq 1;$$

a contradiction. □

Remark 9.48. Let $D \subset \mathbb{C}$ be a nonempty simply connected domain and let $c \in D$. We have introduced two numerical invariants for the pair (D, c). The first is $|f'(c)|$, where f is any Riemann map from D onto $\mathbb{D}$ with $f(c) = 0$. The second is the value at c of $g_D(\cdot, c) + \log|(\cdot - c)|$. The two invariants agree.

Exercises

9.1. Prove that the equivalent forms for the Laplacian given in equations (9.1) and (9.2) are correct.

9.2. Show that a function $z \mapsto f(z)$ on a symmetric region Ω is harmonic (analytic) if and only if the function $z \mapsto \overline{f(\bar{z})}$ is.

9.3. Show that if $\omega = f\, dz + g\, d\bar{z}$ is a continuous differential form on a domain D, then

$${}^*\omega = -\imath\,(f\, dz - g\, d\bar{z}).$$

9.4. We have shown that if u_1 and u_2 are real-valued harmonic functions on D, then

$$u_1\, {}^*du_2 - u_2\, {}^*du_1$$

is a closed form on D and have asserted that it also holds for complex-valued harmonic functions. Prove this assertion.

9.5. Prove the maximum and minimum principles for real-valued harmonic functions:

1. As a general result for real-valued functions that satisfy the MVP.

2. As a consequence of Harnack's inequalities for positive harmonic functions.

9.6. Let K be a compact subset of a domain $D \subseteq \mathbb{C}$, and let u be a positive harmonic function on D.

Show that there exists a constant $c \geq 1$ that depends only on K and D, but not on u, such that

$$\frac{1}{c} \leq \frac{u(z_1)}{u(z_2)} \leq c,$$

for all z_1 and $z_2 \in K$.

9.7. Complete the proof of Theorem 9.14 and show that both possibilities for its conclusion do occur.

9.8. Let u be a continuous real-valued function on a domain D.

Suppose that the partial derivatives $\dfrac{\partial^2 u}{\partial x^2}$ and $\dfrac{\partial^2 u}{\partial y^2}$ exist and satisfy Laplace's equation $\Delta u = 0$ in D. Show that u is harmonic on D.
Hint: Use the notation in the proof of Theorem 9.20. Let $c = a + \imath\, b$ in D. Show first that for all $\epsilon > 0$, the function

$$F(z) = u(z) - v(z) + \epsilon(x - a)^2$$

satisfies the maximum principle in $\{|z - \zeta| \leq r_0\}$.

9.9. Does the area MVP imply harmonicity for continuous functions?

9.10. If u is real-valued and harmonic on $|z| < 1$, continuous on $|z| \leq 1$, and $u(\mathrm{e}^{\imath\theta}) = \cos 2\theta + \sin 2\theta$, find $u\left(\frac{\imath}{2}\right)$.

9.11. Suppose that $u(0) = 1$, where u is harmonic and positive in a neighborhood of $\{z \in \mathbb{C}; |z| \leq 1\}$. Prove that $\frac{1}{7} \leq u\left(\frac{3}{4}\right) \leq 7$.

9.12. Let α be a real number. For $\zeta = \mathrm{e}^{\imath\theta}$ with $\theta \in \mathbb{R}$, let

$$\varphi(\zeta) = \cos\theta + \imath\,\alpha\,\sin\theta.$$

Which of the following assertions are true for all α in $\mathbb{R}$? Which are true for some values of α?

(a) The function $f(z) = \dfrac{1}{2\pi\imath}\displaystyle\int_{|\zeta|=1} \frac{\varphi(\zeta)}{\zeta - z}\,\mathrm{d}\zeta$ is holomorphic for $|z| < 1$.
(b) There exists a function f holomorphic for $|z| < 1$, continuous for $|z| \leq 1$ and satisfying $f(\zeta) = \varphi(\zeta)$ for $|\zeta| = 1$.
(c) There exists a function f holomorphic for $|z| < 1$ such that $\Re f$ is continuous for $|z| \leq 1$ and satisfies $\Re f(\zeta) = \Re\varphi(\zeta)$ for $|\zeta| = 1$.

9.13. Let g be a continuous complex-valued function defined on S^1. Prove that there exists a continuous function f on $\{z \in \mathbb{C};\ |z| \leq 1\}$ with f holomorphic on

$\{z \in \mathbb{C};\ |z| < 1\}$ and $f_{|_{S^1}} = g$ if and only if

$$\int_{|\zeta|=1} g(\zeta)\,\zeta^n\,\mathrm{d}\zeta = 0 \quad \text{for } n = 0, 1, 2, \ldots.$$

9.14. Does there exist a function f holomorphic on $|z| < 1$ such that

$$\lim_{z \to \zeta} f(z) = \zeta + \zeta^{-1} \quad \text{for all } \zeta \text{ with } |\zeta| = 1?$$

9.15. Let $f \in \mathbf{C}^2(D)$. Show that f is holomorphic or anti-holomorphic on D if and only if f and f^2 (the square of f) are harmonic on D.

9.16. Assume h is harmonic in a domain D. Prove that so are h_z, $h_{\bar{z}}$, h_x, and h_y. The rest of this exercise concerns universal bounds for harmonic functions and assume that $|h(z)| \le M$ for all $z \in D$.

- Let $D = U(z_0, R)$ for some $z_0 \in \mathbb{C}$ and some $R > 0$. Show that there exists a universal constant $c > 0$ such that the absolute values of each of the four partial derivatives evaluated at $z = 0$ are bounded by $\dfrac{c\,M}{R}$.
- Construct examples to show that the above estimates do not hold for arbitrary $z \in U(z_0, R)$.
- Let D be arbitrary again. Show that the absolute values of each of the four partial derivatives evaluated at $z \in D$ are bounded by $\dfrac{c\,M}{R - |z|}$, where R is the distance from z to the boundary of D.

9.17. If h is harmonic and $|h|$ is bounded for $0 < |z - z_0| < R$, show that h extends to be harmonic in $|z - z_0| < R$.

9.18. Let u be a nonconstant real-valued harmonic function on a domain D. Show that the set of *critical values* of $\mathrm{d}u = u_x \mathrm{d}x + u_y \mathrm{d}y$ (i.e., the set of points in D at which both u_x and u_y vanish) is discrete.

9.19. Show that if u is real valued and continuous in D, then u is superharmonic in D if and only if for all $\zeta \in D$, and $r_0 > 0$ sufficiently small so that $\mathrm{cl}\, U(\zeta, r_0) \subset D$, u satisfies

$$u(\zeta) \ge \frac{1}{2\pi} \int_0^{2\pi} u(\zeta + r\,\mathrm{e}^{\imath\theta})\,\mathrm{d}\theta \quad \text{for all } 0 \le r < r_0.$$

Conclude that superharmonic functions satisfy the minimum principle.

9.20. Show that if u is real valued and continuous in D, then u is subharmonic in D if and only if for all $\zeta \in D$, and $r_0 > 0$ sufficiently small so that $\mathrm{cl}\, U(\zeta, r_0) \subset D$, u satisfies the *Area Mean Value Inequality*

$$u(\zeta) \le \frac{-1}{2\pi\imath r^2} \iint_{|z-\zeta| \le r} u(z)\,\mathrm{d}z\,\mathrm{d}\bar{z}.$$

9.21. (a) Prove that there are no nonconstant subharmonic functions on $\mathbb{C}$ that are bounded from above.
(b) Generalize the above result to the finitely punctured plane.

9.22. (a) Show that $z \mapsto -\log|z|$ with $0 < |z| < 1$ is the Green's function for the unit disc with singularity at the origin.
(b) What is the Green's function for the unit disc with a singularity at an arbitrary point c of the disc?

9.23. (a) Let D_1 and D_2 be domains with $c \in D_1 \subset D_2$. Show that for all $z \in D_1$,

$$g_{D_1}(z,c) < g_{D_2}(z,c).$$

(b) Let $0 < r < R$ and assume that D is a domain with

$$\{|z| < r\} \subset D \subset \{|z| < R\}.$$

Show that for all $0 < |z| < r$,

$$\log r - \log|z| < g_D(z,0) < \log R - \log|z|.$$

9.24. Let D be a proper simply connected subdomain of $\mathbb{C}$ and $c \in D$. Let $f : D \to \mathbb{D}$ be the Riemann map with $f(c) = 0$ and $f'(c) > 0$. Show that $z \mapsto -\log|f(z)|$ defines the Green's function for D with singularity at c. Is the normalization $f'(c) > 0$ needed?

9.25. Show that the Green's function is a conformal invariant in the following sense: If $f : D_1 \to D_2$ is a conformal map between plane domains, $z \in D_1$, and g is the Green's function on D_2 with singularity at $f(z)$, then $g \circ f$ is the Green's function on D_1 with singularity at z.

9.26. Extend the discussion of the Green's function to include domains in the Riemann sphere $\mathbb{C} \cup \{\infty\}$.

Chapter 10
Zeros of Holomorphic Functions

There are certain (classical families of) functions of a complex variable that mathematicians have studied frequently enough for them to acquire their own names. These are, of course, functions that arise naturally and repeatedly in various mathematical settings. Many of these functions are defined by infinite products. Examples of such *named* functions include Euler's Γ-function, the Riemann ζ-function, and the Euler Φ-function. We will study only the first of these, in Sect. 10.4. There is a long history of synergy between the understanding of such functions and the development of complex analysis. Indeed, motivation for much of the theory and techniques of complex analysis was the desire to understand specific functions. In turn, the understanding of these functions has fed and continues to feed the development of the theory of complex variables.

Holomorphic functions in general, and these classically studied functions in particular, are often understood by their zeros. In this chapter we develop techniques to study the zeros of holomorphic functions. We show that one can always construct a meromorphic function with prescribed zeros and poles. To do so, we begin by developing the theory for infinite products of holomorphic functions in the first section of this chapter. This theory is applied in Sects. 10.2 and 10.6 to study, respectively, meromorphic functions on an arbitrary proper domain in the complex plane with prescribed zeros and poles and bounded analytic functions on the unit disc with prescribed zeros. In Sect. 10.5 we introduce the notion of divisors on a plane domain and discuss some of their properties; they will prove useful in the following section. Recall that we discussed finite Blaschke products in an earlier chapter, as opposed to the infinite Blaschke products that we study here in Sect. 10.6. This in turn leads us to begin the study of bounded analytic functions on $\mathbb{D}$. In Sect. 10.3 we prove Bers's theorem that a plane domain is determined by its ring of holomorphic functions; the beginning of the interplay between the geometry of a plane domain and its ring of holomorphic functions.

R.E. Rodríguez et al., *Complex Analysis: In the Spirit of Lipman Bers*, Graduate Texts in Mathematics 245, DOI 10.1007/978-1-4419-7323-8_10,

10.1 Infinite Products

We begin with some language needed to discuss infinite products. We then develop lemmas that lead to Theorem 10.5, which relates the uniform convergence of certain infinite sums to the uniform convergence (to a holomorphic function) of corresponding infinite products.

Definition 10.1. Let $u_n \in \mathbb{C}$ for each $n \in \mathbb{Z}_{>0}$ and set

$$p_n = (1 + u_1)(1 + u_2) \cdots (1 + u_n).$$

If $\lim_{n\to\infty} p_n$ exists and equals p, we write

$$p = \prod_{n=1}^{\infty} (1 + u_n).$$

We call p_n the *partial product* of the *infinite product* p. We say that the infinite product $\prod_{n=1}^{\infty}(1 + u_n)$ *converges* if $\{p_n\}$ does.

Lemma 10.2. *Let* $\{u_n\}_{n=1}^{\infty} \subset \mathbb{C}$ *and set*

$$p_N = \prod_{n=1}^{N} (1 + u_n) \quad \text{and} \quad p_N^* = \prod_{n=1}^{N} (1 + |u_n|).$$

Then

$$p_N^* \le \mathrm{e}^{|u_1|+\cdots+|u_N|} \quad \text{and} \quad |p_N - 1| \le p_N^* - 1.$$

Proof. We know that $x > 0$ implies that $\mathrm{e}^x \ge 1 + x$. Therefore, $1 + |u_n| \le \mathrm{e}^{|u_n|}$ so that $p_N^* \le \mathrm{e}^{|u_1|+\cdots+|u_N|}$.

The second statement is proved by induction on N. For $N = 1$, $|p_1 - 1| = |u_1| = p_1^* - 1$. For $N \ge 1$,

$$\begin{aligned} |p_{N+1} - 1| &= |p_N(1 + u_{N+1}) - 1| \\ &= |(p_N - 1)(1 + u_{N+1}) + u_{N+1}| \\ &\le |p_N - 1| \cdot |1 + u_{N+1}| + |u_{N+1}| \end{aligned}$$

and by induction this expression is

$$\begin{aligned} &\le (p_N^* - 1) \cdot (1 + |u_{N+1}|) + |u_{N+1}| \\ &= p_{N+1}^* - 1. \end{aligned}$$

□

Theorem 10.3. *If $\{u_n\}$ is a sequence of bounded functions on a set S such that $\sum |u_n|$ converges uniformly on S, then:*

(1) $f(z) = \prod_{n=1}^{\infty}(1 + u_n(z))$ *converges uniformly on S.*

(2) *If $J : \mathbb{Z}_{>0} \to \mathbb{Z}_{>0}$ is any bijection, then*

$$f(z) = \prod_{k=1}^{\infty}(1 + u_{J(k)}(z)).$$

(3) $f(z_0) = 0$ *if and only if* $u_n(z_0) = -1$ *for some* $n \in \mathbb{Z}_{>0}$.

Proof. By uniform convergence of $\sum |u_n|$ on S, there exists c in $\mathbb{R}_{>0}$ such that

$$\sup_{z \in S} \sum |u_n(z)| \le c.$$

Let

$$p_N(z) = \prod_{n=1}^{N}(1 + u_n(z)) \text{ and } q_M(z) = \prod_{k=1}^{M}(1 + u_{J(k)}(z)).$$

We know that

$$|p_N| \le |p_N - 1| + 1 \le p_N^* \le e^{|u_1| + \cdots + |u_N|} \le e^c.$$

Choose ϵ with $0 < \epsilon < \frac{1}{4}$. Then there exists an $N_0 \in \mathbb{Z}_{>0}$ such that

$$\sum_{n=N_0}^{\infty} |u_n(z)| < \epsilon \quad \text{for all } z \in S;$$

in particular, $|u_n(z)| < \epsilon < \frac{1}{4}$ for all $z \in S$ and all $n > N_0$. Choose M_0 such that

$$\{1, 2, \ldots, N_0\} \subset \{J(1), J(2), \ldots, J(M_0)\}.$$

If $M, N > \max\{M_0, N_0\}$, then we can write

$$q_M(z) - p_N(z) = p_N(z) \left(\frac{\prod_1 (1 + u_n(z))}{\prod_2 (1 + u_n(z))} - 1 \right).$$

Here the symbols $\prod_1$ and $\prod_2$ denote products taken over appropriate disjoint indices. For our purposes the important facts about $\prod_1$ and $\prod_2$ are that only indices $n > N_0$ appear and that the indices that appear in the two products are disjoint.

Let us define $\widetilde{u}_n(z)$ by

$$1 + \widetilde{u}_n(z) = \begin{cases} 1 + u_n(z) & \text{if } n \text{ appears in } \prod_1 \\ \dfrac{1}{1 + u_n(z)} & \text{if } n \text{ appears in } \prod_2 \end{cases}$$

and let I be the union of the indexing sets in $\prod_1$ and $\prod_2$. Note that $I \subset \mathbb{Z}_{>0}$.

Now if a and $b \in \mathbb{C}$ and $\delta \in \mathbb{R}_{>0}$ satisfy $|a| < \delta < \frac{1}{2}$ and $\dfrac{1}{1+a} = 1 + b$, then $|b| < 2\delta$. Therefore

$$\begin{aligned} |q_M(z) - p_N(z)| &= |p_N(z)| \left| \prod_{n \in I} (1 + \widetilde{u}_n(z)) - 1 \right| \\ &\le |p_N(z)| \left(\prod_{n \in I} (1 + |\widetilde{u}_n(z)|) - 1 \right) \\ &\le |p_N(z)| \left(\exp\left(\sum_{n \in I} |\widetilde{u}_n(z)| \right) - 1 \right) \\ &\le |p_N(z)| \left(\exp\left(\sum_{n=N_0}^{\infty} |\widetilde{u}_n(z)| \right) - 1 \right) \\ &\le |p_N| \, (e^{2\epsilon} - 1). \end{aligned}$$

We now claim that $e^x - 1 \le 2x$ for $0 \le x \le \frac{1}{2}$. This claim may be verified as follows. Define $F(x) = e^x - 1 - 2x$ for $x \in \mathbb{R}$. Observe that $F(0) = 0$ and $F'(x) = e^x - 2$. Since the (real) exponential function is increasing, $F' \le e^{\frac{1}{2}} - 2$ on $[0, \frac{1}{2}]$. But

$$\begin{aligned} e^{\frac{1}{2}} &= 1 + \frac{1}{2} + \frac{(\frac{1}{2})^2}{2!} + \frac{(\frac{1}{2})^3}{3!} + \cdots \\ &< 1 + \frac{1}{2} + \left(\frac{1}{2}\right)^2 + \cdots = \frac{1}{1 - \frac{1}{2}} = 2, \end{aligned}$$

and thus F is nonpositive on $[0, \frac{1}{2}]$. Therefore

$$|q_M(z) - p_N(z)| \le e^c 4\epsilon \quad \text{for all } z \in S$$

and we can conclude as follows:

1. If we let J be the identity map, then $p_N(z) \to f(z)$ uniformly for $z \in S$.
2. For arbitrary J, we conclude that $q_M(z) \to f(z)$ uniformly for $z \in S$.

3. Since $p_{N_0} = p_M - (p_M - p_{N_0})$, we have that, for sufficiently large M,

$$\begin{aligned}|p_{N_0}| &\leq |p_M| + |p_M - p_{N_0}| \leq |p_M| + (\mathrm{e}^{2\epsilon} - 1)\,|p_{N_0}| \\ &\leq |p_M| + 4\epsilon\,|p_{N_0}|\end{aligned}$$

or, equivalently, that

$$|p_M(z)| \geq (1 - 4\epsilon)\,|p_{N_0}(z)|\,.$$

Therefore $f(z) = 0$ if and only if $p_{N_0}(z) = 0$. □

Theorem 10.4. *Assume* $0 \leq u_n < 1$.

(1) *If* $\sum_{n=1}^{\infty} u_n < \infty$, *then* $0 < \prod_{n=1}^{\infty}(1 - u_n) < \infty$.

(2) *If* $\sum_{n=1}^{\infty} u_n = +\infty$, *then* $\prod_{n=1}^{\infty}(1 - u_n) = 0$.

Proof. The first claim is a consequence of the previous theorem. To prove the second claim, we start with the observation that

$$1 - x \leq \mathrm{e}^{-x} \quad \text{for } 0 \leq x \leq 1.$$

Let $p_N = (1-u_1)(1-u_2)\cdots(1-u_N)$. Since $p_1 \geq p_2 \geq \cdots \geq p_N \geq 0$, $\lim_{N\to\infty} p_N =$ exists. We call it p. Now

$$0 \leq p \leq p_N = \prod_{n=1}^{N}(1 - u_n) \leq \mathrm{e}^{-(u_1+\cdots+u_N)}.$$

Since $\lim_{N\to\infty} \mathrm{e}^{-(u_1+\cdots+u_N)} = 0$, the theorem follows. □

Finally, we can establish when an infinite product is holomorphic.

Theorem 10.5. *Let D be a domain in $\mathbb{C}$ and suppose that $\{f_n\}$ is a sequence in $\mathbf{H}(D)$ with f_n not identically 0 for all n.*

(a) If $\sum_{n=1}^{\infty} |1 - f_n|$ *converges uniformly on compact subsets of D, then* $\prod_{n=1}^{\infty} f_n$ *converges uniformly on compact subsets of D to a function f in $\mathbf{H}(D)$ and*

$$\nu_z(f) = \sum_{n=1}^{\infty} \nu_z(f_n) \quad \textit{for all } z \in D.$$

(b) If $\sum_{n=1}^{\infty} |1 - f_n|$ *diverges pointwise on* D*, then* $\prod_{n=1}^{\infty} f_n$ *converges to* 0 *on* D.

Proof. For part (a), we only have to verify the formula for the order of z. We note that the sum in that formula is finite (i.e., all but finitely many summands are zero). Let $z_0 \in D$ and let $K \subset D$ be a compact set containing a neighborhood of z_0. There is an N in $\mathbb{Z}_{>0}$ such that $|1 - f_n(z)| < \frac{1}{2}$ for all $z \in K$ and all $n \geq N$. Therefore, $f_n(z) \neq 0$ for all $z \in K$ and for all $n \geq N$. Thus

$$\nu_{z_0}(f) = \nu_{z_0}\left(\prod_{n=1}^{N-1} f_n\right) + \nu_{z_0}\left(\prod_{n=N}^{\infty} f_n\right) = \sum_{n=1}^{N-1} \nu_{z_0}(f_n) + 0.$$

Part (b) is easily verified. □

10.2 Holomorphic Functions with Prescribed Zeros

Our goal is to construct a holomorphic function with arbitrarily prescribed zeros (at a discrete set of points in any given domain). To this end we begin by defining *the elementary functions*, first introduced by Weierstrass. We investigate some of their properties and then use them along with Theorem 10.5 to construct the required holomorphic functions.

Definition 10.6. The *Weierstrass elementary functions* are the entire functions E_p, for $p \in \mathbb{Z}_{\geq 0}$, defined as follows. Let $z \in \mathbb{C}$ and set

$$E_0(z) = 1 - z,$$

and, for $p \in \mathbb{Z}_{>0}$,

$$E_p(z) = (1 - z)\exp\left(z + \frac{z^2}{2} + \cdots + \frac{z^p}{p}\right).$$

Note that, for all nonnegative integers p, $E_p(0) = 1$ and $E_p(z) = 0$ if and only if $z = 1$. Furthermore, the unique zero of E_p is simple.

Lemma 10.7. *If* $|z| \leq 1$, *then* $\left|1 - E_p(z)\right| \leq |z|^{p+1}$ *for all nonnegative integers* p.

Proof. The statement is clearly true if $p = 0$.

If $p \geq 1$, we have

$$\begin{aligned} E_p'(z) &= (1 - z)\mathrm{e}^{z + \frac{z^2}{2} + \cdots + \frac{z^p}{p}}[1 + z + \cdots + z^{p-1}] - \mathrm{e}^{z + \frac{z^2}{2} + \cdots + \frac{z^p}{p}} \\ &= -z^p \mathrm{e}^{z + \frac{z^2}{2} + \cdots + \frac{z^p}{p}}. \end{aligned}$$

We therefore conclude that $\nu_0(-E'_p) = p$. Further,

$$-E'_p(z) = z^p e^{z+\frac{z^2}{2}+\cdots+\frac{z^p}{p}} = z^p \sum_{n=0}^{\infty} \frac{1}{n!}\left(z + \frac{z^2}{2} + \cdots + \frac{z^p}{p}\right)^n = \sum_{n\geq p} b_n z^n,$$

with $b_p = 1$ and $b_n > 0$ for all $n \geq p$. Therefore

$$1 - E_p(z) = \sum_{n\geq p} \frac{b_n}{n+1} z^{n+1}.$$

Set

$$\phi(z) = \frac{1 - E_p(z)}{z^{p+1}},$$

and observe that $\phi \in \mathbf{H}(\mathbb{C})$ and that $\phi(z) = \sum_{n\geq 0} a_n z^n$, with $a_n > 0$ for all $n \in \mathbb{Z}_{\geq 0}$. For $|z| \leq 1$, we have

$$|\phi(z)| = \left|\sum_{n\geq 0} a_n z^n\right| \leq \sum_{n\geq 0} a_n |z^n| \leq \sum_{n\geq 0} a_n = \phi(1) = 1;$$

thus $\left|1 - E_p(z)\right| \leq |z|^{p+1}$ for $|z| \leq 1$. □

Theorem 10.8 (Weierstrass Theorem). *Assume that $\{z_n\}$ is a sequence of nonzero complex numbers with $\lim_{n\to\infty} |z_n| = \infty$.*

If $\{p_n\} \subseteq \mathbb{Z}_{\geq 0}$ is a sequence of nonnegative integers with the property that for all positive real numbers r we have

$$\sum_{n=1}^{\infty} \left(\frac{r}{|z_n|}\right)^{1+p_n} < \infty, \tag{10.1}$$

then the infinite product

$$P(z) = \prod_{n=1}^{\infty} E_{p_n}\left(\frac{z}{z_n}\right), \ z \in \mathbb{C}$$

defines an entire function whose zero set is $\{z_1, z_2, \ldots\}$. More precisely, if $z = c$ appears $\nu \geq 0$ times in the above sequence of zeros, then $\nu_c(P) = \nu$.

Furthermore, condition (10.1) is always satisfied for $p_n = n - 1$. Thus any discrete set in $\mathbb{C}$ is the zero set of an entire function.

Proof. We first show that (10.1) holds for $p_n = n - 1$. In this case we have to show convergence of the series $\sum a_n$, with $a_n = \left(\frac{r}{|z_n|}\right)^n$. But $|a_n|^{\frac{1}{n}} \to 0$ as $n \to \infty$, and the root test allows us to conclude that

$$\sum_{n=1}^{\infty} \left(\frac{r}{|z_n|}\right)^n < \infty.$$

Now let $\{p_n\}$ be any sequence of nonnegative integers satisfying condition (10.1) for all $r > 0$; fix $r > 0$ and assume that $|z| \le r$. From Lemma 10.7 we conclude that

$$\left|1 - E_{p_n}\left(\frac{z}{z_n}\right)\right| \le \left|\frac{z}{z_n}\right|^{p_n+1} \le \left(\frac{r}{|z_n|}\right)^{p_n+1}.$$

Therefore we can apply Theorem 10.5 to conclude that $\prod E_{p_n}\left(\frac{z}{z_n}\right)$ converges uniformly on all compact subsets of $\mathbb{C}$ to an entire function that has the required zero set. □

We next prove a generalization of two consequences of the fundamental theorem of algebra, which has been proven previously. The first of these algebraic consequences is that for every finite sequence $\{z_1, \ldots, z_n\}$ of points in the complex plane (that may contain repeated points) there is a polynomial vanishing precisely at the points of that sequence. The second is that every nonzero complex polynomial p has a factorization

$$p(z) = c \prod_{j=1}^{n} (z - z_j) \text{ for all } z \in \mathbb{C},$$

where c is a nonzero constant and $\{z_j\}$ are the zeros of p, repeated according to their multiplicity. We need the analytic tools that were developed to handle infinite sequences.

Theorem 10.9 (Weierstrass Factorization Theorem). *Let f be in $\mathbf{H}(\mathbb{C}) - \{0\}$, and set $k = \nu_0(f)$. Let $\{z_n; n \in I\}$ denote the zeros of f in $\mathbb{C} - \{0\}$, listed according to their multiplicities.*

There exist a $g \in \mathbf{H}(\mathbb{C})$ and a sequence of nonnegative integers $\{p_n; n \in I\}$ such that

$$f(z) = z^k e^{g(z)} \prod_{n \in I} E_{p_n}\left(\frac{z}{z_n}\right)$$

for all z in $\mathbb{C}$.

Proof. Observe that $I \subseteq \mathbb{N}$ may be finite (including the possibility that I is empty) or countable. In the finite case the theorem has already been established. In any case, we can choose any sequence $\{p_n; n \in I\}$ of nonnegative integers such that (10.1) holds for all $r > 0$ and set

$$P(z) = \prod_{n\in I} E_{p_n}\left(\frac{z}{z_n}\right) \text{ and } G(z) = \frac{f(z)}{z^k P(z)}.$$

Then $G \in \mathbf{H}(\mathbb{C})$ and $G(z) \neq 0$ for all $z \in \mathbb{C}$. Since G is a nonvanishing entire function, there is a $g \in \mathbf{H}(\mathbb{C})$ with $e^{g(z)} = G(z)$ for all $z \in \mathbb{C}$. □

Theorem 10.10. *Let D be a proper subdomain of $\widehat{\mathbb{C}}$. Let A be a subset of D that has no limit point in D, and let ν be a function mapping A to $\mathbb{Z}_{>0}$. Then there exists a function $f \in \mathbf{H}(D)$ with $\nu_z(f) = \nu(z)$ for all $z \in A$, whose restriction to $D - A$ has no zeros.*

Proof. To begin, we make the following observations:

1. A is either finite or countable.
2. Without loss of generality, we may assume that $\infty \in D - A$ and that A is nonempty.
3. If A is finite, let $A = \{z_1, \ldots, z_n\}$. Set $\nu_j = \nu(z_j)$, for all $1 \leq j \leq n$, and choose $z_0 \in \mathbb{C} - D$. In this case we set

$$f(z) = \frac{(z - z_1)^{\nu_1} \cdots (z - z_n)^{\nu_n}}{(z - z_0)^{\nu_1 + \cdots + \nu_n}},$$

and note that f is holomorphic on D, and does not vanish on $\widehat{\mathbb{C}} - \{z_1, \ldots, z_n\}$ since $f(\infty) = 1$.

Since $D - A \subset \widehat{\mathbb{C}} - \{z_1, \ldots, z_n\}$, we have thus established the theorem for finite sets A.

To prove the theorem for infinite sets A, let $K = \widehat{\mathbb{C}} - D$. Note that K is a nonempty compact subset of $\mathbb{C}$. Let $\{\alpha_n\}_{n\in\mathbb{N}}$ be a sequence whose terms consist of all $\alpha \in A$, where each α is repeated $\nu(\alpha)$ times.

We first claim that, for each positive integer n, we can choose a $\beta_n \in K$ such that $|\beta_n - \alpha_n| \leq |\beta - \alpha_n|$ for all $\beta \in K$. To see that this can be done, note that, for each positive integer n, the function $z \mapsto l_n(z) = |z - \alpha_n|$ is continuous on K and, therefore, achieves a minimum at some $\beta_n \in K$.

The function f we are seeking (whose existence must be established) is

$$f(z) = \prod_{n=1}^{\infty} E_n\left(\frac{\alpha_n - \beta_n}{z - \beta_n}\right).$$

We show next that the product on the RHS converges in D, by proving that $\sum \left|1 - E_n\left(\frac{\alpha_n - \beta_n}{z - \beta_n}\right)\right|$ converges uniformly on compact subsets of D. For this we first prove that $\lim_{n\to\infty} |\beta_n - \alpha_n| = 0$. If we assume $|\beta_n - \alpha_n| \geq \delta$ for some $\delta > 0$ and infinitely many n, then for some subsequence $\{\alpha_{n_j}\}$ of $\{\alpha_n\}$,

$$|z - \alpha_{n_j}| \geq \delta \text{ for all } z \in K. \tag{10.2}$$

But a subsequence of this subsequence converges to some point α in $\widehat{\mathbb{C}}$. From (10.2) we conclude that $\alpha \notin K$. Thus we arrive at the contradiction that $\alpha \in D$ and is a limit point of A. Next, we put $r_n = 2\,|\alpha_n - \beta_n|$ and observe that $\{r_n\}$ converges to zero. Let K_0 be any nonempty compact subset of D; since K and K_0 are disjoint compact subsets of $\widehat{\mathbb{C}}$, the distance between them must be positive. Therefore, the fact that $r_n \to 0$ implies there is an $N \in \mathbb{Z}_{>0}$ such that $|z - \beta_n| > r_n$ for all $z \in K_0$ and all $n > N$. Thus

$$\left|\frac{\alpha_n - \beta_n}{z - \beta_n}\right| \le \frac{r_n}{2r_n} = \frac{1}{2} \quad \text{for all } n > N \text{ and all } z \in K_0,$$

and hence

$$\left|1 - E_n\left(\frac{\alpha_n - \beta_n}{z - \beta_n}\right)\right| \le \left|\frac{\alpha_n - \beta_n}{z - \beta_n}\right|^{n+1} \le \left(\frac{1}{2}\right)^{n+1}$$

for all $n > N$ and all $z \in K_0$, where the first inequality follows from Lemma 10.7. By Theorem 10.5, the infinite product defining f converges and $f \in \mathbf{H}(D)$. Finally, it follows from Lemma 10.7 that $f(z) = 0$ if and only if $E_n\left(\frac{\alpha_n - \beta_n}{z - \beta_n}\right) = 0$ for some $n \in \mathbb{Z}_{>0}$ if and only if $z = \alpha_n$ for some $n \in \mathbb{Z}_{>0}$. □

As an immediate corollary we obtain the following result.

Theorem 10.11. *If D is a nonempty proper subdomain of $\widehat{\mathbb{C}}$, then $\mathbf{M}(D)$ is the field of fractions of the integral domain $\mathbf{H}(D)$; that is, for every $f \in \mathbf{M}(D)$ there exist g in $\mathbf{H}(D)$ and h in $\mathbf{H}(D) - \{0\}$ such that $f = \frac{g}{h}$.*

10.3 The Ring $\mathbf{H}(D)$

Let D_1 and D_2 be nonempty proper subdomains of $\widehat{\mathbb{C}}$, and assume there exists a conformal map $F : D_1 \to D_2$ between these domains. The map F induces a ring isomorphism $F^* : \mathbf{H}(D_2) \to \mathbf{H}(D_1)$ defined by

$$F^*(f)(z) = f(F(z)), \quad f \in \mathbf{H}(D_2),\ z \in D_1.$$

Similarly, an anti-conformal map F of D_1 onto D_2 induces a ring isomorphism $F^* : \mathbf{H}(D_2) \to \mathbf{H}(D_1)$ defined by

$$F^*(f)(z) = \overline{f(F(z))}, \quad f \in \mathbf{H}(D_2),\ z \in D_1.$$

We have thus defined a map $*$ that sends a conformal or anti-conformal map F between two domains in the extended complex plane to an isomorphism F^* between their rings of holomorphic functions. It should be observed that for the identity map F, F^* is also the identity, and for any F, $(F^{-1})^* = (F^*)^{-1}$.

While if $J : D \to J(D) = \overline{D}$ is the anti-conformal conjugation map ($J(z) = \overline{z}$), then $J^* : \mathbf{H}(\overline{D}) \to \mathbf{H}(D)$ is defined by $J^*(f)(z) = \overline{f(\overline{z})}$, $f \in \mathbf{H}(\overline{D})$, and $z \in D$. We also observe that for any two such maps

$$D_1 \xrightarrow{F_1} D_2 \xrightarrow{F_2} D_3, \quad (F_2 \circ F_1)^* = F_1^* \circ F_2^*.$$

The main purpose of this section is to obtain a theorem due to Bers that is the converse to these facts; essentially that the map $*$ is an isomorphism between appropriate categories. We shall freely use the results that follow from the exercises to this chapter.

Theorem 10.12. *Let D_1 and D_2 be proper subdomains of $\widehat{\mathbb{C}}$. If $\varphi : \mathbf{H}(D_2) \to \mathbf{H}(D_1)$ is a ring isomorphism, then*

$$\varphi(\imath) = \pm\imath.$$

Furthermore:

(a) *If $\varphi(\imath) = \imath$, then there exists a unique conformal map F of D_1 onto D_2 such that $\varphi = F^*$.*
(b) *Similarly, if $\varphi(\imath) = -\imath$, there exists a unique anti-conformal map F of D_1 onto D_2 such that $\varphi = F^*$.*

Proof. The proof proceeds in a number of steps.

(1) It is immediate that $\varphi(\imath) = \pm\imath$, since the two constant functions $\pm\imath$ are the only elements of the two rings $\mathbf{H}(D_j)$ whose squares are the constant function -1. So we start by proving (a).
(2) We now show that we may assume that neither D_1 nor D_2 contain the point ∞. If $D \subset \widehat{\mathbb{C}}$ is any proper domain in the sphere and $\infty \in D$, then there exists some $c \in \mathbb{C}$ that is not in D. The conformal map $T_c : z \mapsto \frac{1}{z-c}$ maps D onto a subdomain of $\mathbb{C}$. Let us assume for the moment that both domains[1] D_1 and D_2 contain ∞ and that $c \notin D_1$ and $d \notin D_2$. In this case,

$$(T_c^*)^{-1} \circ \varphi \circ T_d^* : \mathbf{H}(T_d(D_2)) \to \mathbf{H}(T_c^{-1}(D_1))$$

is a ring isomorphism preserving $\imath$. If we can establish the existence of a conformal map $G : T_c(D_1) \to T_d(D_2)$ such that $G^* = (T_c^*)^{-1} \circ \varphi \circ T_d^*$, then $\varphi = (T_d^{-1} \circ G \circ T_c)^*$.
(3) $\varphi(c) \in \mathbb{C}$ for all constant functions $c \in \mathbf{H}(D_2)$. If c is a complex rational constant (i.e., if $c = a + b\,\imath$ with a and $b \in \mathbb{Q}$), then the assertion follows from (1) and the easily proved fact that $\varphi(x) = x$ for every $x \in \mathbb{Q}$. For a nonrational constant, the conclusion follows from the observation that nonrational

[1] The argument is, of course, simplified if only one of these contains ∞.

constants c are characterized by the existence of the multiplicative inverse of $c - r$ for every rational constant r. Observe that the same argument applied to φ^{-1} shows that φ defines (by restriction) a ring isomorphism of $\mathbb{C}$ onto itself that fixes all complex rational constants.

(4) Let us define, for $c \in D_2$,

$$I_c = \{f \in \mathbf{H}(D_2); f(c) = 0\}$$

and, similarly, for $c \in D_1$,

$$I_c^* = \{f \in \mathbf{H}(D_1); f(c) = 0\}.$$

These are obviously both principal (with generator $z \mapsto z - c$) and maximal ideals in their respective rings. They are the only such ideals in these rings (as c varies over the points in D_2 and D_1, respectively).

(5) We are now ready to define the map $F : D_1 \to D_2$. Let $c \in D_1$. Then I_c^* is a principal maximal ideal in $\mathbf{H}(D_1)$, and it follows that this fact holds if and only if $\varphi^{-1}(I_c^*)$ is a principal maximal ideal in $\mathbf{H}(D_2)$, thus if and only if $\varphi^{-1}(I_c^*) = I_{F(c)}$ for a unique point $F(c) \in D_2$. The map we have defined is injective and surjective. We leave it to the reader to verify these two claims.

(6) The definition of F yields $I_c^* = \varphi(I_{F(c)})$. We want to conclude from this fact that for all $f \in \mathbf{H}(D_2)$ and all $c \in D_1$ we have

$$\varphi(f)(c) = f(F(c)). \tag{10.3}$$

So let $f \in \mathbf{H}(D_2)$ and $c \in D_1$. Then the function $f - f(F(c))$ belongs to $I_{F(c)}$, and it follows that

$$\varphi(f - f(F(c))) = \varphi(f) - \varphi(f(F(c))) \text{ belongs to } I_c^*.$$

Thus

$$\varphi(f)(c) = \varphi(f(F(c)))(c) = \varphi(f(F(c))),$$

and it remains to show that $\varphi(d) = d$ for all $d \in \mathbb{C}$.

(7) Let $\{b_n = F(a_n)\}$ be an infinite and discrete sequence of distinct points in D_2. By the Mittag-Leffler theorem (see Exercise 10.10), there exists $f \in \mathbf{H}(D_2)$ such that $f(b_n) = n$ for all positive integers n. Thus

$$\varphi(f)(a_n) = \varphi(f(F(a_n))) = \varphi(f(b_n)) = \varphi(n) = n,$$

and we conclude that $\{a_n\}$ is an infinite and discrete sequence of distinct points in D_1.

(8) As a consequence of the last assertion, the closure of $F(K)$ is compact in D_2 whenever K is compact in D_1.

(9) For all $\beta \in \mathbb{C}$, there exist $f \in \mathbf{H}(D_2)$ and $c \in D_1$ such that $\varphi(f)(c) = \beta$ and $\varphi(f)$ is univalent in a neighborhood of c. To verify this claim, we note that there certainly exist nonconstant functions in $\mathbf{H}(D_2)$; in particular, it is convenient for future use to consider the identity function on D_2 defined by $I(z) = z$. Since $\varphi(I) \notin \mathbb{C}$ (by (3)), there exists $c \in D_1$ such that $\varphi(I)$ is univalent in a neighborhood of c. If $\varphi(I)(c) = \alpha$, then $f = I + \varphi^{-1}(\alpha - \beta)$ has the required properties.

(10) We already know from (3) that the isomorphism φ defines a ring isomorphism σ of $\mathbb{C}$ onto itself that fixes all complex rational constants. To show that σ is the identity on $\mathbb{C}$, it suffices to show that σ is continuous (see Exercise 2.19(c)), and by the additivity property of σ all we have to show is that it is bounded in a neighborhood of the origin; we will actually show the equivalent condition that σ^{-1} is bounded in a neighborhood of the origin.

To do so, we choose $f \in \mathbf{H}(D_2)$ and $c \in D_1$ such that $\varphi(f)(c) = 0$ and $\varphi(f)$ is univalent in a neighborhood $V \subset D_1$ of c. We may and do assume that the closure K of V in D_1 is compact. Since $\varphi(f)(V)$ is an open set and $0 \in \varphi(f)(V)$, there exists $\delta > 0$ such that the disc $U(0, \delta) \subseteq \varphi(f)(V)$. If σ^{-1} were not bounded in a neighborhood of the origin, we could choose a sequence of distinct complex numbers $\{\alpha_n\}$ with each $|\alpha_n| < \delta$, $\alpha_n \to 0$, and $\sigma^{-1}(\alpha_n) \to \infty$. Thus there would be a sequence $\{c_n\} \subset U$ such that $\varphi(f)(c_n) = \alpha_n$ for all positive integers n. If we set $d_n = F(c_n)$, then the sequence $\{d_n\}$ is contained in the compact set $F(K)$, and thus $\{f(d_n)\}$ is a bounded sequence, but

$$f(d_n) = f(F(c_n)) = \sigma^{-1}(\varphi(f)(c_n)) = \sigma^{-1}(\alpha_n) \to \infty.$$

(11) The fact that $\sigma = \varphi|_{\mathbb{C}}$ is the identity map on $\mathbb{C}$ finally implies that Equation (10.3) holds:

$$\varphi(f)(c) = f(F(c))$$

for all $f \in \mathbf{H}(D_2)$ and all $c \in D_1$. It is quite easy to identify the function F we have defined with an analytic function on D_1: apply the last equation to the function I and conclude that $\varphi(I) = F$. Thus F is in fact in $\mathbf{H}(D_1)$ and is conformal onto D_2.

(12) The uniqueness of F is now clear.

(13) To establish (b), we follow φ by the map

$$(J^*)^{-1} : \mathbf{H}(D_1) \to \mathbf{H}(J(D_1)),$$

where J is the anti-conformal involution, observe that

$$((J^*)^{-1} \circ \varphi)(\imath) = \imath$$

and apply the previous case. □

Corollary 10.13. *Let D_1 and D_2 be proper subdomains of $\widehat{\mathbb{C}}$, and let $\varphi : \mathbf{H}(D_2) \to \mathbf{H}(D_1)$ be a ring isomorphism. Then φ is either a $\mathbb{C}$-linear or a conjugate-$\mathbb{C}$-linear algebra isomorphism.*

10.4 Euler's Γ-Function

In this section we introduce an important function among whose remarkable properties is the fact that it extends the factorial function on the integers to an entire function. Our development is brisker than in previous sections.

10.4.1 Basic Properties

Define, for $z \in \mathbb{C}$,

$$G(z) = \prod_{n=1}^{\infty} \left(1 + \frac{z}{n}\right) e^{-\frac{z}{n}}.$$

The infinite product converges to an entire function with a simple zero at each negative integer. We claim that the entire function h defined for all $z \in \mathbb{C}$ by

$$h(z) = z\,G(z)\,G(-z)$$

satisfies

$$h(z) = \frac{\sin \pi z}{\pi}. \tag{10.4}$$

Simple calculations show that

$$h(z) = z \prod_{n=1}^{\infty} \left(1 - \frac{z^2}{n^2}\right),$$

and hence, using (7.3) of Chap. 7 for the last equality,

$$\frac{h'(z)}{h(z)} = \frac{\mathrm{d}}{\mathrm{d}z} \log h(z) = \frac{1}{z} - \sum_{n=1}^{\infty} \frac{2z}{n^2 - z^2} = \pi \cot \pi z.$$

It follows that $h(z) = c \sin \pi z$ for some nonzero constant c. To evaluate c, we note that

$$\lim_{z\to 0}\frac{\sin \pi z}{z} = \pi\,, \text{ and } \lim_{z\to 0}\frac{h(z)}{z} = 1,$$

and thus we conclude that $c = \dfrac{1}{\pi}$.

The function $z \mapsto G(z-1)$ is entire and has a simple zero at each nonpositive integer and no other zeros. It follows that

$$G(z-1) = z\mathrm{e}^{\gamma(z)}G(z) \tag{10.5}$$

for some entire function γ. We proceed to determine this function. Differentiating logarithmically both sides of the last equation, we obtain

$$\sum_{n=1}^{\infty}\left(\frac{1}{z-1+n}-\frac{1}{n}\right) = \frac{1}{z} + \gamma'(z) + \sum_{n=1}^{\infty}\left(\frac{1}{z+n}-\frac{1}{n}\right).$$

Since

$$\begin{aligned}\sum_{n=1}^{\infty}\left(\frac{1}{z-1+n}-\frac{1}{n}\right) &= \sum_{n=0}^{\infty}\left(\frac{1}{z+n}-\frac{1}{n+1}\right)\\ &= \frac{1}{z} - 1 + \sum_{n=1}^{\infty}\left(\frac{1}{z+n}-\frac{1}{n+1}\right),\end{aligned}$$

we conclude that

$$\gamma'(z) = \sum_{n=1}^{\infty}\left(\frac{1}{n}-\frac{1}{n+1}\right) - 1 = 0.$$

Hence the function γ is constant; it is known as *Euler's constant*. To determine γ, we return to our function G and observe that if we set $z = 1$ in (10.5) we obtain that $1 = G(0) = \mathrm{e}^{\gamma}G(1)$, and hence

$$\mathrm{e}^{-\gamma} = G(1) = \prod_{n=1}^{\infty}\left(1+\frac{1}{n}\right)\mathrm{e}^{-\frac{1}{n}}.$$

Thus

$$\mathrm{e}^{-\gamma} = \lim_{n\to\infty}(n+1)\mathrm{e}^{-(1+\frac{1}{2}+\cdots+\frac{1}{n})},$$

and

$$-\gamma = \lim_{n\to\infty}\left[\log(n+1) - \left(1+\frac{1}{2}+\cdots+\frac{1}{n}\right)\right].$$

Since $\lim\limits_{n\to\infty}(\log(n+1) - \log n) = \lim\limits_{n\to\infty}\log\left(1+\dfrac{1}{n}\right) = 0$, we obtain

$$\gamma = \lim_{n\to\infty}\left[\left(1+\frac{1}{2}+\cdots+\frac{1}{n}\right)-\log n\right]. \tag{10.6}$$

Next set, for $z \in \mathbb{C}$,

$$H(z) = \mathrm{e}^{\gamma z}G(z),$$

and compute that

$$H(z-1) = \mathrm{e}^{\gamma z}\mathrm{e}^{-\gamma}G(z-1) = \mathrm{e}^{\gamma z}zG(z) = zH(z).$$

We can now introduce

Definition 10.14. *Euler's* Γ*-function* is defined by

$$\Gamma(z) = \frac{1}{zH(z)} = \frac{\mathrm{e}^{-\gamma z}}{z}\prod_{n=1}^{\infty}\left(1+\frac{z}{n}\right)^{-1}\mathrm{e}^{\frac{z}{n}}, \quad z \in \mathbb{C}. \tag{10.7}$$

Note that Γ is a meromorphic function on $\mathbb{C}$, with simple poles at $z = 0, -1, -2, \ldots$, and that it has no zeros. The Γ-function satisfies a number of useful functional equations. We derive some of these, which will lead us to (10.12). The first of these functional equations is

$$\Gamma(z+1) = \frac{1}{(z+1)H(z+1)} = \frac{1}{H(z)} = z\Gamma(z). \tag{10.8}$$

Further, it follows from (10.4) that

$$\Gamma(z)\Gamma(1-z) = \frac{\pi}{\sin \pi z}. \tag{10.9}$$

A simple calculation shows that

$$\Gamma(1) = \mathrm{e}^{-\gamma}\prod_{n=1}^{\infty}\left(1+\frac{1}{n}\right)^{-1}\mathrm{e}^{\frac{1}{n}} = 1.$$

Together with (10.8), this implies that

$$\Gamma(n) = (n-1)!, \quad \text{for all } n \in \mathbb{Z}_{>0}.$$

Also, it follows from (10.9) that $\left(\Gamma\left(\frac{1}{2}\right)\right)^2 = \dfrac{\pi}{\sin\frac{\pi}{2}} = \pi$, and hence

$$\Gamma\left(\frac{1}{2}\right) = \sqrt{\pi},$$

since $\Gamma\left(\frac{1}{2}\right) > 0$ from (10.7).

We derive some other properties of Euler's Γ-function that we will need. We start with a calculation, from (10.7):

$$
\begin{aligned}
\frac{\mathrm{d}}{\mathrm{d}z}\frac{\Gamma'(z)}{\Gamma(z)} &= \frac{\mathrm{d}}{\mathrm{d}z}\frac{\mathrm{d}}{\mathrm{d}z}(\log(\Gamma(z)) \\
&= \frac{\mathrm{d}}{\mathrm{d}z}\left(-\gamma - \frac{1}{z} - \sum_{n=1}^{\infty}\left(\frac{1}{z+n} - \frac{1}{n}\right)\right) \\
&= \sum_{n=0}^{\infty}\left(\frac{1}{z+n}\right)^2. \qquad (10.10)
\end{aligned}
$$

Both functions

$$z \mapsto \Gamma(2z) \text{ and } z \mapsto \Gamma(z)\Gamma\left(z+\frac{1}{2}\right)$$

have simple poles precisely at the points $0, -1, -2, \ldots$ and $-\frac{1}{2}, -\frac{3}{2}, \ldots$. The ratio of the two functions is hence entire without zeros. The next calculation will show more:

$$
\begin{aligned}
\frac{\mathrm{d}}{\mathrm{d}z}\left(\frac{\Gamma'(z)}{\Gamma(z)}\right) + \frac{\mathrm{d}}{\mathrm{d}z}\left(\frac{\Gamma'(z+\frac{1}{2})}{\Gamma(z+\frac{1}{2})}\right) &= \sum_{n=0}^{\infty}\frac{1}{(z+n)^2} + \sum_{n=0}^{\infty}\frac{1}{(z+n+\frac{1}{2})^2} \\
&= 4\left(\sum_{n=0}^{\infty}\frac{1}{(2z+2n)^2} + \sum_{n=0}^{\infty}\frac{1}{(2z+2n+1)^2}\right) \\
&= 4\left(\sum_{m=0}^{\infty}\frac{1}{(2z+m)^2}\right) \\
&= 2\frac{\mathrm{d}}{\mathrm{d}z}\left(\frac{\Gamma'(2z)}{\Gamma(2z)}\right).
\end{aligned}
$$

Therefore, there exists a constant A such that

$$2\frac{\Gamma'(2z)}{\Gamma(2z)} = \frac{\Gamma'(z)}{\Gamma(z)} + \frac{\Gamma'(z+\frac{1}{2})}{\Gamma(z+\frac{1}{2})} - A$$

or, equivalently,

$$\frac{\mathrm{d}}{\mathrm{d}z}\log\Gamma(2z) = \frac{\mathrm{d}}{\mathrm{d}z}\log\left[\Gamma(z)\Gamma\left(z+\frac{1}{2}\right)\right] - A.$$

Thus there exists a constant B such that

$$\operatorname{Log}\Gamma(2z) = \operatorname{Log}\left[\Gamma(z)\Gamma\left(z+\frac{1}{2}\right)\right] - Az - B$$

or, equivalently,

$$\Gamma(2z)\,\mathrm{e}^{Az+B} = \Gamma(z)\,\Gamma\left(z+\frac{1}{2}\right). \tag{10.11}$$

Next work backward to determine A and B. Setting $z = \frac{1}{2}$ in (10.11), we obtain $1\cdot\mathrm{e}^{\frac{1}{2}A+B} = \sqrt{\pi}\cdot 1$; that is, $\frac{1}{2}A + B = \frac{1}{2}\log\pi$. Setting $z = 1$ in (10.11), we obtain $\mathrm{e}^{A+B} = \frac{1}{2}\sqrt{\pi}$; that is, $A + B = \frac{1}{2}\log\pi - \log 2$. Thus $A = -2\log 2$ and $B = \frac{1}{2}(\log\pi) + \log 2$, and we have established

10.4.1.1 Legendre's Duplication Formula

$$\sqrt{\pi}\,\Gamma(2z) = 2^{2z-1}\,\Gamma(z)\,\Gamma\left(z+\frac{1}{2}\right). \tag{10.12}$$

10.4.2 Estimates for $\Gamma(z)$

The estimate of $\Gamma(z)$ for large values of $|z|$ that is found in this section is known as *Stirling's formula*. To derive this formula, we first express the partial sums $\sum_{k=0}^{n}\left(\frac{1}{z+k}\right)^2$ of $\frac{\mathrm{d}}{\mathrm{d}z}\frac{\Gamma'(z)}{\Gamma(z)}$ in (10.10) as a *convenient* line integral. View $z = x + \imath y$ as a (fixed) parameter and $\zeta = \xi + \imath\,\eta$ as a variable, and define

$$\Phi(\zeta) = \frac{\pi\cot\pi\zeta}{(z+\zeta)^2}, \quad \text{for } \zeta\in\mathbb{C}.$$

The function Φ has singularities at $\zeta = -z$ and at $\zeta\in\mathbb{Z}$; if $z\notin\mathbb{Z}$, Φ has a double pole at $-z$ and simple poles at the integers. Let Y be a positive real number, n be a nonnegative integer, and K be the rectangle in the ζ-plane described by $-Y\le\eta\le Y$ and $0\le\xi\le n+\frac{1}{2}$; see Fig. 10.1. Then the residue theorem yields the next result.

Lemma 10.15. *If $\Re z > 0$, then*

$$\frac{1}{2\pi\imath}\,\text{pr. v.}\int_{\partial K}\Phi(\zeta)\,\mathrm{d}\zeta = -\frac{1}{2z^2} + \sum_{\nu=0}^{n}\frac{1}{(z+\nu)^2},$$

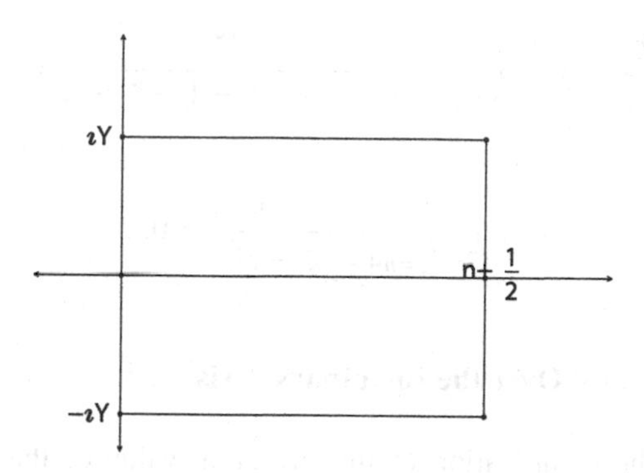

Fig. 10.1 The rectangle K

where as usual ∂K is positively oriented.

We plan to let $Y \to \infty$ and $n \to \infty$. We thus have to study several line integrals, as follows.

10.4.2.1 The Integral Over the Horizontal Sides: $\eta = \pm Y$

On the horizontal sides $\eta = \pm Y$, $\cot \pi\zeta$ converges uniformly to $\pm\imath$ as Y goes to ∞. Thus $\dfrac{\cot \pi\zeta}{(z+\zeta)^2}$ converges to 0 on each of the line segments $\xi \geq 0, \eta = \pm Y$ as $Y \to \infty$. We need to show that

$$\lim_{n\to\infty} \lim_{Y\to\infty} \int_0^{n+\frac{1}{2}} \frac{\cot \pi(\xi \pm \imath Y)}{(z+\xi+\imath Y)^2} \mathrm{d}\xi = 0.$$

Since we are able to control the speed with which Y and n approach infinity, this presents a small challenge; for example, we can set $Y = n^2$ and then let $n \to \infty$.

10.4.2.2 The Integral Over the Vertical Side $\xi = n + \frac{1}{2}$

On the vertical line, $\xi = n + \frac{1}{2}$, $\cot\, \pi\zeta$ is bounded since cot is a periodic function. Thus we conclude that for some constant c,

$$\left| \int_{\xi=n+\frac{1}{2}} \Phi(\zeta)\mathrm{d}\zeta \right| \leq c \int_{\xi=n+\frac{1}{2}} \frac{\mathrm{d}\eta}{|\zeta+z|^2}.$$

On $\xi = n + \frac{1}{2}$, we have $\overline{\zeta} = n + \frac{1}{2} - \imath\,\eta = 2n + 1 - \zeta$. We then use residue calculus (as in case 1 of Sect. 6.6) to conclude that

$$\frac{1}{\imath}\int_{\xi=n+\frac{1}{2}}\frac{\mathrm{d}\zeta}{|\zeta+z|^2}=\frac{1}{\imath}\int_{\xi=n+\frac{1}{2}}\frac{\mathrm{d}\zeta}{(\zeta+z)(2n+1-\zeta+\overline{z})}=\frac{2\pi}{2n+1+2x}.$$

Therefore,

$$\lim_{n\to\infty}\int_{\xi=n+\frac{1}{2}}\frac{\mathrm{d}\eta}{|\zeta+z|^2}=0.$$

10.4.2.3 The Integral Over the Imaginary Axis

We now turn to the computation of the principal value of the integral over the imaginary axis, which may be written as follows.

$$\frac{1}{2}\int_0^\infty \cot\pi\imath\eta\left[\frac{1}{(\imath\eta+z)^2}-\frac{1}{(\imath\eta-z)^2}\right]\mathrm{d}\eta=-\int_0^\infty \cot h\pi\eta\frac{2\eta z}{(\eta^2+z^2)^2}\mathrm{d}\eta.$$

We now return to the main task. We let Y and n tend to ∞ in Lemma 10.15 to conclude that

$$\begin{aligned}\frac{\mathrm{d}}{\mathrm{d}z}\frac{\Gamma'(z)}{\Gamma(z)}&=\sum_{n=0}^\infty\frac{1}{(z+n)^2}\\&=\frac{1}{2z^2}+\text{pr. v.}\int_{+\infty}^{-\infty}\Phi(\zeta)\,\mathrm{d}\zeta\\&=\frac{1}{2z^2}+\int_0^\infty \cot h\pi\eta\frac{2\eta z}{(\eta^2+z^2)^2}\,\mathrm{d}\eta.\end{aligned}$$

Replacing $\cot h\pi\eta$ by $1+\dfrac{2}{\mathrm{e}^{2\pi\eta}-1}$ in the above expression and noting that

$$\int_0^\infty\frac{2\eta z}{(\eta^2+z^2)^2}\,\mathrm{d}\eta=\frac{1}{z},$$

we obtain

$$\begin{aligned}\frac{\mathrm{d}}{\mathrm{d}z}\frac{\Gamma'(z)}{\Gamma(z)}&=\frac{1}{2z^2}+\int_0^\infty\left(1+\frac{2}{\mathrm{e}^{2\pi\eta}-1}\right)\frac{2\eta z}{(\eta^2+z^2)^2}\,\mathrm{d}\eta\\&=\frac{1}{z}+\frac{1}{2z^2}+\int_0^\infty\frac{4\eta z}{(\eta^2+z^2)^2}\frac{\mathrm{d}\eta}{\mathrm{e}^{2\pi\eta}-1}.\end{aligned}$$

Note that we have restricted our study to values of z with positive real parts. We can integrate with respect to z and change the order of integration in the last integral to conclude that

$$\frac{\Gamma'(z)}{\Gamma(z)}=\widetilde{C}+\operatorname{Log}z-\frac{1}{2z}-\int_0^\infty\frac{2\eta}{\eta^2+z^2}\frac{\mathrm{d}\eta}{\mathrm{e}^{2\pi\eta}-1}.$$

Using integration by parts, we see that

$$-\int_0^\infty \frac{2\eta}{\eta^2+z^2}\frac{\mathrm{d}\eta}{\mathrm{e}^{2\pi\eta}-1} = \frac{1}{\pi}\int_0^\infty \frac{z^2-\eta^2}{(\eta^2+z^2)^2}\log(1-\mathrm{e}^{-2\pi\eta})\,\mathrm{d}\eta;$$

therefore

$$\frac{\Gamma'(z)}{\Gamma(z)} = \widetilde{C} + \operatorname{Log} z - \frac{1}{2z} + \frac{1}{\pi}\int_0^\infty \frac{z^2-\eta^2}{(\eta^2+z^2)^2}\log(1-\mathrm{e}^{-2\pi\eta})\,\mathrm{d}\eta,$$

and we conclude that

$$\operatorname{Log}\Gamma(z) = C' + Cz + \left(z-\frac{1}{2}\right)\operatorname{Log} z + J(z), \tag{10.13}$$

where

$$J(z) = \frac{1}{\pi}\int_0^\infty \frac{z}{(\eta^2+z^2)^2}\log\frac{1}{1-\mathrm{e}^{-2\pi\eta}}\,\mathrm{d}\eta.$$

If $z \to \infty$ and z stays away from $\imath\mathbb{R}$, then $J(z) \to 0$. We have almost established the next result.

10.4.3 *The Formulae for the Function*

Theorem 10.16 (Stirling's Formula). *For* $\Re z > 0$,

$$\Gamma(z) = \sqrt{2\pi}\, z^{z-\frac{1}{2}}\, \mathrm{e}^{-z}\, \mathrm{e}^{J(z)}. \tag{10.14}$$

Proof. We know from (10.13) that

$$\Gamma(z) = \mathrm{e}^{C'+Cz}\, z^{z-\frac{1}{2}}\, \mathrm{e}^{J(z)}, \tag{10.15}$$

and we only need to determine the constants C' and C, which we do using (10.15) and the two functional equations (10.8) and (10.9) already derived for Γ. Replacing $\Gamma(z)$ by the RHS of (10.15) (and also $\Gamma(z+1)$ by the corresponding value) in $\Gamma(z+1) = z\Gamma(z)$, we obtain

$$C = -(z+\frac{1}{2})\operatorname{Log}\left(1+\frac{1}{z}\right) + J(z) - J(z+1),$$

and letting z tend to ∞ we conclude that $C = -1$. To obtain C' one can proceed in a similar manner, replacing $\Gamma(z)$ and $\Gamma(1-z)$ by the corresponding RHS of (10.15) in $\Gamma(z)\Gamma(1-z) = \dfrac{\pi}{\sin \pi z}$, with $z = \frac{1}{2} + \imath y$. We leave the details to the reader (Exercise 10.9). □

Corollary 10.17.

$$\lim_{n\to\infty} \frac{n!}{\sqrt{2\pi n}\left(\frac{n}{e}\right)^n} = 1.$$

Proof. Note that

$$(n+1)^{n+\frac{1}{2}} = n^{n+\frac{1}{2}} \left(1+\frac{1}{n}\right)^n \left(1+\frac{1}{n}\right)^{\frac{1}{2}},$$

and therefore

$$\lim_{n\to\infty} \frac{(n+1)^{n+\frac{1}{2}}}{e\, n^{n+\frac{1}{2}}} = 1. \tag{10.16}$$

Applying Stirling's formula (10.14) with $z = n+1$, we obtain

$$\Gamma(n+1) = \sqrt{2\pi}(n+1)^{n+\frac{1}{2}} \mathrm{e}^{-(n+1)} \mathrm{e}^{J(n+1)}.$$

Since we already know that $\lim_{n\to i\infty} J(n+1) = 0$, the claim is proved. □

With some further work we can prove the following integral expression for the Γ-function.

$$\Gamma(z) = \int_0^\infty \mathrm{e}^{-t}\, t^{z-1}\, \mathrm{d}t, \quad \text{for } \Re z > 0. \tag{10.17}$$

Proof. Denote the RHS of (10.17) by $F(z)$. It certainly defines a holomorphic function whenever the integral converges, that is, in the right half-plane $\Re z > 0$. Since

$$F(z+1) = \int_0^\infty \mathrm{e}^{-t} \cdot t^z \mathrm{d}t = z \int_0^\infty \mathrm{e}^{-t} \cdot t^{z-1} \mathrm{d}t = zF(z) \;\; \text{for } \Re z > 0,$$

we see that F and Γ satisfy the same functional equation (10.8). Thus

$$\frac{F(z+1)}{\Gamma(z+1)} = \frac{F(z)}{\Gamma(z)} \;\; \text{for } \Re z > 0,$$

and the function F can be extended to be defined on all of $\mathbb{C}$. For z with $1 \le \Re z \le 2$,

$$|F(z)| \le \int_0^\infty \mathrm{e}^{-t} \cdot t^{\Re z - 1} \mathrm{d}t = F(\Re z),$$

and since F is a continuous function on the closed interval $[1,2]$, it is bounded there. The last estimate shows that F is bounded on the strip $1 \le \Re z \le 2$. We need a lower bound for $|\Gamma|$ on the same strip; it suffices to determine how negative can $\log|\Gamma| = \Re \log \Gamma$ become. From (10.13), we see that

$$\log|\Gamma(z)| = \frac{1}{2}\log 2\pi - \Re z + \left(\Re z - \frac{1}{2}\right)\log|z| - \Im z \operatorname{Arg} z + \Re J(z).$$

Only the term $-\Im z$ Arg z approaches $\pm\infty$ as $\Im z$ approaches $+\infty$. This term is comparable to $-\frac{\pi}{2}|\Im z|$. Thus as $\Im z$ goes to $\pm\infty$, $\left|\frac{F}{\Gamma}\right|$ grows at most like a constant times $\exp\left(-\frac{\pi}{2}|\Im z|\right)$. Since $\frac{F}{\Gamma}$ is periodic, it is a function of $W = e^{2\pi\imath z}$ (on the punctured plane $\mathbb{C}-\{0\}$); it has an isolated singularity at $W = 0$. As W approaches 0, $\frac{F}{\Gamma}$ grows at most as $|W|^{-\frac{1}{2}}$, and as W approaches ∞, $\frac{F}{\Gamma}$ grows at most as $|W|^{\frac{1}{2}}$. Hence the singularities at both ends are removable and $\frac{F}{\Gamma}$ is constant. Since $F(1) = 1 = \Gamma(1)$, this constant is equal to 1. □

10.5 Divisors and the Field of Meromorphic Functions

It is convenient to introduce the following.

Definition 10.18. Let D be a nonempty plane domain. A *divisor* on D is a formal expression:

$$\prod_i z_i^{\nu_i},$$

where the $\{z_i\}$ form a discrete set in D and the ν_i are integers. We will also write a divisor $\mathcal{D}$ as

$$\mathcal{D} = \prod_{z\in D} z^{\nu_z(\mathcal{D})},$$

with the understanding that $\nu_z(\mathcal{D}) \in \mathbb{Z}$ for all $z \in D$ and $\nu_z(\mathcal{D}) = 0$ for all z not in a discrete subset of D (that depends on the divisor, of course). If $\nu_z = 0$ for all $z \in D$, the divisor is denoted by I.

There is a commutative multiplication law for divisors, where I is the unit element, and if $\mathcal{D}_1 = \prod_{z\in D} z^{\nu_z(\mathcal{D}_1)}$ and $\mathcal{D}_2 = \prod_{z\in D} z^{\nu_z(\mathcal{D}_2)}$ are two divisors, then

$$\mathcal{D}_1 \cdot \mathcal{D}_2 = \prod_{z\in D} z^{\nu_z(\mathcal{D}_1)+\nu_z(\mathcal{D}_2)}.$$

The set of all divisors on D with this operation becomes a commutative group, denoted by $\operatorname{Div}(D)$.

An interesting and important subset of $\operatorname{Div}(D)$ is obtained from the divisors naturally associated to meromorphic functions on D. To each not identically zero meromorphic function f on D we can associate its divisor (f) defined by

$$(f) = \prod_{z \in D} z^{\nu_z(f)},$$

where $\nu_z(f)$ denotes the order of the function f at the point z in D (see Sect. 3.5). In particular, to any constant (nonzero) function in D, we associate the divisor I. A divisor of a nonzero meromorphic function in D is called a *principal divisor*. The set of principal divisors is a subgroup of $\mathrm{Div}(D)$, and the function $(\cdot)$ that associates to each nonzero meromorphic function f on D its divisor (f) is a homomorphism from the multiplicative group $\mathbf{M}(D) - \{0\}$ to $\mathrm{Div}(D)$, whose image is the subgroup of principal divisors. Much more is true as shown next.

Theorem 10.19. *Let D be a nonempty domain in $\mathbb{C}$. The map*

$$(\cdot) : \mathbf{M}(D) - \{0\} \to \mathrm{Div}(D)$$

is a surjective homomorphism, whose kernel is the set of nowhere vanishing holomorphic functions on D.

Proof. Let $\mathcal{D} \in \mathrm{Div}(D)$. Since I is clearly in the image of the map $(\cdot)$, we assume $\mathcal{D} \neq I$. We can then write $\mathcal{D} = \mathcal{D}_1\, \mathcal{D}_2^{-1}$, where $\mathcal{D}_1$ and $\mathcal{D}_2$ are relatively prime integral divisors (see Exercise 10.5), and use Theorem 10.10 to conclude the proof. □

10.6 Infinite Blaschke Products

Let

$$\mathcal{D} = \prod_i z_i^{\nu_i}$$

be an integral divisor (a divisor such that $\nu_i \geq 0$ for all i, see Exercise 10.5) on the unit disc $\mathbb{D}$. It is clear that the definitions of Sect. 8.5 associate a Blaschke product $B_{\mathcal{D}}$ to the divisor $\mathcal{D}$. We end this book with some applications of Blaschke products to the study of bounded analytic functions on $\mathbb{D}$.

Theorem 10.20. *Let $I \neq \mathcal{D} = \prod_i a_i^{\nu_i}$ be an integral divisor on the unit disc $\mathbb{D}$. The Blaschke product $B_{\mathcal{D}} = \prod_i B_{a_i}^{\nu_i}$ converges to a nonconstant bounded analytic function on $\mathbb{D}$ if and only if*

$$\sum_i \nu_i\,(1 - |a_i|) < \infty. \tag{10.18}$$

If $\sum_i \nu_i\,(1 - |a_i|) = \infty$, then the Blaschke product converges to the constant zero function on the disc.

Proof. Without loss of generality, we assume that $a_i \neq 0$ for all i. It is convenient to write the divisor $\mathcal{D} = \prod_j a_j$, with each a_i appearing in this product ν_i times.

Assume (10.18) translated to $\sum_j (1 - |a_j|) < \infty$. By Theorem 10.5, it suffices to show that $\sum_j \left| 1 + \frac{|a_j|}{a_j} \frac{z - a_j}{1 - \bar{a}_j z} \right|$ converges uniformly on compact subsets $|z| \leq r$ of $\mathbb{D}$, with $0 < r < 1$. For $|z| = r$, we estimate

$$\begin{aligned} \left| 1 + \frac{|a_j|}{a_j} \frac{z - a_j}{1 - \bar{a}_j z} \right| &= \left| \frac{a_j - |a_j|^2 z + |a_j| z - |a_j| a_j}{a_j (1 - \bar{a}_j z)} \right| \\ &= \left| \frac{a_j (1 - |a_j|) + |a_j| z (1 - |a_j|)}{a_j (1 - \bar{a}_j z)} \right| \\ &= \left| \frac{(1 - |a_j|) \left(1 + \frac{|a_j|}{a_j} z \right)}{1 - \bar{a}_j z} \right| \\ &\leq \frac{(1 - |a_j|)(1 + r)}{1 - r}. \end{aligned}$$

By the maximum principle for holomorphic functions, the same estimate is valid for $|z| \leq r$. Hence on this compact set

$$\sum_j |1 - B_{a_j}(z)| \leq \frac{1 + r}{1 - r} \sum_j (1 - |a_j|)$$

is finite, and the infinite Blaschke product converges there. The infinite product

$$B_{\mathcal{D}} = \prod_i B_{a_i}^{\nu_i}$$

is bounded by 1 on $\mathbb{D}$, because each factor is. We already know that this bounded analytic function is nonconstant, since $\nu_{a_i}(B_{\mathcal{D}}) = \nu_i$.

If $\sum_j (1 - |a_j|) = +\infty$, the infinite product converges to the zero function by the second part of Theorem 10.4. □

Let F be a nowhere vanishing holomorphic function defined on a neighborhood of $\mathbb{D}$. Then $\log |F|$ is harmonic there, and the mean value property for such functions tells us that

$$\log |F(0)| = \frac{1}{2\pi} \int_0^{2\pi} \log \left| F(re^{\imath\theta}) \right| d\theta, \tag{10.19}$$

for $0 \leq r \leq 1$.

An important generalization of this simple formula that relates the values of F on the boundary of $\mathbb{D}$ to its value at the origin is the next result.

Theorem 10.21 (Jensen's Formula). *Let f be a holomorphic function in a neighborhood V of $\mathbb{D}$, and assume that $f(0) \neq 0$. For fixed $0 < r < 1$, let $\{a_1, \ldots, a_k\}$ denote the zeros of f in the disc $U(0, r)$ listed with multiplicity, and assume that f does not vanish on the circle $\{|z| = r\}$. Then*

$$\log|f(0)| + \sum_{j=1}^{k} \log\left|\frac{r}{a_j}\right| = \frac{1}{2\pi}\int_0^{2\pi} \log|f(re^{\imath\theta})| d\theta.$$

Proof. For $a \in \mathbb{D}$, the modified Blaschke factor

$$\widehat{B}_a(z) = \frac{z-a}{1-\bar{a}z}, \quad z \in V,$$

has been encountered many times before. Its properties readily imply that

$$F(z) = \frac{f(z)}{\prod_{j=1}^{k} \widehat{B}_{\frac{a_j}{r}}\left(\frac{z}{r}\right)}, \quad z \in V,$$

defines a function that is holomorphic and free of zeros in V. Thus, by (10.19),

$$\log|f(0)| - \sum_{j=1}^{k} \log\left|\widehat{B}_{\frac{a_j}{r}}(0)\right| = \frac{1}{2\pi}\int_0^{2\pi} \left[\log|f(re^{\imath\theta})| - \sum_{j=1}^{k} \log\left|\widehat{B}_{\frac{a_j}{r}}(e^{\imath\theta})\right|\right] d\theta.$$

Evaluating

$$\widehat{B}_{\frac{a_j}{r}}(0) = -\frac{a_j}{r} \quad \text{and} \quad \left|\widehat{B}_{\frac{a_j}{r}}(e^{\imath\theta})\right| = 1$$

leads us to the required formula. □

As a consequence we have the following reflection of the subharmonicity of $\log|f|$.[2]

[2]In more general definitions of subharmonic functions than used in this text, they are allowed to assume the value $-\infty$.

Corollary 10.22 (Jensen's Inequality). *Under the hypothesis of the theorem,*

$$\log|f(0)| \leq \frac{1}{2\pi}\int_0^{2\pi} \log\left|f(re^{\imath\theta})\right| d\theta.$$

Proof. For each j, $\log\left|\frac{r}{a_j}\right| \geq 0$. □

Theorem 10.23. *Let $\mathcal{D} = \prod_i a_i^{\nu_i}$ be an integral divisor on the unit disc $\mathbb{D}$. There exists a bounded analytic function f on $\mathbb{D}$ with divisor equal to $\mathcal{D}$ if and only if $\sum_i \nu_i\,(1-|a_i|) < \infty$. Furthermore, the function f is then of the form $f = B_{\mathcal{D}}\,F$, with F holomorphic, free of zeros, and bounded on $\mathbb{D}$.*

Proof. Let $f \in \mathbf{H}(\mathbb{D})$ be a bounded function with $(f) = \mathcal{D}$; choose $M \in \mathbb{R}_{>0}$ such that $|f| \leq M$.

Assume for the moment that $a_j \neq 0$ for all j. We can certainly find a sequence of numbers $0 < r_i < 1$ with $\lim_{i\to\infty} r_i = 1$ and $\left|a_j\right| \neq r_i$ for all i and j. By Jensen's formula applied with $r = r_i$,

$$\log|f(0)| + \sum_{|a_j|<r} \nu_j \log\left|\frac{r}{a_j}\right| = \frac{1}{2\pi}\int_0^{2\pi} \log\left|f(re^{\imath\theta})\right| d\theta \leq \log M.$$

Letting $i \to +\infty$ (thus $r_i \to 1$), we conclude that

$$\sum_j \nu_j(1-\left|a_j\right|) \leq \sum_j \nu_j \log\frac{1}{\left|a_j\right|} \leq \log M - \log|f(0)|.$$

If $\nu = \nu_{\mathcal{D}}(0) > 0$ (i.e., if $f(0) = 0$), set $g(z) = \frac{f(z)}{z^\nu}$. Then g is a bounded analytic function on $\mathbb{D}$ with $g(0) \neq 0$, and the previous argument may be applied to g to conclude that in this case we also have $\sum_j \nu_j(1-\left|a_j\right|) < \infty$.

Conversely, if $\sum_i \nu_i\,(1-|a_i|) < \infty$, it follows from Theorem 10.20 that $B_{\mathcal{D}}$ is a bounded analytic function on $\mathbb{D}$ satisfying $(B_{\mathcal{D}}) = \mathcal{D}$. Furthermore, if f is analytic and bounded on $\mathbb{D}$ with $(f) = \mathcal{D}$, then $F = \frac{f}{B_{\mathcal{D}}} \in \mathbf{H}(\mathbb{D})$ is zero-free and bounded on $\mathbb{D}$. □

Exercises

10.1. Calculate $\prod_{n=2}^{\infty}\left(1-\frac{1}{2^n}\right)$.

10.2. Let $\{a_n\}$ be a sequence in $\mathbb{C}_{\neq -1}$. The infinite product $\prod_{n=1}^{\infty}(1+a_n)$ is said to be *absolutely convergent* if the corresponding series $\sum_{n=1}^{\infty}\operatorname{Log}(1+a_n)$ converges absolutely.

1. Show that $\prod_{n=1}^{\infty}(1+a_n)$ converges absolutely if and only if $\sum_{n=0}^{\infty}|a_n|$ converges.
2. Show that the value of an absolutely convergent product does not change if the factors are reordered.
3. Find examples that show that the convergence of the series $\sum_{n=0}^{\infty}a_n$ is neither necessary nor sufficient for the convergence of the infinite product $\prod_{n=1}^{\infty}(1+a_n)$.

10.3. Show that $\Gamma(\frac{1}{6}) = 2^{-\imath}(\frac{3}{\pi})^{\frac{1}{2}}\cdot(\Gamma(\frac{1}{3}))^2$.

10.4. Find the residues of Γ at the poles $z=-n$, $n\geq 1$.

10.5. A divisor $\mathcal{D}$ on a domain D is *integral* if $\nu_z(\mathcal{D})\geq 0$ for all $z\in D$. Define appropriately the *greatest common divisor* $\gcd(\mathcal{D}_1,\mathcal{D}_2)$ and the *least common multiple* $\operatorname{lcm}(\mathcal{D}_1,\mathcal{D}_2)$ of two integral divisors $\mathcal{D}_1$ and $\mathcal{D}_2$. Show that both the gcd and the lcm exist, and obtain formulae for them. The integral divisors $\mathcal{D}_1$ and $\mathcal{D}_2$ are *relatively prime* if their gcd is the empty set.

10.6. Show that two nonzero meromorphic functions f and g in D give rise to the same divisor if and only if there exists h in $\mathbf{H}(D)$ that does not vanish in D and such that $f=gh$.

10.7. Show that a principal divisor on D is integral if and only if it is the divisor of an analytic function on D.

10.8. Show that if f and g are analytic functions in D (where at least one is not the zero function), then there exists a function $h\in\mathbf{H}(D)$ such that h is a greatest common divisor for f and g. That is, h divides f and g (in $\mathbf{H}(D)$), and it is divisible by every holomorphic function dividing both f and g.

If neither f nor g are the zero function, show that

$$(h)=\gcd((f),(g)).$$

Hint: Apply Theorem 10.10.

10.9. Replace $\Gamma(z)$ and $\Gamma(1-z)$ by the corresponding RHS of (10.15) in $\Gamma(z)\Gamma(1-z) = \dfrac{\pi}{\sin \pi z}$, with $z = \frac{1}{2} + \imath\, y$ in the proof of Stirling's formula (Theorem 10.16) to obtain C' in a similar manner to that used to obtain C. Give full details.

10.10. Prove the following form of the Mittag-Leffler Theorem. Let $\{a_n\}$ be a discrete sequence of distinct points in a domain $D \subset \mathbb{C}$, and let $\{b_n\}$ be an arbitrary sequence of complex numbers. Then there exists a holomorphic function f on D such that $f(a_n) = b_n$ for all n.

Let $D \subseteq \mathbb{C}$ be a domain with $\mathbf{H}(D)$ its ring of holomorphic functions. The following six exercises study the ideal structure of this ring; in particular, they establish that every finitely generated ideal in $\mathbf{H}(D)$ is a principal ideal.

10.11. Show that every nonempty collection of holomorphic functions in D, except for the set consisting of the single function zero, has a greatest common divisor.

10.12. Show that if h is a greatest common divisor for $f_1, \dots, f_n$ in $\mathbf{H}(D)$, then there exist $g_1, \dots, g_n$ in $\mathbf{H}(D)$ such that

$$f_1\, g_1 + \cdots + f_n\, g_n = h.$$

Hint: First consider the case when $n = 2$ and $h = 1$.

10.13. Let $f_1, \dots, f_n$ be holomorphic functions in D and consider the ideal I generated by them in $\mathbf{H}(D)$:

$$I = \{f_1\, g_1 + \cdots + f_n\, g_n\,;\, g_j \in \mathbf{H}(D)\}.$$

Prove that I is a principal ideal; that is, there exists a function f in $\mathbf{H}(D)$ such that

$$I = \{f\, g\,;\, g \in \mathbf{H}(D)\}.$$

10.14. Show that there exist nonprincipal maximal ideals in $\mathbf{H}(D)$.

10.15. Let $\varphi : \mathbf{H}(D) \to \mathbb{C}$ be a $\mathbb{C}$-algebra homomorphism. Show that there exists a unique $c \in D$ such that

$$\varphi(f) = f(c) \quad \text{for all } f \in \mathbf{H}(D).$$

10.16. Characterize the principal maximal ideals in the ring of holomorphic functions on a plane domain. Do the concepts of "principal maximal" and "maximal principal" ideals coincide?

10.17. Complete the proof of Theorem 10.12 by supplying all missing details.

Bibliographical Notes

The references required for proofs and definitions were included in the body of the book. The purpose of these notes is to list a number of basic books, not necessarily the latest edition or reprint, on complex analysis that the authors of this volume have found useful and interesting. Although we tried to write a book following a path that Bers might have chosen, certain portions show the influences of the authors in our list, particularly those of Lars Ahlfors and Henri Cartan. This is an incomplete list, arranged as four categories, reflecting the tastes and limited knowledge of the authors.

1. *Undergraduate texts*: The texts by Churchill and Brown [6], Derick [8], Fisher [9], and Silverman [33] are each very appropriate for an undergraduate course that is centered around applications and uses the Ahlfors and Bers approach to complex integration and analytic continuation along a path; Palka [27] is a very thorough and careful undergraduate text that is also frequently used in graduate courses, and uses the Ahlfors and Bers approach to complex integration and analytic continuation along a path; Marsden [23] is more theoretical while Bak–Newman [2] is the only undergraduate text on the subject that ends up with a treatment of the prime number theorem.
2. *Graduate texts*: Ahlfors [1] is among the outstanding mathematics books in any field. It clearly influenced almost all books on the subject that followed, including, of course, our own. Another classical book is the lovely treatment by Nevanlinna–Paatero [26]; Narasimhan [24] is an outstanding very modern treatment. The beautiful text by Cartan [5] starts with a treatment of formal power series; it is reflected in and is an inspiration for our chapter on convergent power series. The first of the two Hille volumes [13] is a standard introduction to the subject; the second [14] deals with many interesting special topics. The book by Heins [12] covers many prerequisites currently dealt with in other courses and some advanced topics. Conway's book [7] is very concise yet complete. It includes the big Picard theorem. Berenstein and Gay [3] is in tune with more recent, modern developments in complex variables. The Green–Krantz text [11] is a treatment of complex variables as an outgrowth of real multivariate calculus.

R.E. Rodríguez et al., *Complex Analysis: In the Spirit of Lipman Bers*, Graduate Texts in Mathematics 245, DOI 10.1007/978-1-4419-7323-8,

The Remmert volume [28] deals with some number theoretic topics that we have omitted and contains short biographical notes on some of the principal contributors to classical function theory. It also contains a discussion of the origins of some of the basic concepts in function theory. The text by Boas [4] includes a discussion of Abel summability.

Another standard reference is Lang [21]. Lang and Bers probably influenced each other's view of complex variables as a result of long discussions in the Columbia mathematics department fifth floor lounge while Lang was writing his book and Bers was teaching Complex Variables I and II during the 1966–1967 academic year.

The first chapter of Hörmander [15] serves as a thorough review for those who have learned one variable complex analysis and are interested in seeing how it leads naturally to a study of several variables. Rudin [30] is an integrated treatment of real and complex analysis, while the very recent volume by Lax-Zalcman [22] applies complex analysis to provide efficient proofs of real variable theorems. The text by Needham [25] is intended mainly for beginners, undergraduates perhaps. It is full of theory and applications, and all the material is presented from a wonderful geometric perspective. The volume by Stein–Shakarchi [34] treats complex analysis as part of the wider analytic universe.

3. *Problem books*: There are paperback books that contain problem sets and their solutions. Classical is a five-volume set by Konrad Knopp. Whenever Bers taught the yearlong graduate complex analysis course at Columbia, he taught without a text and simply told his students to read and work through all of the problems in Knopp's books on the theory of functions [16–20] (Bers translated one of these five volumes from the German [18]). Knopp's problem books are still eminently relevant and useful. A paperback by Shakarchi [32] includes solutions to all of the undergraduate level problems from Lang's graduate text (problems from the first eight chapters) and solutions to selected problems from the more advanced chapters.
4. *History*: Although not primarily a book on the history of mathematics, [28] belongs on this list. Roy [29] is a scholarly and well-researched treatment of sources in the development of infinite series and products. His appendix to our first chapter is indicative of the breadth of the book. The Freitag–Busam text [10] also contains interesting historical remarks. A volume by Sandifer [31] contains a thorough history of part of the work of a mathematical giant.

References

1. Ahlfors, L.V.: Complex Analysis, 3rd edn. McGraw-Hill, New York (1979)
2. Bak, J., Newman, D.J.: Complex Analysis. Springer, Berlin (1982)
3. Berenstein, C.A., Gay, R.: Complex variables, an introduction. In: Graduate Texts in Mathematics, vol. 125. Springer, Berlin (1991)
4. Boas, R.P.: Invitation to Complex Analysis. Random House, New York (1987)
5. Cartan, H.: Elementary Theory of Analytic Functions of One or Several Complex Variables. Addison-Wesley, Reading (1963)
6. Churchill, R.V., Brown, J.W.: Complex Analysis and Applications, 5th edn. McGraw-Hill, New York (1990)
7. Conway, J.B.: Functions of One Complex Variable, 2nd edn. Springer, Berlin (1978)
8. Derrick, W.R.: Complex Analysis and Applications, 2nd edn. Wadsworth International Group, New York (1982)
9. Fisher, S.D.: Complex Variables, 2nd edn. Dover Publications, New York (1999)
10. Freitag, E., Busam, R.: Complex Analysis. Springer, Berlin (2005)
11. Greene, R.E., Krantz, S.G.: Function Theory of one Complex Variable. John Wiley & Sons Inc., New York (1997)
12. Heins, M.: Complex Function Theory. Academic Press, New York (1968)
13. Hille, E.: Analytic Function Theory, vol. I. Blaisdell, New York (1959)
14. Hille, E.: Analytic Function Theory, vol. II. Blaisdell, Waltham (1962)
15. Hörmander, L.: An Introduction to Complex Analysis in Several Variables. Van Nostrand, Princeton (1966)
16. Knopp, K.: Theory of Functions I. Elements of the General Theory of Analytic Functions. Dover Publications, New York (1945)
17. Knopp, K.: Theory of Functions II. Applications and Continuation of the General Theory. Dover Publications, New York (1947)
18. Knopp, K.: Problem Book in the Theory of Functions: Problems in the Elementary Theory of Functions, , vol. 1. Dover Publications, New York (1948) (Translated by Lipman Bers)
19. Knopp, K.: Elements of the Theory of Functions. Dover Publications Inc., New York (1953) (Translated by Frederick Bagemihl)
20. Knopp, K.: Problem Book in the Theory of Functions: Problems in the Advanced Theory of Functions, vol. II. Dover Publications, New York, NY (1953) (Translated by F. Bagemihl)
21. Lang, S.: Complex analysis, 4th edn. In: Graduate Texts in Mathematics, vol. 103. Springer, Berlin (1999)
22. Lax, P.D., Zalcman, L.: Complex Proofs of Real Theorem. American Mathematical Society University Lecture Series, AMS, New York (2012)
23. Marsden, J.E.: Basic Complex Analysis. W. H. Freeman and Company, New York (1973)

R.E. Rodríguez et al., *Complex Analysis: In the Spirit of Lipman Bers*, Graduate Texts in Mathematics 245, DOI 10.1007/978-1-4419-7323-8,

24. Narasimhan, R.: Complex analysis in One Variable. Verlag, Birkhäuser (1985)
25. Needham, T.: Visual Complex Analysis. Oxford University Press, Oxford (2004)
26. Nevanlinna, R., Paatero, V.: Introduction to Complex Analysis. Addison-Wesley, New York (1964)
27. Palka, B.: An Introduction to Complex Function Theory. Springer, Berlin (1991)
28. Remmert, R.: Theory of Complex Functions. Springer, Berlin (1991) (Translated by R. B. Burckel)
29. Roy, R.: Sources in the Development of Mathematics. Cambridge University Press, Cambridge (2011)
30. Rudin, W.: Real and Complex Analysis, 3rd edn. McGraw-Hill, New York (1987)
31. Sandifer, C.E.: The Early Mathematics of Leonhard Euler. The Mathematical Association of America, Washington (2007)
32. Shakarchi, R.: Problems and Solutions for Complex Analysis. Springer, Berlin (1999)
33. Silverman, R.A.: Complex Analysis with Applications. Prentice-Hall (1974)
34. Stein, E.M., Shakarchi, R.: Complex Analysis. Princeton Lectures in Analysis. Princeton University Press, Princeton (2003)

Index

R.E. Rodríguez et al., *Complex Analysis: In the Spirit of Lipman Bers*, Graduate Texts in Mathematics 245, DOI 10.1007/978-1-4419-7323-8,

T

U

V

W

Z